The IMA Volumes in Mathematics and its Applications

Volume 137

Series Editors
Douglas N. Arnold   Fadil Santosa

Springer
New York
Berlin
Heidelberg
Hong Kong
London
Milan
Paris
Tokyo

Institute for Mathematics and its Applications (IMA)

The **Institute for Mathematics and its Applications** was established by a grant from the National Science Foundation to the University of Minnesota in 1982. The primary mission of the IMA is to foster research of a truly interdisciplinary nature, establishing links between mathematics of the highest caliber and important scientific and technological problems from other disciplines and industry. To this end, the IMA organizes a wide variety of programs, ranging from short intense workshops in areas of exceptional interest and opportunity to extensive thematic programs lasting a year. IMA Volumes are used to communicate results of these programs that we believe are of particular value to the broader scientific community.

The full list of IMA books can be found at the Web site of the Institute for Mathematics and its Applications:

http://www.ima.umn.edu/springer/full-list-volumes.html.

Douglas N. Arnold, Director of the IMA

* * * * * * * * * *

IMA ANNUAL PROGRAMS

1982–1983	Statistical and Continuum Approaches to Phase Transition
1983–1984	Mathematical Models for the Economics of Decentralized Resource Allocation
1984–1985	Continuum Physics and Partial Differential Equations
1985–1986	Stochastic Differential Equations and Their Applications
1986–1987	Scientific Computation
1987–1988	Applied Combinatorics
1988–1989	Nonlinear Waves
1989–1990	Dynamical Systems and Their Applications
1990–1991	Phase Transitions and Free Boundaries
1991–1992	Applied Linear Algebra
1992–1993	Control Theory and its Applications
1993–1994	Emerging Applications of Probability
1994–1995	Waves and Scattering
1995–1996	Mathematical Methods in Material Science
1996–1997	Mathematics of High Performance Computing
1997–1998	Emerging Applications of Dynamical Systems
1998–1999	Mathematics in Biology

Continued at the back

Christopher B. Croke Irena Lasiecka
Gunther Uhlmann Michael S. Vogelius
Editors

Geometric Methods in Inverse Problems and PDE Control

With 11 Illustrations

Springer

Christopher B. Croke
Department of Mathematics
University of Pennsylvania
Philadelphia, PA 19104-6395
USA
ccroke@math.upenn.edu

Irena Lasiecka
Department of Mathematics
University of Virginia
Charlottesville, VA 22903
USA
il2v@weyl.math.virginia.edu

Gunther Uhlmann
Department of Mathematics
University of Washington
Seattle, WA 98195-4350
USA
gunther@math.washington.edu

Michael S. Vogelius
Department of Mathematics
Rutgers University
New Brunswick, NJ 08903
USA
vogelius@math.rutgers.edu

Series Editors:
Douglas N. Arnold
Fadil Santosa
Institute for Mathematics and
its Applications
University of Minnesota
Minneapolis, MN 55455
USA
http://www.ima.umn.edu

Mathematics Subject Classification (2000): 35-06, 53-06, 93-06

Library of Congress Cataloging-in-Publication Data

ISBN 0-387-40529-1 Printed on acid-free paper.
Geometric methods in inverse problems and PDE control / [edited by] Christopher B. Croke . . . [et al.]
p. cm. — (IMA volumes mathematic ; v. 137)
Includes index.
ISBN 0-387-40529-1 (acid-free paper)
1. Inverse problems (Differential equations) 2. Differential equations, Partial. 3. Control theory. I. Croke, Christopher B. II. IMA volumes in mathematics and its applications ;v. 137.
QA377.G4585 2003
515′.357—dc21 2003054414

Printed in the United States of America.

9 8 7 6 5 4 3 2 1 SPIN 10944400

Camera-ready copy provided by the IMA.

www.springer-ny.com

Springer-Verlag New York Berlin Heidelberg
A member of BertelsmannSpringer Science+Business Media GmbH

FOREWORD

This IMA Volume in Mathematics and its Applications

GEOMETRIC METHODS IN INVERSE PROBLEMS AND PDE CONTROL

contains a selection of articles presented at 2001 IMA Summer Program with the same title.

We would like to thank Christopher B. Croke (University of Pennsylvania), Irena Lasiecka (University of Virginia), Gunther Uhlmann (University of Washington), and Michael S. Vogelius (Rutgers University) for their excellent work as organizers of the two-week summer workshop and for editing the volume.

We also take this opportunity to thank the National Science Foundation for their support of the IMA.

Series Editors

Douglas N. Arnold, Director of the IMA

Fadil Santosa, Deputy Director of the IMA

PREFACE

This volume contains a selected number of articles based on lectures delivered at the IMA 2001 Summer Program on "Geometric Methods in Inverse Problems and PDE Control." The focus of this program was some common techniques used in the study of inverse coefficient problems and control problems for partial differential equations, with particular emphasis on their strong relation to fundamental problems of geometry. Inverse coefficient problems for partial differential equations arise in many application areas, for instance in medical imaging, nondestructive testing, and geophysical prospecting. Control problems involving partial differential equations may arise from the need to optimize a given performance criterion, e.g., to dampen out undesirable vibrations of a structure , or more generally, to obtain a prescribed behaviour of the dynamics. For both types of problems a central issue becomes the determination of boundary conditions that will lead to a desired state of the solution of the associated partial differential equation. As just one example of the strong relation to geometric problems let us mention that a combination of unique continuation results (for the wave equation) with the so-called boundary control method positively solves the problem of determining a metric of a Riemannian manifold (with boundary) from the dynamic Dirichlet-to-Neumann data boundary map.

The article by Katchalov, Kurylev and Lassas provides new insights into the boundary control method. The success of this method depends on a new version of Holmgren's uniqueness theorem for the wave equation. In his paper Tataru surveys the key results on Carleman estimates that are required to prove such a unique continuation result.

An important issue in control theory is to establish sufficient conditions for controllability and stabilization – in particular, conditions expressed in terms of the geometry of the physical domain. The article by Gulliver, Lasiecka, Littman and Triggiani contains an extensive account of the use of differential geometric methods in control.

As explained in the article by Uhlmann, the Dirichlet-to-Neumann map for the wave equation determines the boundary distance function (the minimal travel time along geodesics connecting points on the boundary of a Riemannian manifold). A natural question to ask is if one can determine the metric from this data alone. This question is at the center of the boundary rigidity problem studied in Riemannian geometry. The paper by Croke considers in detail this problem.

Sharafutdinov surveys the linearization of the boundary rigidity problem, namely the integral geometry problem of determining a tensor field from its integral along geodesics; he also surveys some recent results for the local boundary rigidity problem (when only Riemannian metrics close to a particular one are considered).

Many results concerning inverse problems as well as exact controllability are based on the study of propagation of singularities for the underlying partial differential equation. The theory of Fourier integral operators, Lagrangian distributions and more generally, microlocal analysis, have been crucial for the study of certain aspects of these singularities. Taylor develops in his paper an approach to obtain delicate regularity estimates for Fourier integral operators.

Microlocal analysis is also used in Uhlmann's paper to show that the Dirichlet-to-Neumann map for the wave equation uniquely determines the scattering relation (or lens map). Roughly speaking this map encodes information of all travel times, not just the ones associated with minimizing geodesics. An approach to find the Riemannian metric from the scattering relation is also described in detail.

The paper by Bryan and Vogelius and sections 5 and 6 of Uhlmann's paper consider elliptic inverse boundary value problems. Bryan and Vogelius survey the problem of crack imaging based on electrical measurements at the boundary of the medium. The measurements are a finite set of points on the graph of the Dirichlet-to-Neumann map. Most results in this area are based on a detailed study of the global behaviour of the level curves (surfaces) of the corresponding potentials. Uhlmann describes how to use Carleman estimates to obtain information about the conductivity from measurements of the Dirichlet-to-Neumann map on particular subsets of the boundary.

Berryman, Borcea, Papanicolaou and Tsogka consider the important issue of the relative merit of single frequency vs. time domain imaging. Using a linear sampling (or MUSIC) approach to determine potential source locations, they present evidence that the time domain approach performs much better in the presence of significant spatial heterogeneity.

The articles by Zelditch and Ballmann are on inverse spectral problems, motivated by the famous question posed by Mark Kac: can one hear the shape of a drum? Zelditch finds new spectral invariants which he uses to discover a class of domains in the plane, with analytic boundaries, which are determined by their Dirichlet spectrum. Ballmann, by unification and extension of work of Sunada and of Gordon and Schueth, develops general methods for the construction of non-isometric, isospectral manifolds.

As organizers we would like to thank the IMA for hosting this very successful workshop, and we wish to express our hope (and belief) that it and its proceedings may serve to encourage interaction between researchers in differential geometry, inverse problems and control of PDE's.

Christopher B. Croke (University of Pennsylvania)
Irena Lasiecka (University of Virginia)
Gunther Uhlmann (University of Washington)
Michael S.Vogelius (Rutgers University)

CONTENTS

Foreword v

Preface vii

On the construction of isospectral manifolds 1
Werner Ballmann

Statistical stability and time-reversal imaging in random media 15
James G. Berryman, Liliana Borcea, George C. Papanicolaou, and Chrysoula Tsogka

A review of selected works on crack identification 25
Kurt Bryan and Michael S. Vogelius

Rigidity theorems in Riemannian geometry 47
Christopher B. Croke

The case for differential geometry in the control of single and coupled PDEs: the structural acoustic chamber 73
R. Gulliver, I. Lasiecka, W. Littman, and R. Triggiani

Energy measurements and equivalence of boundary data for inverse problems on non-compact manifolds 183
A. Katchalov, Y. Kurylev, and M. Lassas

Ray transform and some rigidity problems for Riemannian metrics 215
Vladimir Sharafutdinov

Unique continuation problems for partial differential equations 239
Daniel Tataru

Remarks on Fourier integral operators 257
Michael Taylor

The Cauchy data and the scattering relation 263
Gunther Uhlmann

Inverse resonance problem for $\mathbb{Z}_2$-symmetric analytic obstacles in the plane 289
Steve Zelditch

List of workshop participants 323

ON THE CONSTRUCTION OF ISOSPECTRAL MANIFOLDS

WERNER BALLMANN*

Abstract. I discuss two general methods, namely Sunada's method and a more recent method, due to Gordon and Schueth, for the construction of isospectral closed Riemannian manifolds. I prove a theorem unifying and extending the two methods and apply it to obtain isospectral metrics on $S^2 \times S^3$.

Key words. Laplace operator, spectrum.

AMS(MOS) subject classifications. 58J53.

Introduction. Let M be a closed Riemannian manifold. Let $\Delta = \Delta_M$ be the Laplace operator of M. Then the spectrum of M consists of the sequence

$$0 = \lambda_0 \leq \lambda_1 \leq \lambda_2 \leq \ldots$$

of eigenvalues of Δ, where we count each eigenvalue as often as its multiplicity requires. The spectrum of M is related to the geometry and topology of M. For example, the asymptotic expansion of the heat kernel of Weyl and Minakshisundaram-Pleijel shows that dimension, volume and mean scalar curvature are invariants of M determined by its spectrum, see [BGM].

Two closed Riemannian manifolds M and M' are called *isospectral* if the corresponding sequences of eigenvalues coincide. First examples of isospectral and not isometric manifolds were discovered by J. Milnor, flat tori in dimension 16 [Mil]. M. F. Vignéras found surfaces and 3-manifolds with curvature -1 [Vi, 1980] and A. Ikeda manifolds with curvature $+1$ [Ik1, 1983]. C. Gordon and E. Wilson discovered the first examples of isospectral deformations [GWi, 1984].

In this paper I discuss two general methods for the construction of isospectral, but not isometric Riemannian manifolds, namely Sunada's method and a more recent method, due to Gordon and Schueth. In Theorem 3.4 I unify and extend these two methods, see Remark 3.5. I apply Theorem 3.4 to obtain isospectral metrics on $S^2 \times S^3$.

Some of my discussion follows historical lines, but I do not try to give a historical account of the matter. I should also point out that not all known pairs of isospectral manifolds fall under the methods in the form I discuss them.

Acknowledgment. I would like to thank Dorothee Schueth for many clarifying remarks and elucidating discussions and the referee for pointing out some errors and suggesting that Part 2 of Remark 3.5 be included.

---

*Mathematisches Institut, Universität Bonn, Beringstrasse 1, 53115 Bonn, Germany (ballmann@math.uni-bonn.de). Partially supported by SFB256 (U Bonn).

1. Sunada's method. The first general method for establishing isospectrality is due to T. Sunada [Sun]. It is concerned with subcovers of a given Riemannian manifold. Examples arising from this method have the same local geometry, they can be distinguished only by global invariants.

In the first part of my discussion, I follow [Ik2]. Let $\pi : N \to M := \Gamma\backslash N$ be a finite normal Riemannian covering with group Γ of deck transformations. If $f : M \to \mathbb{R}$ is a smooth function, then

$$\Delta_N(f \circ \pi) = \Delta_M(f) \circ \pi.$$

In particular, if $\lambda \geq 0$ and

$$V_\lambda := \{f : N \to \mathbb{R} \mid \Delta_N f = \lambda f\},$$

then the multiplicity of λ as an eigenvalue of Δ_M is equal to the dimension of

$$V_\lambda^\Gamma := \{f \in V_\lambda \mid f = f \circ g \text{ for all } g \in \Gamma\}.$$

Suppose that $p : N \to B = G\backslash N$ is a further finite normal Riemannian covering with group $G \supset \Gamma$. Let R be the natural representation of G on V_λ on the right, $R_g(f) = f \circ g$. Then V_λ^Γ is Γ-invariant, and the orthogonal projection $P : V_\lambda \to V_\lambda^\Gamma$ is given by

$$Pf = \frac{1}{|\Gamma|}\sum_{g\in\Gamma} f \circ g = \frac{1}{|\Gamma|}\sum_{g\in\Gamma} R_g(f).$$

Therefore,

$$\dim V_\lambda^\Gamma = \operatorname{tr} P = \frac{1}{|\Gamma|}\sum_{g\in\Gamma} \operatorname{tr} R_g.$$

If $h \in \Gamma$ is conjugate to $g \in G$ in G, $h = kgk^{-1}$ for some $k \in G$, then $R_h = R_k R_g R_k^{-1}$ and hence $\operatorname{tr} R_h = \operatorname{tr} R_g$. It follows that

$$\dim V_\lambda^\Gamma = \frac{1}{|\Gamma|}\sum_{[g]} \big|[g] \cap \Gamma\big| \cdot \operatorname{tr} R_g,$$

where the summation is over the conjugacy classes $[g]$ of G. Hence for N and G fixed, the spectrum of $M = \Gamma\backslash N$ only depends on the numbers $|[g] \cap \Gamma|$.

THEOREM 1.1 (Sunada [Sun]). *Let Γ and Γ' be subgroups of G and set $M = \Gamma\backslash N$ and $M' = \Gamma'\backslash N$. Then if each conjugacy class of G meets Γ and Γ' in the same number of elements, then M and M' are isospectral.* □

This method of constructing isospectral Riemannian manifolds is very powerful. For example, Sunada showed that for certain genera $g \geq 2$, there is a set $\mathcal{I}$ of positive dimension in Teichmüller space $\mathcal{T}_g$ such that any surface in $\mathcal{I}$ has an isospectral, non-isometric companion in $\mathcal{T}_g$.

C. Gordon and D. DeTurck generalized Sunada's method to the case where N is possibly noncompact, where G is a closed subgroup of the isometry group of N, and Γ and Γ' are subgroups of G which act properly discontinuously and freely on N with compact quotients $M = \Gamma\backslash N$ and $M' = \Gamma'\backslash N$ [DTG]. Note that the original method of Sunada does not give rise to isospectral deformations since G is a discrete subgroup of the isometry group of N. However, as one of the applications of their extension Gordon and DeTurck recovered most of the isospectral deformations from [GWi].

P. Bérard observed that Sunada's condition and the corresponding condition in [DTG] is equivalent to the condition that the quasi-regular representations of G on $L^2(\Gamma\backslash G)$ and $L^2(\Gamma'\backslash G)$ are equivalent [Be1, Be2]. He also extended Sunada's method to the case where the actions of Γ and Γ' are not free.

One of the highlights among the applications of Sunada's method, in a version yet slightly extending Bérard's extension, is the construction of isospectral plane domains by Gordon, Webb, and Wolpert [GWW]. Surveys of and historical comments on these developments are contained in [Go3, Sch1].

For the sake of clarity and simplicity I assume throughout that G is a finite group of covering transformations. I now discuss Bérard's observation in this simple context.

LEMMA 1.2. *Sunada's assumption is equivalent to the condition that the quasi-regular representations ρ and ρ' of G on $L^2(\Gamma\backslash G)$ and $L^2(\Gamma'\backslash G)$, respectively, are equivalent.*

Proof. Recall that a (finite dimensional) complex representation ρ of a finite group G is characterized by its character $c = c(g) = \operatorname{tr}\rho(g)$. Hence the quasi-regular representations ρ and ρ' of G are equivalent if and only if their characters c and c' coincide. Identify

$$L^2(\Gamma\backslash G) = \{f \in L^2(G) \mid f(kh) = f(h) \text{ for all } k \in \Gamma \text{ and } h \in G\},$$ [1]

and let $P : L^2(G) \to L^2(\Gamma\backslash G)$ be the orthogonal projection. Then $\operatorname{tr}(\rho_g) = \operatorname{tr}(R_g \circ P)$ for all $g \in G$, where R denotes the right regular representation of G on $L^2(G)$. For $f \in L^2(G)$ and $h \in G$,

$$((R_g P)(f))(h) = \frac{1}{|\Gamma|} \sum_{k\in\Gamma} f(k^{-1}hg).$$

An orthonormal basis of $L^2(G)$ is given by the delta functions δ_h, $h \in G$. Hence

[1] I use the normalized counting measures on G and quotients of G.

$$\operatorname{tr}(R_g \circ P) = \frac{1}{|\Gamma|} \sum_{h \in G, k \in \Gamma} \delta_h(k^{-1}hg) = \frac{|[g] \cap \Gamma|}{|\Gamma|} \cdot |C_G(g)|,$$

where $C_G(g)$ denotes the centralizer of g in G. The lemma follows. □

As an application, we recover Sunada's theorem in the following form:

THEOREM 1.3. *Let Γ and Γ' be subgroups of G and set $M = \Gamma\backslash N$ and $M' = \Gamma'\backslash N$. Then if the quasi-regular representations of G on $L^2(\Gamma\backslash G)$ and $L^2(\Gamma'\backslash G)$ are equivalent, then M and M' are isospectral.* □

My understanding used to be that the method of Sunada dealt with subcoverings of a given manifold, here the manifold N. However, one may also look at M and M' as covering spaces of $B = G\backslash N$ (under the assumptions of Sunada). This point of view is enlightening when compared to my formulation of the method of Gordon and Schueth in Theorem 3.4. A corresponding direct proof of Theorem 1.3 runs as follows:

View $N \to B$ as a (left) principal G-bundle. Consider the associated fiber bundle $G \times_G N$, where the latter consists of equivalence classes $[g, x]$ with $g \in G$ and $x \in N$ and where, for $h \in G$, $[gh, x] = [g, hx]$. Define a G-action on $G \times_G N$ by $h[g, x] = [hg, x]$. Then the map $G \times_G N \to N$, $[g, x] \to gx$, is a G-equivariant diffeomorphism, hence

$$M = \Gamma\backslash N = \Gamma\backslash(G \times_G N) = (\Gamma\backslash G) \times_G N.$$

In other words, the choice of a point $x \in N$ over a point $z \in B$ identifies the fiber of $\pi : M \to B$ over z with $\Gamma\backslash G$. With respect to this identification, the restriction of a function $f : M \to \mathbb{C}$ to the fiber of π over z is a function on $\Gamma\backslash G$. In this sense,

$$L^2(M) = L^2(\Gamma\backslash N) = L^2\big(L^2(\Gamma\backslash G) \times_G N\big).$$

That is, we view functions on M as sections of the associated Hermitian vector bundle $W = L^2(\Gamma\backslash G) \times_G N \to B$. Note that the principal bundle $p : N \to B$ is a covering, hence has a unique G-connection and this connection is flat. The associated connection on W is also flat and, under the above identification, its Laplacian corresponds to the Laplacian on functions on M. Hence the spectrum of the Laplacian of M coincides with the spectrum of the Laplacian of the vector bundle W. Now W is the vector bundle over B induced by the quasi-regular representation of G on $L^2(\Gamma\backslash G)$, hence Theorem 1.3 follows.

The difference to the argument of Bérard is the additional intermediate step of showing that the Laplacian of M is unitarily equivalent to the Laplacian of $W \to B$. Bérard gives a direct unitary equivalence of the Laplacians of M and M', more precisely, of the corresponding Sobolev spaces.[2]

[2]For that reason, Bérard does not need to assume that the action of G is free, that is, that B is a manifold. Note however that the above arguments also extend to the case where B is a Riemannian orbifold.

REMARK 1.4. The same arguments as above show that any differential operator on B lifts to differential operators on M and M' with the same spectrum. For example, the Hodge Laplacians on forms on M and M' have the same spectrum.

2. The method of Gordon and Schueth. The first pair of manifolds — with boundary though — which are isospectral but not locally isometric is due to Z. Szabo [Sza]. Motivated by his work, C. Gordon constructed the first closed examples and came up with the following general criterion, compare [Go1, Go2].

THEOREM 2.1. *Let M and M' be closed Riemannian manifolds. Suppose that a torus T acts freely and isometrically on M and M'. Suppose furthermore that the following hold:*

1. *The orbits of the action are totally geodesic.*
2. *For any subtorus S of T of codimension at most one, the quotient manifolds $S\backslash M$ and $S\backslash M'$ are isospectral.*

Then M and M' are isospectral.

Gordon's criterion has turned out to be a very powerful tool in the construction of isospectral manifolds; for a more elaborate historical account I refer to [Sch1].

Let T be a torus and $M \to B$ be a principal bundle with structure group T, where B is a closed and connected Riemannian manifold. Fix a left invariant metric g_T on T. Then for any principal connection ω on M, there is a unique Riemannian metric g_ω on M such that the projection $M \to B$ is a Riemannian submersion with totally geodesic fibers isometric to T and horizontal distribution equal to $\ker \omega$, see Theorem 9.59 in [Bes].

In [Sch1], Dorothee Schueth used Theorem 2.1 to derive the following very useful more special criterion, see Theorem 1.6 in [Sch1].

THEOREM 2.2. *Let ω and ω' be principal connections on M. Suppose that for each ζ in the dual of the Lie algebra $\mathfrak{t}$ of T there is an automorphism $\Phi_\zeta : M \to M$ such that*

1. *Φ_ζ factors over an isometry of B.*
2. *$\zeta \circ \omega' \circ \Phi_{\zeta *} = \zeta \circ \omega$.*

Then (M, g_ω) and $(M, g_{\omega'})$ are isospectral.

In her proof, Schueth showed that for any closed subgroup S of T of codimension at most one, the quotient manifolds $(M, g_\omega)/S$ and $(M, g_{\omega'})/S$ are isometric. In one of her applications, Schueth obtained pairs of isospectral metrics on $S^2 \times T^2$.

One of my aims is to extend the criteria of Gordon and Schueth and to explain why tori play such a prominent role in their criteria.

3. An extension. Let M and B be closed Riemannian manifolds and $\pi : M \to B$ be a Riemannian submersion. Let F_z denote the fiber of π over $z \in B$.

Let $c : [0,1] \to B$ be a piecewise smooth curve from $z_0 = c(0)$ to $z_1 = c(1)$. Then the horizontal displacement $h_c : F_{z_0} \to F_{z_1}$ is a diffeomorphism. Recall the following well known fact, see Theorem 9.56 in [Bes].

LEMMA 3.1. *The fibers of π are totally geodesic if and only if, for each piecewise smooth curve $c : [0,1] \to B$, h_c is an isometry.*

From now on assume that B is connected and that the fibers of π are totally geodesic. Then all fibers of π are isometric. Fix an origin $o \in B$ and set $F := F_o$. Denote the induced metric on F by g_F.

Let $H \subset \mathrm{Iso}(F)$ be the holonomy group of π at o; that is, H consists of all isometries h_c, where $c : [0,1] \to B$ is a piecewise smooth curve with $c(0) = c(1) = o$.

Fix a closed subgroup $G \subset \mathrm{Iso}(F)$ containing H and let $N \to B$ be the principal bundle with structure group G and fibers

$$N_z := \{h_c \circ g \mid g \in G\}, \quad z \in B,$$

where $c : [0,1] \to B$ is some piecewise smooth curve from o to z. Since $H \subset G$, N_z does not depend on the choice of c. Composition on the right defines the right action of G on N. Horizontal displacement defines a principal connection ω on N: Let $c : [0,1] \to B$ be piecewise smooth and $x \in N_{c(0)}$. Let $h_t : F_{c(0)} \to F_{c(t)}$ be the horizontal displacement along $c|[0,t]$. Then $h_t \circ x$, $0 \le t \le 1$, is the horizontal lift of c to N starting in x.

Let $\mu : G \to \mathrm{Iso}(F)$ be the inclusion. Then the canonical map

$$N \times_\mu F \to M, \quad [x,p] \mapsto x(p),$$

is a diffeomorphism. Moreover, the horizontal distribution induced by ω coincides with the original horizontal distribution of π. In this sense, the Riemannian metric on M is induced from the connection on N and the Riemannian metrics on F and B.

Denote the Laplace operator of F by Δ_F. Since F is closed, $L^2(F)$ is the Hilbert space sum of the eigenspaces of Δ_F. Fix a Hilbert space decomposition

$$L^2(F) = \hat{\bigoplus}_{\alpha \in A} V_\alpha$$

into pairwise orthogonal and G-invariant subspaces such that, for each α, the induced representation $\rho_\alpha : G \to U(V_\alpha)$ is irreducible. Here $U(V_\alpha)$ denotes the group of unitary transformations of V_α. Then each V_α is contained in an eigenspace of Δ_F. In particular, V_α is a finite dimensional Euclidean space. Set

$$W_\alpha := N \times_{\rho_\alpha} V_\alpha .$$

Since V_α is Euclidean and ρ_α is unitary, W_α carries a canonical Hermitian metric. The connection ω on N induces a Hermitian covariant derivative ∇^α on W_α. Denote the corresponding connection Laplacian by Δ_α. In terms of a local orthonormal frame $Y_1, \dots, Y_n$ on B,

$$\Delta_\alpha = -\sum \left(\nabla^\alpha_{Y_j} \nabla^\alpha_{Y_j} - \nabla^\alpha_{\nabla^B_{Y_j} Y_j} \right),$$

where ∇^B denotes the Levi–Civita connection of B.

Recall that V_α consists of functions $f : F \to \mathbb{C}$. Hence

$$\big(\sigma : z \mapsto [k(z), v(z)]\big) \mapsto \big(f : y \mapsto v(z)(k(z)^{-1}(y))\big),$$

where $z = \pi(y)$, is a linear map from sections of W_α to functions on M. Therefore, sections of W_α are in canonical one–to–one correspondence with functions $f : M \to \mathbb{C}$ such that $f \circ h_c \in V_\alpha$ for all piecewise smooth curves $c : [0,1] \to B$ with $c(0) = o$. In particular, via this correspondence

$$L^2(M) = \hat{\oplus}_\alpha L^2(W_\alpha)$$

is a decomposition into pairwise perpendicular closed subspaces.

Let $(U_1, \dots, U_k, X_1, \dots, X_n)$ be a local orthonormal frame of M, where $(U_1, \dots, U_k)$ is a local orthonormal vertical frame and $(X_1, \dots, X_n)$ is the horizontal lift of a local orthonormal frame of B. Then

$$\Delta = -\sum (U_i^2 - \nabla_{U_i} U_i) - \sum (X_j^2 - \nabla_{X_j} X_j)\,. \tag{3.1}$$

Since the fibers are totally geodesic, the vertical part

$$\Delta_{\mathcal{V}} := -\sum (U_i^2 - \nabla_{U_i} U_i) \tag{3.2}$$

of Δ corresponds to the Laplace operator of the fibers. Call

$$\Delta_{\mathcal{H}} := -\sum (X_j^2 - \nabla_{X_j} X_j) \tag{3.3}$$

the horizontal part of Δ.

LEMMA 3.2. *If a smooth section σ of W_α corresponds to a smooth function f on M, then $\Delta_\alpha \sigma$ corresponds to $\Delta_{\mathcal{H}} f$.*

Proof. Let $z_0 \in B$ and $y \in F_{z_0}$ be a point over z_0. Choose $\varepsilon > 0$ smaller than the injectivity radius of B at z_0. Choose a smooth curve $c : [0, 1/2] \to B$ from o to z_0. For $q \in B_\varepsilon(z_0)$ let $c_z : [0,1] \to B$ be the piecewise smooth curve with $c|[0,1/2]) = c$ and such that $c|[1/2,1]$ is the unique shortest geodesic from z_0 to z. Let $\tilde{c}_z$ be the horizontal lift of c_z with $\tilde{c}_z(1/2) = y$.

Let $v_1, \dots, v_s$ be an orthonormal basis of V_α. Then for $z \in B_\varepsilon(z_0)$, the vectors $\sigma_i(z) := [h_{c_z}, v_i]$, $1 \le i \le s$, are an orthonormal basis of the fiber of W_α over z. Moreover, the local frame $(\sigma_1, \dots, \sigma_s)$ is smooth and parallel along the curves c_z. Hence locally, σ can be expressed as a linear combination

$$\sigma(z) = \sum \varphi_i(z)\, \sigma_i(z)$$

with smooth coefficient functions $\varphi_1, \dots, \varphi_s$ and

$$(\Delta_\alpha \sigma)(z_0) = \sum (\Delta_B \varphi_i)(z_0)\, \sigma_i(z_0),$$

where Δ_B denotes the Laplace operator of B.

By the definition of c_z, $y_0 = h_{c_z}^{-1}(\tilde{c}_z(1)) \in F$ is independent of $z \in B_\varepsilon(z_0)$. Hence the function f corresponding to σ satisfies

$$f(\tilde{c}_z(1)) = \sum \varphi_i(z)\, v_i(y_0).$$

Therefore,

$$(\Delta_{\mathcal{H}} f)(y) = \sum (\Delta_B \varphi_i)(z_0)\, v_i(y_0).$$

Hence $\Delta_\alpha \sigma$ corresponds to $\Delta_{\mathcal{H}} f$. □

For each α, there is an eigenvalue λ_α of Δ_F such that V_α is contained in the corresponding eigenspace of Δ_F. Therefore,

$$\Delta f = \lambda_\alpha f + \Delta_{\mathcal{H}} f$$

for any smooth function f such that $f \circ h_c \in V_\alpha$ for all piecewise smooth curves $c : [0,1] \to B$ with $c(0) = o$. Now list the eigenvalues of Δ_α in increasing order and include any eigenvalue as often as its multiplicity,

$$\lambda_{\alpha,0} \leq \lambda_{\alpha,1} \leq \lambda_{\alpha,2} \leq \dots$$

Then by the above, the spectrum of Δ is given by

$$\lambda_\alpha + \lambda_{\alpha,k}\,, \quad \alpha \in A\,,\ k \geq 0\,,$$

where each eigenvalue occurs as often as its multiplicity demands.

My aim is a criterion for isospectrality of Riemannian manifolds. Let $\pi : M \to B$ and $\pi' : M' \to B'$ be Riemannian submersions as above with totally geodesic fibers and B and B' connected. Fix decompositions

$$L^2(F) = \hat{\oplus}_{\alpha \in A} V_\alpha \quad \text{and} \quad L^2(F') = \hat{\oplus}_{\alpha \in A} V'_\alpha$$

as above (and with the same set A of indices). The following criterion is a variation of the general criterion of Gordon in Theorem 2.1.

THEOREM 3.3. *Suppose that for each $\alpha \in A$,*

1. $\lambda_\alpha = \lambda'_\alpha$ and

2. Δ_α and Δ'_α are isospectral.

Then M and M' are isospectral.

Let B be a closed and connected Riemannian manifold and $N \to B$ be a principal bundle with structure group G, where G is a compact Lie group. Let F be a closed Riemannian manifold and $\mu : G \to \mathrm{Iso}(F)$ be a representation. Set $M := N \times_\mu F$.

Given a principal connection ω on N, there is precisely one Riemannian metric g_ω on M such that the natural projection $\pi : M \to B$ is a Riemannian submersion with totally geodesic fibers isometric to F and horizontal distribution induced by ω, see Theorem 9.59 in [Bes]. Extending Theorem 2.2, I obtain the following criterion for manifolds to be isospectral.

THEOREM 3.4. *Let F and F' be closed Riemannian manifolds and $\mu : G \to \mathrm{Iso}(F)$ and $\mu' : G \to \mathrm{Iso}(F')$ be morphisms. Assume that there is a unitary isomorphism $\tau : L^2(F) \to L^2(F')$ intertwining the corresponding right representations of G and the Laplacians of F and F'. Let ω and ω' be principal connections on N. Suppose that for each irreducible unitary representation ρ of G there is an automorphism $\Phi_\rho : N \to N$ such that*

1. Φ_ρ factors over an isometry of B.

2. $\rho_* \circ \omega' \circ \Phi_{\rho *} = \rho_* \circ \omega$.

Then $(M = N \times_\mu F, g_\omega)$ and $(M' = N \times_{\mu'} F', g_{\omega'})$ are isospectral.

My proof of Theorem 3.4 is different from the proof of Theorem 2.2 in [Sch1]. I use a correspondence between the Laplacian of M and the Laplacians of certain vector bundles.

REMARKS 3.5. 1) In Theorems 2.2 and 3.4, suppose that $F = F'$ with the same G-action. Then it would be overkill to assume that there is an automorphism $\Phi : N \to N$ which factors over an isometry $\phi : B \to B$ such that $\omega' \circ \Phi_* = \omega$. This would imply that (M, g_ω) and $(M', g_{\omega'})$ are isometric. One of the highlights in the work of Schueth is the construction of non-trivial examples where Theorem 2.2 applies, see Chapter 2 of [Sch1] and Section 4 below.

Theorem 3.4 explains why tori are crucial in the constructions of isospectral manifolds of Gordon and Schueth: If $F = F'$ with the same G-action and if G has an irreducible unitary representation ρ with injective differential, then $\omega' \circ \Phi_{\rho *} = \omega$ by Assumption 2 of Theorem 3.4, and then Φ_ρ is an isometry between (M, g_ω) and $(M', g_{\omega'})$.

Actually Assumption 2 in Theorem 3.4 can be relaxed somewhat: Only the irreducible unitary representations ρ of G which occur in the natural representation of G on $L^2(F)$ need to be considered.

2) This is a good place to look back at the proof of Sunada's theorem (in Bérard's version Theorem 1.3) at the end of Section 1: There $N \to B = G\backslash N$ is a Riemannian covering. The fibers $F = \Gamma\backslash G$ and $F' = \Gamma'\backslash G$ are finite sets of the same cardinality, but, in the interesting cases, with inequivalent G-actions. By assumption, there is a unitary isomorphism $L^2(F) \to L^2(F')$ intertwining the right representations of G. The Laplacians on F and F' are trivial since F and F' are finite sets. Hence the assumption on the Laplacians in Theorem 3.4 is trivially satisfied. Since $N \to B$ is a covering, there is only one G-connection on this G-principal bundle, hence $\omega = \omega'$ and we may choose $\Phi_\rho = \mathrm{id}_N$ for all irreducible unitary representations ρ of G to satisfy the assumptions of Theorem 3.4. Hence Theorem 3.4 unifies Sunada's method with the method of Gordon and Schueth.

Proof of Theorem 3.4. Let $\tau : L^2(F) \to L^2(F')$ be a unitary isomorphism intertwining the induced representations of G and the Laplacians of F and F'. As above, fix a Hilbert space decomposition $L^2(F) = \hat{\bigoplus}_{\alpha \in A} V_\alpha$ into pairwise orthogonal subspaces such that, for each α, the representation $\rho_\alpha : G \to U(V_\alpha)$ is irreducible. Set $V'_\alpha = \tau(V_\alpha)$.

For each $\alpha \in A$, there exist a G–equivariant bundle isomorphism $\Phi_\alpha : N \to N$ satisfying Assumptions 1 and 2 of Theorem 3.4 with $\rho = \rho_\alpha$. Moreover,

$$W_\alpha := N \times_{\rho_\alpha} V_\alpha \quad \text{and} \quad W'_\alpha := N \times_{\rho_\alpha} V'_\alpha$$

are Hermitian vector bundles over B. The map

$$\Psi_\alpha : W_\alpha \to W'_\alpha , \quad \Psi_\alpha([x,v]) = [\Phi_\alpha(x), \tau(v)] ,$$

is well defined, hence a vector bundle isomorphism which preserves the metrics of the fibers. Now by Assumption 1, Ψ_α factors over an isometry of the base manifolds, hence Ψ_α induces a unitary transformation $L^2(W_\alpha) \to L^2(W'_\alpha)$.

Denote the covariant derivatives on W_α and W'_α induced by ω and ω' by ∇_α and ∇'_α, respectively. Now ∇_α and ∇'_α only depend on $\rho_{\alpha *} \circ \omega$ and $\rho_{\alpha *} \circ \omega'$, respectively. Hence by Assumption 2, $\Psi^*_\alpha(\nabla'_\alpha) = \nabla_\alpha$. It follows that Ψ_α intertwines the Laplacians Δ_α and Δ'_α. Hence by Theorem 3.3, (M, g_ω) and $(M', g_{\omega'})$ are isospectral. □

The main point in the proof of Theorem 3.4 is the decomposition of the Laplacian into vertical and horizontal part, $\Delta = \Delta_{\mathcal{V}} + \Delta_{\mathcal{H}}$, as in (3.1). There is a similar formula in the case where the metric in the vertical direction is rescaled by a positive function ψ on B,

$$g_\psi := \psi^2 g_{\mathcal{V}} + g_{\mathcal{H}} ,$$

where $g_{\mathcal{V}}$ and $g_{\mathcal{H}}$ denote the restrictions of the original metric on M to the vertical and horizontal distribution. The corresponding Laplacian Δ_ψ is given by

$$\Delta_\psi = \frac{1}{\psi^2} \Delta_{\mathcal{V}} + \Delta_{\mathcal{H}} - kZ ,$$

where $k = \dim F$ and Z is the horizontal lift of $\operatorname{grad} \psi$.

Note that for the rescaled metric, the fibers are totally umbilic submanifolds with second fundamental form $S(U,V) = -\langle U, V \rangle Z$ and the horizontal displacements are homotheties. Vice versa, suppose the horizontal displacements of a Riemannian submersion with compact fibers are homotheties. Then rescaling vertically so that the fibers are of constant volume one obtains a Riemannian submersion with totally geodesic fibers.

Theorem 3.4 extends to this situation: The data are connections ω and ω' as in Theorem 3.4 and functions ψ and ψ' on B. Rescaling the metrics g_ω and $g_{\omega'}$ in the vertical directions by ψ and ψ' one obtain metrics $g_{\psi,\omega}$ and $g_{\psi',\omega'}$ on M and M', respectively. Then $(M, g_{\psi,\omega})$ and $(M', g_{\psi',\omega'})$ are isospectral provided Assumptions 1 and 2 of Theorem 3.4 hold and, in addition, $\psi' \circ \phi_\rho = \psi$ for all ρ. Compare Theorem 4.3 in [Sch1] for a special version (where I subsume the rescaling of the metric g_B on the base there into g_B here).

4. An application. In this work I discuss just one application of Theorem 3.4, others are possible. In my application, I rely on results from Chapter 2 in [Sch1].

THEOREM 4.1. *Suppose that F has dimension ≥ 3 and constant scalar curvature and that T^2 acts effectively on F. Then $M = S^2 \times F$ admits isospectral Riemannian metrics for which the projection $M \to S^2$ is a Riemannian submersion with totally geodesic fibers isometric to F and base S^2 with the standard metric.*

In particular, there is a pair of isospectral metrics on $S^2 \times S^3$. At the time of its discovery in the spring of 2000, this was the lowest dimensional known simply connected closed Riemannian manifold which admits a pair of isospectral metrics.

In the proof of Theorem 4.1, I use ideas and results from Chapter 2 of [Sch1]. Let $B = S^2 = \{z \in \mathbb{R}^3 \mid |z| = 1\}$ and $N = S^2 \times T^2$, where T^2 acts on N by right multiplication on the factor T^2. Let $F = F'$ be a compact Riemannian manifold of dimension ≥ 3 and constant scalar curvature and $\mu = \mu' : T^2 \times F \to F$ be an effective isometric action. In particular, $M = M' := N \times_\mu F$ is diffeomorphic to $S^2 \times F$.

Consider the following two pairs of symmetric matrices,

$$c_1 = c_1' = \begin{pmatrix} -1 & 0 & 0 \\ 0 & 0 & 0 \\ 0 & 0 & 1 \end{pmatrix}, \quad c_2 = \begin{pmatrix} 0 & 1 & 0 \\ 1 & 0 & 1 \\ 0 & 1 & 0 \end{pmatrix}, \quad c_2' = \begin{pmatrix} 0 & & \sqrt{2} \\ 0 & 0 & 0 \\ \sqrt{2} & 0 & 0 \end{pmatrix}.$$

The remarkable property of these matrices is that for any pair $\zeta = (\zeta_1, \zeta_2)$ of real numbers, the matrices $\zeta_1 c_1 + \zeta_2 c_2$ and $\zeta_1 c_1' + \zeta_2 c_2'$ have the same eigenvalues, that is, there is a matrix $A_\zeta \in SO(3)$ such that

$$A_\zeta(\zeta_1 c_1 + \zeta_2 c_2)A_\zeta^{-1} = \zeta_1 c_1' + \zeta_2 c_2',$$

but A_ζ cannot be chosen independently of ζ, see Proposition 2.4 in [Sch1].

Define 1–forms $\lambda = (\lambda_1, \lambda_2)$ and $\lambda' = (\lambda_1', \lambda_2')$ on S^2 with values in the Lie algebra $\mathfrak{t}^2 \cong \mathbb{R}^2$ of T^2 by

$$\lambda_i(z)(Z) = \langle c_i(z) \times z, Z\rangle \quad \text{and} \quad \lambda_i'(z)(Z) = \langle c_i'(z) \times z, Z\rangle, \tag{4.1}$$

where $\langle .,. \rangle$ denotes the inner product of $\mathbb{R}^3$. Note that the dependence on the foot point $z \in S^2$ is quadratic. Let $\zeta = (\zeta_1, \zeta_2)$ be a pair of real numbers. Then A_ζ as above is an isometry of S^2. Considering ζ as an element of the dual of $\mathfrak{t}^2$,

$$\zeta \circ \lambda' \circ A_{\zeta *} = \zeta \circ \lambda$$

by the above property of A_ζ.

Define connection forms ω and ω' on N with values in $\mathfrak{t}^2$ by

$$\omega((Z, V)) = \lambda(Z) + V \quad \text{and} \quad \omega'((Z, V)) = \lambda'(Z) + V,$$

where vector fields on N are written in the form (Z, V) with Z tangent to $B = S^2$ and V tangent to T^2. Then for any element ζ in the dual of $\mathfrak{t}^2$,

$$\zeta \circ \omega' \circ \Phi_{\zeta *} = \zeta \circ \omega,$$

where $\Phi_\zeta = (A_\zeta, \mathrm{id})$. It follows that the connection forms ω and ω' satisfy the assumptions of Theorem 3.4. Hence (M, g_ω) and $(M, g_{\omega'})$ are isospectral.

It remains to show that (M, g_ω) and $(M, g_{\omega'})$ are not isometric. To that end I follow [Sch1] and study the loci of maximal scalar curvature. I start with some more general considerations.

Let M and B be Riemannian manifolds and $\pi : M \to B$ be a Riemannian submersion with totally geodesic fibers. Let s_M, s_B and $s_\mathcal{V}$ be the scalar curvature of M, B and fibers of π, respectively. Then in $y \in M$,

$$s_M(y) = s_B(\pi(y)) + s_\mathcal{V}(y) - \frac{1}{4} \sum_{i,j} \left\| [Y_i, Y_j]_\mathcal{V}(y) \right\|^2, \tag{4.2}$$

where $Y_1, \ldots, Y_n$ is a local orthonormal frame about y of the horizontal distribution $\mathcal{H}$ and where the index $\mathcal{V}$ indicates the vertical component, see Corollary 9.37 in [Bes].

Now I consider the case where $N = B \times T^2$ and $M = N \times_\mu F$. I also assume that ω is of the form $\omega((Z, V)) = \lambda(Z) + V$, where λ is a 1–form on B with values in $\mathfrak{t}^2$.

For a left invariant vector field V on T^2, let V^* be the vector field on F defined by

$$V^*(p) = \partial_t(e^{tV}(p))|_{t=0}.$$

Then the horizontal distribution on M induced by ω is given by the following rule:

$$(Z, W) \text{ is horizontal} \iff W(z, p) = \{-\lambda_z(Z(z, p))\}^*(p) \text{ for all } (z, p).$$

In short: (Z, W) is horizontal if and only if $W = -\lambda(Z)^*$.

In order to apply (4.2), one needs to compute the vertical part of the Lie bracket of horizontal lifts of vector fields X and Y on B,

$$[(X, -\lambda(X)^*), (Y, -\lambda(Y)^*)]_\mathcal{V}.$$

Then X, Y and λ only depend on $z \in B$. Since T^2 is abelian and $\lambda(X)$ and $\lambda(Y)$ are left invariant vector fields on T^2, $[\lambda(X)^*, \lambda(Y)^*] = 0$. By the definition of the horizontal distribution, the vertical part of $([X, Y], 0)$ is $\lambda([X, Y])^*$. Therefore,

$$\begin{aligned}
[(X, -\lambda(X)^*),&(Y, -\lambda(Y)^*)]_\mathcal{V} \\
&= (0, \{-X(\lambda(Y)) + Y(\lambda(X)) + \lambda([X, Y])\}^*) \\
&= (0, \{-d\lambda(X, Y)\}^*).
\end{aligned}$$

Conclusion: If (z,p) is a point in M, if $(Z_1,\ldots,Z_n)$ is a local orthonormal frame about z in B and $((Z_1,-\lambda(Z_1)^*),\ldots,(Z_n,-\lambda(Z_n)^*))$ is its horizontal lift to M, then

$$\sum_{i,j}\left\|[(Z_i,-\lambda(Z_i)^*),(Z_j,-\lambda(Z_j)^*)]_{\mathcal{V}}(z,p)\right\|^2$$

$$=\sum_{i,j}\left\|\{(d\lambda(z))(Z_i(z),Z_j(z))\}^*(p)\right\|^2.$$

I now return to the case in question, where $B=S^2$. Let F^* be the set of points in F with a twodimensional T^2–orbit. That is $p\in F^*$ if and only if $\{V^*(p)\mid V\in\mathfrak{t}^2\}$ is twodimensional. It is immediate from the slice theorem that $F\setminus F^*$ has codimension at least two in F. Now $B=S^2$, hence up to order, there is only one summand $\left\|\{(d\lambda)_z(Z_i(z),Z_j(z))\}^*(p)\right\|^2$. For $p\in F^*$ it vanishes precisely in the points (z,p) where $d\lambda(z)=0$. Now for λ, λ' as in (4.1),

$$\{z\in S^2\mid d\lambda(z)=0\}=\{z\in S^2\mid z_1=-z_3 \text{ or } z=(\pm1/\sqrt{2},0,\pm1/\sqrt{2})\}$$

and

$$\{z\in S^2\mid d\lambda'(z)=0\}=\{(0,\pm1,0)\},$$

see Proposition 2.10 in [Sch1]. It follows from (4.2) and the fact that s_B and $s_{\mathcal{V}}$ are constant that in the first case, the locus where the scalar curvature attains its maximum has components of codimension one, in the second case the locus has codimension two. Hence (M,g_ω) and $(M,g_{\omega'})$ are not isometric.

5. Concluding remarks. Independently of this work and at about the same time, Gordon extended the work from [Sch1] in a different direction. Building on Gordon's extension, Schueth obtained isospectral metrics on fivedimensional spheres and also on $S^2\times S^3$, see [Sch2]. An interpretation of Schueth's formulation of Gordon's extension is as follows.

THEOREM 5.1. *Let M and M' be closed Riemannian manifolds. Suppose a compact Lie group G acts isometrically on M and M'. Assume that, for each irreducible unitary representation ρ of G, there is a G-equivariant diffeomorphism $F_\rho: M\to M'$ such that*

1. *F_ρ is isometric modulo $\ker_\rho$-orbits.*
2. *F_ρ preserves volume.*

Then M and M' are isospectral.

Sketch of proof. For each unitary representation ρ of G, let V_ρ and V'_ρ be the corresponding isotypical components of $L^2(M)$ and $L^2(M')$, respectively. Since G acts by isometries, these subspaces are invariant under the Laplacians Δ and Δ'. Assumptions (1) and (2) can be used to show that there is a G-equivariant unitary isomorphism $\tau: V_\rho\to V'_\rho$ intertwining Δ and Δ'. □

For a somewhat different argument see [Sch2]. I haven't worked out the details of the extension of this result to manifolds of the form $M \times_G F$ as in Theorem 3.4. To that end, it is probably more convenient to follow the original arguments of Bérard in [Be1, Be2] more closely and to show directly that the Laplacians of manifolds of this kind, under the appropriate assumptions, are unitarily equivalent.

REFERENCES

[BGM] M. Berger, P. Gauduchon, and E. Mazet. *Le Spectre d'une variété Riemannienne.* Springer LNM 194, Berlin [a.o.], 1971.

[Be1] P. Bérard. Transplantation et isospectralité I. *Math. Annalen* **292** (1992), 547–559.

[Be2] P. Bérard. Transplantation et isospectralité II. *J. London. Math. Soc.* **48** (1993), 565–576.

[Bes] A.L. Besse. *Einstein manifolds.* Springer–Verlag 1987.

[DTG] D. DeTurck and C. Gordon. Isospectral Riemannian metrics and potentials. With an appendix by Kyung Bai Lee. *Comm. Pure Appl. Math.* **42** (1989), 1067–1095.

[Go1] C.S. Gordon. Isospectral closed Riemannian manifolds which are not locally isometric. *J. Differential Geometry* **37** (1993), 639–649.

[Go2] C.S. Gordon. Isospectral closed Riemannian manifolds which are not locally isometric: II. In: *Geometry of the Spectrum* (R. Brooks, C. Gordon, and P. Perry, eds.), Contemp. Math. **173** (1994), 121–131.

[Go3] C.S. Gordon. Survey of isospectral manifolds. Dillen, Franki J.E. (ed.) et al., *Handbook of differential geometry.* Volume I. Amsterdam: North-Holland. pp. 747–778 (2000).

[Go4] C.S. Gordon. Isospectral deformations of metrics on spheres. *Inventiones Math.* **145** (2001), 317–331.

[GWW] C.S. Gordon, D. Webb, and S. Wolpert. Isospectral plane domains and surfaces via Riemannian orbifolds. *Inventiones Math.* **110** (1992), 1–22.

[GWi] C.S. Gordon and E. Wilson. Isospectral deformations of compact solvmanifolds. *J. Differential Geometry* **19** (1984), 241–256.

[Ik1] A. Ikeda. On spherical space forms which are isospectral but not isometric. *J. Math. Soc. Japan* **35** (1983), 437–444.

[Ik2] A. Ikeda. On space forms of real Grassmann manifolds which are isospectral but not isometric. *Kodai Math. J.* **20** (1997), 1–7.

[Mil] J. Milnor. Eigenvalues of the Laplace operators on certain manifolds. *Proc. Nat. Acad. Sci. USA* **51** (1964), 542.

[Sch1] D. Schueth. Isospectral manifolds with different local geometries. *J. reine angew. Math.* **534** (2001), 41–94.

[Sch2] D. Schueth. Isospectral metrics on five-dimensional spheres. *J. Differential Geometry* **58** (2001), 87–111.

[Sun] T. Sunada. Riemannian coverings and isospectral manifolds. *Annals of Math.* **121** (1985), 169–186.

[Sza] Z.I. Szabo. Locally non–isometric yet super isospectral spaces. *GAFA* **9** (1999), 185–214.

[Vi] M.F. Vignéras. Variétés riemanniennes isospectrales et non isométriques. *Annals of Math.* **112** (1980), 21–32.

STATISTICAL STABILITY AND TIME-REVERSAL IMAGING IN RANDOM MEDIA

JAMES G. BERRYMAN*, LILIANA BORCEA†,
GEORGE C. PAPANICOLAOU‡, AND CHRYSOULA TSOGKA§

Abstract. Localization of targets imbedded in a heterogeneous background medium is a common problem in seismic, ultrasonic, and electromagnetic imaging problems. The best imaging techniques make direct use of the eigenfunctions and eigenvalues of the array response matrix, as recent work on time-reversal acoustics has shown. Of the various imaging functionals studied, one that is representative of a preferred class is a time-domain generalization of MUSIC (MUltiple SIgnal Classification), which is a well-known linear subspace method normally applied only in the frequency domain. Since statistical stability is not characteristic of the frequency domain, a transform back to the time domain after first diagonalizing the array data in the frequency domain takes optimum advantage of both the time-domain stability and the frequency-domain orthogonality of the relevant eigenfunctions.

1. Introduction. There have been many approaches to estimating target location using seismic, ultrasonic, and electromagnetic imaging methods. Some of the most popular ones in recent years continue to be matched-field processing (Bucker, 1976; Jensen *et al.*, 1994), MUSIC (MUltiple SIgnal Classification) (Schmidt, 1979; Johnson, 1982; Schmidt, 1986; Biondi and Kostov, 1989), and other linear subspace methods (Johnson, 1982; Johnson and DeGraaf, 1982; Cheney, 2001). When the targets are imbedded in heterogeneous media so that significant multiple scattering occurs in the background medium during wave propagation between array and target, the randomness has a different character than that usually envisioned in these traditional analyses. Yet there are a great many applications (Fink, 1997; 1999; Fink *et al.*, 2000; Fink, 2001; Fink and Prada, 2001; ter Haar, 2001) ranging from the biomedical to ocean acoustics to nondestructive evaluation, where imaging is important and where sources of randomness not associated with the targets to be imaged can play havoc with the traditional methods. Time-reversal acoustics (Fink *et al.*, 1989; Jackson and Dowling, 1991; Prada and Fink, 1991) offers part of the answer to these difficult imaging questions, and some significant improvements over these methods for imaging in random media are summarized here.

We have found that methods designed to work well for finding targets in homogeneous media do not necessarily work very well for targets

*Lawrence Livermore National Laboratories, P.O. Box 808 L-200, Livermore, CA 94551-9900 (berryman1@llnl.gov).

†Computational and Applied Mathematics, Rice University, MS 134, 6100 Main Street, Houston, TX 77005-1892 (borcea@caam.rice.edu).

‡Department of Mathematics, Stanford University, Stanford, CA 94305 (papanico@math.stanford.edu).

§CNRS/LMA, 31 Chemin Joseph Aiguier, 13402 Marseille cedex 20, France (tsogka@lma.cnrs-mrs.fr).

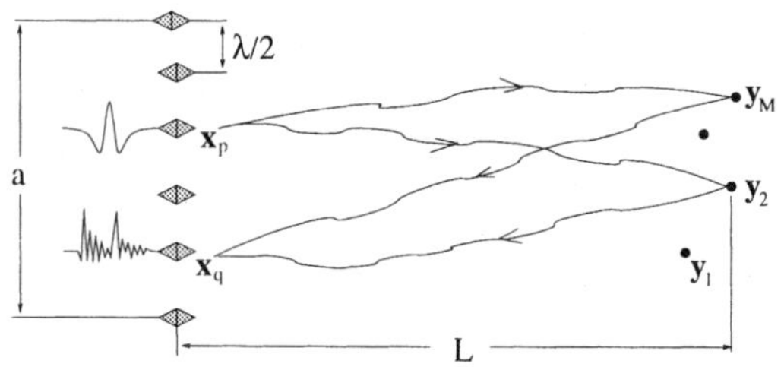

FIG. 1. *Array probing of a randomly inhomogeneous medium containing M small scatterers.*

imbedded in random media. In particular, the fact that the linear subspace methods are normally applied in the frequency domain combined with the fact that statistically stable methods are normally found only in the time domain, forces us to seek different imaging strategies in the random media imaging problems of interest to us here. We find that a set of imaging functionals having the desired characteristics exists, and furthermore that the properties of this set can be completely understood when the time-domain self-averaging — that gives rise to the required statistical stability of the target images — is taken properly into account. We can largely eliminate the undesirable features of the frequency domain methods by making a transform back to the time domain after first diagonalizing sensor array data. While the frequency domain analysis takes optimum advantage of eigenfunction orthogonality of the array data, a transform to the time-domain takes optimum advantage of wave self-averaging which then leads to the statistical stability we require for reliable and repeatable imaging in random media.

We first introduce the imaging problem in the next section. Then we summarize our technical approach. Examples of the cross-range (or bearing) estimates obtained with these methods are presented and then combined with range information from time-delay data to obtain our best estimates and images of target location. The final section provides some pointers to recent related advances in inverse problems and summarizes our conclusions about the imaging methods discussed.

2. Imaging problem. Our analysis assumes that the array has N transducers located at spatial positions $\mathbf{x}_p$, for $p = 1, \ldots, N$. (See Fig. 1.) When used in active mode, the array probes the unknown acoustic medium containing M small scatterers by emitting pulses and recording the time traces of the back-scattered echoes. We call the resulting data set the multistatic array response (or transfer) matrix

$$P(t) = (P_{pq}(t)), \tag{1}$$

where p and q both range over all the array elements. For our simulations, we consider a linear array where two adjacent point transducers are a distance $\lambda/2$ apart, with λ being the carrier (central) wavelength of the probing pulses. Such an arrangement ensures that the collection of transducers behaves like an array having aperture $a = (N-1)\lambda/2$ and not like separate entities, while keeping the interference among the transducers at a minimum (Steinberg, 1983). Our goal is to detect and then localize all M of the targets in the random medium, if possible.

For the numerical examples considered here, we treat ultrasonic imaging problems. Our simulations assume that $\lambda \leq \ell << a = (N-1)\lambda/2 << L$, where λ is the central wavelength, ℓ is a characteristic length scale of the inhomogeneity (like a correlation length), a is the array aperture, and L is the approximate distance to the targets from the array. This is the regime where multipathing, or multiple scattering, is significant even when the standard deviation of sound speed fluctuations is only a few percent. Values used in the codes are $\lambda = 0.5$ mm, $a = 2.5$ mm, and a background wave speed of $c_0 = 1.5$ km/s. More details concerning the simulations may be found in Borcea *et al.* (2002).

Typical array processing methods assume that the targets are far away from the array and, therefore, they look like points. Similarly, the propagation medium is assumed homogeneous and so the observed wavefronts scattered by the targets look like plane waves at the array. Array noise has usually been treated as due either to diffuse sources of white noise coming simultaneously from all directions, or to isolated "noise" having the same types of source characteristics as the targets of interest. But in random media with significant multiple scattering, the resulting "noise" cannot be successfully treated in these traditional ways.

Real-space time-reversal processing of the array response data involves an iterative procedure: sending a signal, recording and storing the scattered return signal, time-reversing and then rebroadcasting the stored signal, with subsequent repetitions. This procedure amounts to using the power method for finding the singular vector of the data matrix having the largest singular value. Alternatively, when the full response/transfer matrix has been measured for a multistatic active array, the resulting data matrix can be analyzed directly by Singular Value Decomposition (SVD) to determine not only the singular vector having the largest singular value, but all singular vectors and singular values — simultaneously (Prada and Fink, 1994; Prada *et al.*, 1996; Mordant *et al.*, 1999).

Imaging is always done using a fictitious medium for the simulated backpropagation that produces these images since the real medium is not known. Its large-scale features could be estimated from other information, such as geological data obtained by seismic methods. For example, migration methods (Claerbout, 1976; Aki and Richards, 1980; Bleistein *et al.*, 2001) can be used, where very large arrays — much larger than those we contemplate using here — are required. However, the small-scale random inhomogeneities are not known and cannot be effectively estimated, so the simplest thing is to ignore them when imaging, and then to use methods that are statistically stable and therefore insensitive to the exact character of these small inhomogeneities.

3. Technical approach. In our simulations, the array response matrix $\widehat{P}(\omega)$ [see definition in (1)] in the frequency domain is symmetric but not Hermitian. In general (as for array elements with nonlinear response characteristics due to clipping at high amplitudes and noise masking at very low amplitudes), it is neither Hermitian nor symmetric, but with slight modifications our methods apply to this case as well. The eigenvectors of $\widehat{P}(\omega)\widehat{P}^H(\omega)$ having unit norm are denoted by $\widehat{\mathbf{U}}_r(\omega)$, for $r = 1, \ldots, N$. The eigenvalues of $\widehat{P}(\omega)\widehat{P}^H(\omega)$ are $\sigma_r^2(\omega)$, with $\sigma_r(\omega)$ being the singular values of $\widehat{P}(\omega)$. The significant singular vectors $\widehat{\mathbf{U}}_r(\omega)$ [*i.e.*, those in the range of $\widehat{P}(\omega)$] have singular values $\sigma_r(\omega) > 0$ for $1 \le r \le M$, where M is either the number of targets, or the size of the array (N) — whichever is smaller. We assume that the number of targets is smaller than the array size N, so that M is in fact the number of distinguishable targets; this assumption is required by the imaging methods we employ (such as MUSIC) as will become clear while presenting the method.

The notation used here is the same as in Borcea *et al.* (2002). We denote by $\widehat{\mathbf{g}}_0(\mathbf{y}, \omega)$ the deterministic source vector observed at the array for a source located at $\mathbf{y}^s$. Then, $\widehat{\mathbf{g}}_0(\mathbf{y}, \omega)$ is given by

$$\widehat{\mathbf{g}}_0(\mathbf{y}^s, \omega) = \begin{pmatrix} \widehat{G}_0(\mathbf{y}^s, \mathbf{x}_1, \omega) \\ \widehat{G}_0(\mathbf{y}^s, \mathbf{x}_2, \omega) \\ \vdots \\ \widehat{G}_0(\mathbf{y}^s, \mathbf{x}_N, \omega) \end{pmatrix}, \tag{2}$$

where $\widehat{G}_0(\mathbf{y}^s, \mathbf{x}_j, \omega)$ is the deterministic two-point Green's function, and $\mathbf{x}_j$ is the location of the j-th array element.

We also define the projection $\mathcal{P}_N\widehat{\mathbf{g}}_0(\mathbf{y}, \omega)$ of $\widehat{\mathbf{g}}_0(\mathbf{y}^s, \omega)$ onto the null-space of $\widehat{P}\widehat{P}^H(\omega)$ by

$$\begin{aligned} \mathcal{P}_N\widehat{\mathbf{g}}_0(\mathbf{y}^s, \omega) &= \widehat{\mathbf{g}}_o(\mathbf{y}^s, \omega) \\ &\quad - \sum_{r=1}^{M} \left[\widehat{\mathbf{U}}_r^H(\omega)\widehat{\mathbf{g}}_o(\mathbf{y}^s, \omega)\right] \widehat{\mathbf{U}}_r(\omega), \end{aligned} \tag{3}$$

for each frequency in the support of the probing pulse $\widehat{f}(\omega)$.

The method we describe here is a time domain variant of MUSIC (Schmidt, 1979; 1986; Lev-Ari and Devaney, 2000; Cheney, 2001) which we label *DOA*, because it gives very stable estimates of the direction of arrival. Frequency domain MUSIC takes a replica (or trial) vector, which is the impulse response or Green's function for a point source at some point in the space, and dots this vector into an observed singular vector at the array. With appropriate normalization, this dot product acts like a direction cosine of the angle between the replica vector and the data vector. If the sum of the squares of these direction cosines is very close to unity, then it is correct to presume that the source point of that replica vector is in fact a target location since it lies entirely in the range of the array response matrix. Crudely speaking, imaging is accomplished by plotting $1/[1-\cos^2(\cdot)]$, which will have a strong peak when the replica source point is close to the target location.

We form the sum

$$\mathcal{G}^{(j)}(\mathbf{y}^s) = \sum_{p=1}^{N} \left| \mathcal{F}_p^{(j)}(\mathbf{y}^s, t_p(\mathbf{y}^s)) \right|^2, \tag{4}$$

with

$$\begin{aligned} \mathcal{F}^{(j)}(\mathbf{y}^s, t) &= \int e^{-i\omega t} \sigma_j(\omega) \widehat{\mathbf{g}}_o(\mathbf{y}^s, \omega) d\omega \\ &- \int e^{-i\omega t} \sigma_j(\omega) \sum_{r=1}^{M} \left[\widehat{\mathbf{U}}_r^H(\omega) \widehat{\mathbf{g}}_o(\mathbf{y}^s, \omega) \right] \widehat{\mathbf{U}}_r(\omega) d\omega, \end{aligned} \tag{5}$$

and display the objective functional

$$\mathcal{R}_{\mathrm{DOA}}(\mathbf{y}^s) = \sum_{j=1}^{M} \frac{\min_{\mathbf{y}^s} \mathcal{G}^{(j)}(\mathbf{y}^s)}{\mathcal{G}^{(j)}(\mathbf{y}^s)}, \tag{6}$$

for points $\mathbf{y}^s$ in the target domain.

The arrival time $t_p(\mathbf{y}^s)$ is the deterministic travel time from the p-th transducer to the search point,

$$t_p(\mathbf{y}^s) = \frac{|\,\mathbf{x}_p - \mathbf{y}^s\,|}{c_0}. \tag{7}$$

4. Examples and range estimation. The set-up for our numerical simulations is shown in Fig. 2. We have computed array response matrices numerically for the two-target configuration illustrated. The probing pulse we use is given by

$$f(t) = -2\pi^2\nu^2 \left(t - \frac{1}{\nu} \right) e^{-\pi^2\nu^2 (t-\frac{1}{\nu})^2}, \tag{8}$$

with central frequency $\nu = 3$MHz, so $\widehat{f}(\omega)$ with $\omega = 2\pi\nu$ is supported over the band of frequencies $0.159 - 7.958$ MHz and the carrier wavelength is $\lambda = 0.5$mm. In this simulation we have two unequal targets, one being bigger and therefore more reflective than the other. The larger target is located at $\mathbf{y}_1 = (8\lambda, 21\lambda)$ and the smaller one at $\mathbf{y}_2 = (4\lambda, 22\lambda)$. The array has aperture $a = 9.5\lambda$ and the transducers are located at

$$\mathbf{x}_p = \left(7\lambda + (p-1)\frac{\lambda}{2}, 4\lambda \right), \qquad \text{for } p = 1, \cdots, 20. \tag{9}$$

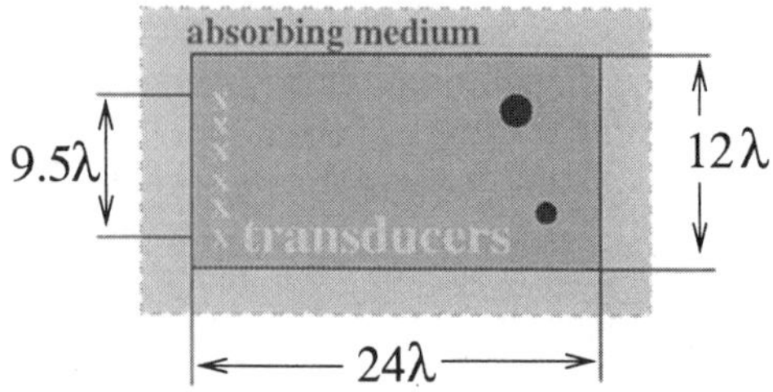

FIG. 2. *The set-up for numerical simulations having two targets, one bigger than the other. The larger target is located at* $y_1 = (8\lambda, 21\lambda)$, *and the smaller one at* $y_2 = (4\lambda, 22\lambda)$. *As indicated, the physical aperture is* 9.5λ.

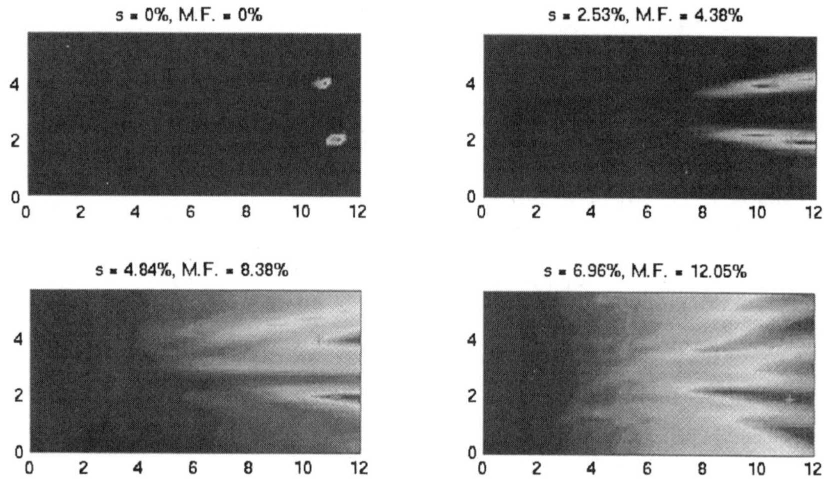

FIG. 3. *The MUSIC central frequency estimate of the location of two targets in random media with different strength of the fluctuations of the sound speed. The exact location of the targets is denoted by green stars (in the color version). The standard deviation s and maximum fluctuations (M.F.) are indicated on the top of each view. The horizontal axis is the range in mm and the vertical axis is the cross-range in mm.*

The ambient medium is either homogeneous or random. Characteristics of the random media are: (1) sound speed $c(\mathbf{x})$ has constant mean $c_0 = 1.5$km/s, (2) correlation length $l = 0.3$mm, and (3) standard deviation s between 1% and 5%. The correlation function for the medium is assumed to be Gaussian.

Examples for frequency-domain MUSIC with two targets are displayed in Fig. 3. It is clear from this Figure that no range information is obtained from frequency-domain objective functionals, and even the cross-range information is often quite haphazard in random media. Lack of statistical stability prevents these imaging approaches from being useful in random media with significant multipathing as considered here. When the real-

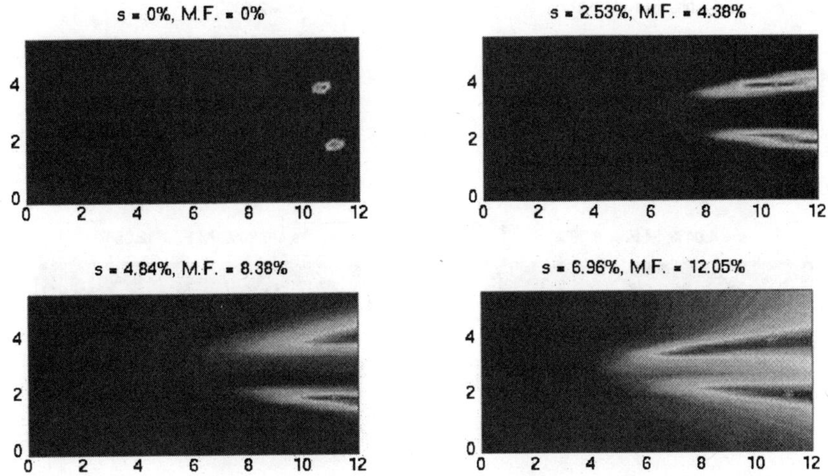

FIG. 4. *The DOA estimate [DOA estimate (6)] of the location of two targets in random media with different strengths of the fluctuations of the sound speed. The exact location of the target is denoted by the green star (in the color version). The standard deviation s and maximum fluctuations (M.F.) are indicated on the top of each view. The horizontal axis is the range in mm and the vertical axis is the cross-range in mm.*

ization of the random medium is changed, the images obtained typically change also — which is what we mean by the phrase "lack of statistical stability" for these methods. Note that this approach works well for homogeneous media, but quickly breaks down when randomness of the velocity field is important.

Examples for time-domain MUSIC with two targets are displayed in Fig. 4. The cross-range results show dramatic improvement over results using other methods (Berryman *et al.*, 2002). Range information is still not to be found here, due to loss of coherence in the random medium; we cannot get exact cancellation at the targets in this situation whereas coherent refocusing is possible in homogeneous media. But the statistical stability of the universal "comet tails" — which was also anticipated by recent theoretical analyses (Blomgren *et al.*, 2002) — is now easily observed. The images are necessarily shown for specific realizations, but the results do not change significantly when the underlying realization of the random medium is changed. This fact has been repeatedly shown in our simulations, and is the main operational characteristic of statistically stable methods.

Target localization also requires an estimate of the range. In the far field, only the arrival time information is useful for this purpose. Arrival time information is present in the singular vectors and can also be averaged (for the same random medium) using multiple copies in the array response matrix for random media — see Borcea *et al.* (2002) — to obtain very stable estimates of arrival times. We will now combine this approach with the time-domain methods to obtain well-localized images of the targets.

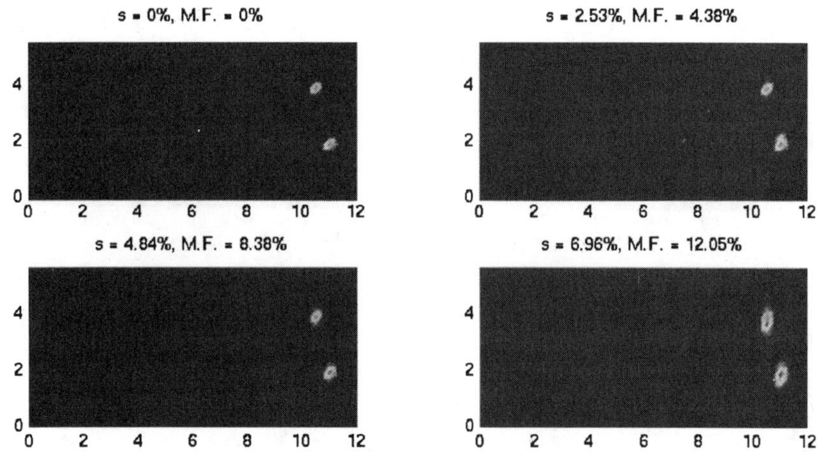

FIG. 5. *The SAT estimate for two targets.*

For each search point $\mathbf{y}^s$, we compute the objective functional

$$\mathcal{R}_{\mathrm{SAT}}(\mathbf{y}^s) = \sum_{j=1}^{M} \frac{\min_{\mathbf{y}^s} \mathcal{G}^{(j)}_{\mathrm{SAT}}(\mathbf{y}^s)}{\mathcal{G}^{(j)}_{\mathrm{SAT}}(\mathbf{y}^s)}, \tag{10}$$

where

$$\mathcal{G}^{(j)}_{\mathrm{SAT}}(\mathbf{y}^s) = \sum_{p=1}^{N} \left| \mathcal{F}^{(j)}_p(\mathbf{y}^s, t_p(\mathbf{y}^s)) \right|^2 \left[\tau^{(j)}_p - t_p(\mathbf{y}^s) \right]^2 . \tag{11}$$

Here $\mathcal{F}^{(j)}(\mathbf{y}^s, t)$ is defined by (5), $t_p(\mathbf{y}^s)$, for $p = 1, \ldots, N$, are the deterministic arrival times given by (7) and $\tau^{(j)}_p$, for $p = 1, \ldots, N$, and $j = 1, \ldots, M$, are the computed arrival times. We call (10) the Subspace Arrival Time (SAT) estimator.

Examples of SAT (or time-domain MUSIC with arrival time estimates from the averaged singular vectors) for two targets are displayed in Fig. 5. This method is statistically stable and gives good estimates of the target locations. These localization results have degraded the least of all those considered (Borcea *et al.*, 2002; Berryman *et al.*, 2002) at the highest values of the random fluctuations.

5. Summary and conclusions. The methods presented here, although differing substantially in detail, have much in common with adjoint field methods that have been used recently in both acoustical imaging (Natterer and Wübbeling, 2001) and in electromagnetic tomography (Dorn *et al.*, 1999; 2002). The use of singular value decomposition of the data matrix is also very close in spirit to various methods for evaluating the resolution (Berryman, 2000) and the sensitivity (Dorn *et al.*, 2002) of tomographic reconstruction methods.

For imaging applications in randomly inhomogeneous acoustical media, the results summarized in this paper lead us to the following conclusions: (1) Single frequency methods (including MUSIC) are not statistically stable, and therefore cannot be used without modification in the presence of significant amounts of spatial heterogeneity in the acoustic wave speed distribution. (2) In contrast, time domain methods are statistically stable for any objective functional having the characteristic that the random Green's functions appear in Hermitian conjugate pairs of gg^* (Borcea *et al.*, 2002), because large random phases cancel precisely. This result has been shown here to be true for DOA, and is expected to be true more generally. (3) The DOA gives only cross-range information. Range information must be obtained separately.

To isolate the targets in random media, we need either multiple views (using multiple arrays) so we can triangulate, or we need to extract a direct measure of range from the data. In the SAT examples shown here, we used arrival time for the range estimation.

Acknowledgments. We thank Biondo Biondi, Jim Candy, Dave Chambers, and Bill Curry for helpful comments on the work and the presentation. Work of J.G.B. performed under the auspices of the U.S. Department of Energy by the University of California Lawrence Livermore National Laboratory under contract No. W-7405-ENG-48 and supported specifically by the LLNL Laboratory Directed Research and Development Program and Environmental Management Sciences Program. The work of L.B. was partially supported by the National Science Foundation under grant number DMS-9971209 and by DOE DE-FG03-00ER25424. The work of G.C.P. was supported by grants AFOSR F49620-01-1-0465, NSF DMS-9971972, DOE DE-FG03-00ER25424, and ONR N00014-02-1-0088.

REFERENCES

Aki K. and Richard P.G., 1980, *Quantitative Seismology: Theory and Methods*, Vol. II, Freeman, New York.

Berryman J.G., 2000, Analysis of approximate inverses in tomography I. Resolution analysis of common inverses: *Optimization Eng.*, **1**, 87–115.

Berryman J.G., Borcea L., Papanicolaou G.C., and Tsogka C., 2002, Statistically stable ultrasonic imaging in random media: *J. Acoust. Soc. Am.*, **112**, 1509–1522.

Biondi B.L. and Kostov C., 1989, High-resolution velocity spectra using eigenstructure methods: *Geophysics*, **54**, 832–842.

Bleistein N., Cohen J.K., and Stockwell J.W., Jr., 2001, *Mathematics of Multidimensional Seismic Imaging, Migration, and Inversion*, Springer, New York.

Blomgren P., Papanicolaou G.C., and Zhou H., 2002, Super-resolution in time-reversal acoustics: *J. Acoust. Soc. Am.*, **111**, 238–248.

Borcea L., Papanicolaou G.C., Tsogka C., and Berryman J.G., 2002, Imaging and time reversal in random media: *Inverse Problems*, **18**, 1247–1279.

BUCKER H.P., 1976, Use of calculated sound field and matched-field detection to locate sound sources in shallow water: *J. Acoust. Soc. Am.*, **59**, 368–373.

CHENEY M., 2001, The linear sampling method and the MUSIC algorithm: *Inverse Problems* **17**, 591–596.

CLAERBOUT J.F., 1976, *Fundamentals of Geophysical Data Processing with Applications to Petroleum Prospecting*, McGraw-Hill, New York.

DORN O., BERTETE-AGUIRRE H., BERRYMAN J.G., AND PAPANICOLAOU G.C., 1999, A nonlinear inversion method for 3D-electromagnetic imaging using adjoint fields: *Inverse Problems*, **15**, 1523–1558.

DORN O., BERTETE-AGUIRRE H., BERRYMAN J.G., AND PAPANICOLAOU G.C., 2002, Sensitivity analysis of a nonlinear inversion method for 3D electromagnetic imaging in anisotropic media: *Inverse Problems*, **18**, 285–317.

FINK M., 1997, Time reversal acoustics: *Physics Today*, **50**(3), 34–40.

FINK M., 1999, Time-reversed acoustics: *Scientific American*, **281**(11), 91–97.

FINK M., 2001, Chaos and time-reversed acoustics: *Physica Scripta*, **T90**, 268–277.

FINK M., CASSEREAU D., DERODE A., PRADA C., ROUX P., TANTER M., THOMAS J.-L., AND WU F., 2000, Time-reversed acoustics: *Rep. Prog. Phys.*, **63**, 1933–1995.

FINK M. AND PRADA C., 2001, Acoustic time-reversal mirrors: *Inverse Problems*, **17**, R1–R38.

FINK M., PRADA C., AND WU F., 1989, Self-focusing in inhomogeneous media with time reversal acoustics mirrors: in McAvoy, B.R., Ed., Proc. IEEE Ultrason. Symp. 1989, Vol. 2, pp. 681–686.

JACKSON D.R. AND DOWLING D.R., 1991, Phase conjugation in underwater acoustics: *J. Acoust. Soc. Am.*, **89**, 171–181.

JENSEN F.B., KUPERMAN W.A., PORTER M.B., AND SCHMIDT H., 1994, *Computational Ocean Acoustics*, AIP Press, New York.

JOHNSON D.H., 1982, The application of spectral estimation methods to bearing estimation problems: *Proc. IEEE*, **70**, 1018–1028.

JOHNSON D.H. AND DEGRAAF S.R., 1982, Improving the resolution of bearing in passive sonar arrays by eigenvalue analysis: *IEEE Trans. Acoustics, Speech, Signal Proc.*, **ASSP-30**, 638–647.

LEV-ARI H. AND DEVANEY A.J., 2000, The time-reversal technique reinterpreted: Subspace-based signal processing for multi-static target location: *IEEE Sensor Array and Multichannel Signal Processing Workshop, Cambridge, MA, March, 2000*, pp. 509–513.

MORDANT N., PRADA C., AND FINK M., 1999, Highly resolved detection and selective focusing in a waveguide using the D.O.R.T. method: *J. Acoust. Soc. Am.*, **105**, 2634–2642.

NATTERER F. AND WÜBBELING F., *Mathematical Methods in Image Reconstruction*, SIAM, Philadelphia, Pennsylvania, 2001, p. 137.

PRADA C. AND FINK M., 1994, Eigenmodes of the time reversal operator: A solution to selective focusing in multiple-target media: *Wave Motion*, **20**, 151–163.

PRADA C., MANNEVILLE F., SPOLIANSKY D., AND FINK M., 1996, Decompostition of the time reversal operator: Detection and selective focusing on two scatterers: *J. Acoust. Soc. Am.*, **99**, 2067–2076.

PRADA C., WU F., AND FINK M., 1991, The iterative time reversal mirror: A solution to self-focusing in the pulse echo mode: *J. Acoust. Soc. Am.*, **90**, 1119–1129.

SCHMIDT R.O., 1979, Multiple emitter location and signal parameter estimation: in Proc. RADC Spectrum Estimation Workshop, Rome, New York, October, 1979, Rome Air Development Center, 243–258.

SCHMIDT R.O., 1986, Multiple emitter location and signal parameter estimation: *IEEE Trans. Antennas Propag.*, **AP-34**, 276–281.

STEINBERG B.D., 1983, *Microwave Imaging and Large Antenna Arrays*, Wiley, New York.

TER HAAR G., 2001, Acoustic surgery: *Physics Today*, **54** (12), 29–34.

A REVIEW OF SELECTED WORKS ON CRACK IDENTIFICATION

KURT BRYAN* AND MICHAEL S. VOGELIUS†

Abstract. We give a short survey of some of the results obtained within the last 10 years or so concerning crack identification using impedance imaging techniques. We touch upon uniqueness results, continuous dependence results, and computational algorithms.

1. The forward problem. Consider first the two-dimensional forward problem, which has certain special features not found in higher dimensions. Let Ω be a bounded, simply connected domain in $\mathbb{R}^2$ with smooth boundary and $\gamma : \bar{\Omega} \to \mathbb{R}$ a bounded function with $\inf_\Omega \gamma \geq \delta > 0$; for the moment we assume that γ is real-analytic, although this restriction will later be relaxed. The domain Ω represents the object in which we wish to detect cracks and γ is the *reference* or *background* conductivity, considered known a priori.

Unless otherwise noted, we define a *crack* in Ω as a curve σ contained in Ω which can be parameterized by a twice continuously differentiable map from $[0,1] \to \Omega$ with non-vanishing derivative; we also require that σ does not self-intersect. We use $\Sigma = \cup_{k=1}^n \{\sigma_k\}$ to denote a collection of n cracks. We assume that the cracks are pairwise disjoint. Note that Σ may be empty.

Let ϕ denote an applied electrical potential on $\partial\Omega$. If we assume that the collection of cracks Σ is *perfectly insulating* (completely blocking the flow of electrical current) then the electrical potential v inside Ω satisfies

$$\begin{aligned} \nabla \cdot (\gamma \nabla v) &= 0 \qquad && \text{in } \Omega \setminus \Sigma, \\ \gamma \frac{\partial v}{\partial \nu} &= 0 && \text{on } \Sigma \\ v &= \phi && \text{on } \partial\Omega \end{aligned} \tag{1}$$

where ν is a unit normal vector field on Σ, consistently oriented on each crack (e.g., if a crack σ is parameterized by $c : [0,1] \to \mathbb{R}^2$, then take $\nu(c(t)) = (c'(t))^{\perp}/|c'(t)|$ where $\perp$ denotes a counter-clockwise rotation through an angle $\pi/2$.) The inverse problem of interest is to determine Σ from one or more pairs of boundary-voltage and current data, $(\phi, \gamma\frac{\partial v}{\partial \nu})$.

The boundary value problem (1) with insulating cracks is equivalent to a problem involving perfectly conducting cracks, and this framework is slightly preferable for the analysis that follows. By a *perfectly conducting* crack we mean a crack which maintains constant electrical potential along

---

*Rose-Hulman Institute of Technology, Terre Haute, IN 47803, USA.

†Rutgers University, New Brunswick, NJ 08903, USA. This work was partially supported by the National Science Foundation under grant DMS-00-72556.

its length. The boundary value problem (1) can be transformed into an equivalent problem for perfectly conducting cracks by considering the "γ-harmonic conjugate" of v. Specifically, let u be a function related to v by $(\nabla u)^{\perp} = \gamma \nabla v$. It is not difficult to verify the existence of u, given that v satisfies (1) with $\gamma \frac{\partial v}{\partial \nu} = 0$ on Σ. The function u is determined only up to an additive constant.

If v is a solution to (1) then the function u satisfies the boundary value problem

$$
\begin{aligned}
\nabla \cdot (\gamma^{-1} \nabla u) &= 0 && \text{in } \Omega \setminus \Sigma, \\
u &= c_k && \text{on } \sigma_k \\
\gamma^{-1} \frac{\partial u}{\partial \nu} &= \psi && \text{on } \partial\Omega
\end{aligned}
\tag{2}
$$

where $\psi = \frac{\partial \phi}{\partial s}$, ν is an outward unit normal vector field on $\partial\Omega$, and $s = (\nu)^{\perp}$. The constants c_k are determined by

$$
c_k = -\int_p^{\sigma_k} \gamma \frac{\partial v}{\partial \nu}\, ds + u(p)
$$

where $\nu = -s^{\perp}$, p is a fixed point on $\partial\Omega$ (or in Ω) and the integration is carried out along any path connecting p to a point in σ_k.

Note that knowledge of pairs $(\phi, \gamma \frac{\partial v}{\partial \nu}|_{\partial\Omega})$ for the perfectly insulating problem is entirely equivalent to knowledge of pairs $(u|_{\partial\Omega}, \psi)$ for the perfectly conducting problem.

The boundary value problem (2) can also be cast in an energy minimization form, specifically, u is the minimizer of

$$
\begin{aligned}
Q(w) &= \frac{1}{2} \int_{\Omega} \gamma^{-1} |\nabla w|^2\, dx - \int_{\partial\Omega} \psi w\, ds \qquad \text{over} \\
K &= H^1(\Omega) \cap \{w = \text{constant on each } \sigma_k\}.
\end{aligned}
\tag{3}
$$

This minimizer, as well as the solution to (2), is uniquely determined modulo an additive constant.

Note that the constant values assumed by u on the cracks are determined as part of the minimization process. We also notice that:

REMARK 1.1. *Let Γ be a simple closed curve in $\Omega \setminus \Sigma$. Then*

$$
\int_{\Gamma} \gamma^{-1} \frac{\partial u}{\partial \nu}\, ds = \int_{\Gamma} \frac{\partial v}{\partial s}\, ds = 0.
\tag{4}
$$

In particular, we may select Γ so that it encloses a single crack. Therefore, for the energy minimizing solutions no crack acts as a source or sink of current. This criterion also serves to uniquely characterize the constants $\{c_k\}_{k=1}^{n}$. See [26] for a more detailed discussion.

However, one could also consider the non-physical boundary value and inverse problem obtained by allowing the constants c_k in (2) to assume arbitrary specified values (which don't necessarily minimize the above energy.) In this case individual cracks will act as net sources or sinks for current. This has important implications for the inverse problem.

In three or more dimensions the forward problem is again governed by Equation (1) in the case of perfectly insulating cracks, or Equation (2) for conducting cracks. A crack in a three-dimensional conductor is defined to be a suitably smooth (e.g., C^2) hypersurface which does not self-intersect. The main difference between the two and higher dimensional cases is that in the latter there is no duality between the perfectly conducting and the perfectly insulating problems.

2. The inverse problem.

2.1. Uniqueness results in the two dimensional case. The first uniqueness result concerning the determination of cracks inside a conductor was proved in [35]. It was demonstrated that with the Dirichlet data corresponding to two input current fluxes of a specified form one could uniquely determine the precise shape and location of a single conductive crack inside a conductor with real-analytic background conductivity; an analogous result was also proved for insulating cracks. In [35] it was also shown that in general two sets of measurements are required to determine a single crack. These results were generalized in [25] to show that $n+1$ input fluxes of a specified form and the resulting Dirichlet data uniquely determine a collection of n conductive cracks. Although we remarked that this result (or rather: a similar result) holds in the case in which the crack constants are arbitrarily specified, our proof in fact required that the constants assumed by the potential on the cracks be the "energy-minimizing" constants in the variational form of the problem, (3).

The results of [25] were improved, simultaneously and independently, in [9] and [36]. In [36] it is shown that with real-analytic background conductivity one can determine a collection of any number of conducting cracks using only two inputs fluxes of a specified form and the corresponding Dirichlet data. The same result is proven in [9] with much weaker assumptions about the background conductivity γ and the smoothness of the cracks. For the "two measurement" uniqueness result it would appear quite essential that the constants assumed on the cracks be "energy minimizing". The authors also provide analogous results for insulating cracks.

In all of the identifiability proofs an essential ingredient is the fact that the potential functions in $\Omega \setminus \Sigma$ which are induced by appropriate input current fluxes do not have "too many" (or any) critical points in $\Omega \setminus \Sigma$. The verification of this fact typically involves a detailed analysis of the equipotential curves of the potential function.

To illustrate the central ideas in the above papers we will, for simplicity, consider the perfectly conducting case with $\gamma \equiv 1$, so that the function

u in (2) is harmonic in $\Omega \setminus \Sigma$. We will consider input fluxes of the form $\psi_i = \delta_{P_0} - \delta_{P_i}$, for $i = 1, 2$, where P_0, P_1, and P_2 are distinct points on $\partial\Omega$, and where δ_{P_i} denotes a delta function on $\partial\Omega$ at P_i (physically a "point" input source of current at P_i.) Note that for this type of input flux the solution to (2) will not be an $H^1(\Omega)$ function, and so not obtained as the minimizer of (3). We rather interpret the solution u as a weak solution to (2), smooth except at the delta function current input, where u has a singularity of the form $\pm \ln |r|/\pi$, where r denotes distance to the input current point. It is also worth noting here that the solution u is continuous in Ω but will typically have $r^{1/2}$ type singularities at the ends of the cracks (see [35].) We then have the following uniqueness result.

THEOREM 2.1. *Let Σ and $\tilde{\Sigma}$ be two collections of cracks in Ω. Let u_1, u_2 (resp., $\tilde{u}_1, \tilde{u}_2$) be the functions which satisfy the boundary value problem (2) with $\gamma = 1$ in $\Omega \setminus \Sigma$ (resp., $\Omega \setminus \tilde{\Sigma}$) with input fluxes ψ_1, ψ_2. Let Γ be any open portion of $\partial\Omega$. If $u_1 = \tilde{u}_1$ and $u_2 = \tilde{u}_2$ on Γ then $\Sigma = \tilde{\Sigma}$.*

One of the main tools we need is a detailed analysis of the equipotential curves of the function u and how such curves can be extended. This is the focus of the following two lemmas.

In [35] the authors prove

LEMMA 2.1. *Let u satisfy $\triangle u = 0$ in $\Omega \setminus \Sigma$ with u constant on each crack σ_k. Let ρ be a nonempty analytic curve in Ω with $\rho \cap \Sigma = \emptyset$ along which u is constant. Then there exists an analytic (open) curve ρ' with $\rho \subset \rho'$ such that*

- *u is constant on ρ',*
- *one endpoint of ρ' lies on $\partial\Omega$ or on σ_j for some j,*
- *the other endpoint of ρ' lies on $\partial\Omega$ or on σ_k for some k, $k \neq j$.*

The proof of Lemma 2.1 is similar to that of the next lemma, proved in [25]. Neither proof requires that the constants c_k assumed by u on the cracks be the energy minimizing constants.

LEMMA 2.2. *Let u satisfy $\triangle u = 0$ in $\Omega \setminus \Sigma$ with u constant on each crack σ_k. Let ρ be a nonempty analytic curve in Ω with $\rho \cap \Sigma = \emptyset$ along which u is constant. Let x^* be some point in ρ at which $\nabla u(x^*) = 0$. Then there exists an analytic curve ρ' which has x^* as an interior point such that*

- *$\rho' \cap \rho = x^*$,*
- *u is constant on ρ'.*

Sketch of Proof. We can expand u in a Taylor series in r in polar coordinates near x^* to obtain $u(x) = u(x^*) + r^N(a \sin(N\theta) + b\cos(N\theta) + rA(r,\theta))$ for $N \geq 2$, $(r,\theta) \in [0,\epsilon] \times [0, 2\pi]$, and A bounded. Here we have used that $\frac{\partial u}{\partial r}(0,\theta) = 0$, due to the fact that $\nabla u(x^*) = 0$. We may assume, via a rotational change of coordinates, that ρ is tangential to the (half) line $\theta = 0$, so that $b = 0$ and $a \neq 0$, i.e., $u(x) = u(x^*) + r^N(a\sin(N\theta) + rA(r,\theta))$. The function $u(r,\theta)$ is analytic in (r,θ) near $r = 0$, and has an analytic extension to $[-\epsilon,\epsilon] \times [0,2\pi]$ (given by $u(r,\theta) = u(-r,\theta+\pi)$ for r negative). Therefore the function $A(r,\theta)$ also has an analytic extension to $[-\epsilon,\epsilon] \times [0,2\pi]$. We then have

$$u(r,\theta) = u(x^*) + r^N(a\sin(N\theta) + rA(r,\theta))$$

for $(r,\theta) \in [-\epsilon,\epsilon] \times [0,2\pi]$ with $N \geq 2$. The function $F(r,\theta) = a\sin(N\theta) + rA(r,\theta)$ satisfies $F(0,\pi/N) = 0$ and $\frac{\partial}{\partial\theta}F(0,\pi/N) = -aN$, and so by the implicit function theorem we can find a unique analytic function $\theta(r)$ such that $\theta(0) = \pi/N$ and $\{(r,\theta) : F(r,\theta) = 0\}$ coincides with $\{r,\theta(r)\}$ in some neighborhood of $(0,\pi/N)$. The curve $(r\cos(\theta(r)), r\sin(\theta(r))) + x^*$ satisfies the requirements in the statement of the Lemma. □

The key fact noted in some variation in both [9] and [36] which allows us to prove a two-measurement uniqueness result is this:

LEMMA 2.3. *Let u satisfy the boundary value problem (2) with $\gamma = 1$. Let $\sigma \in \Sigma$ and let ρ be a curve on which u is constant. Suppose that $\rho \cap \sigma = x^*$ with x^* an endpoint for ρ. Then ρ "can be extended", i.e., there is some curve ρ', with $\rho' \cap \rho \setminus \{x^*\} = \emptyset$, such that*

- *$\rho' \cap \sigma = y^*$, for some point $y^* \in \sigma$,*
- *u is constant on ρ'.*

Sketch of Proof (following the idea in [9]). We may assume that $u = 0$ on σ. Let C be a simple C^2 closed curve which encloses σ but no other crack. We can find a conformal change of coordinates in which σ is mapped to $\partial B_1(0)$ and C is mapped to $\partial B_R(0)$ for $R > 1$. We will still use u to denote the potential function in the new coordinates. Then u is still harmonic and we have $u = 0$ on $\partial B_1(0)$; moreover, u is smooth up to $\partial B_1(0)$ and we can continue u as a harmonic function into the annulus $B_1(0) \setminus B_{1/R}(0)$ by defining $u(z) = -u(\bar{z}^{-1})$. From Remark 1.1 we can see that

$$\int_{C'} \frac{\partial u}{\partial n}\, ds = 0 \tag{5}$$

for any closed curve C' contained in the annular region $B_R(0) \setminus B_{1/R}(0)$, and in particular for $C' = \partial B_1(0)$. In the new coordinates ρ intersects $\partial B_1(0)$ at some point and at this point we must have $\frac{\partial u}{\partial n} = 0$. From Equation (5) we conclude that $\frac{\partial u}{\partial n}$ (which is continuous on $\partial B_1(0)$) must vanish at some other point y on $\partial B_1(0)$, corresponding to some point $y^* \in \sigma$. We then have $\nabla u(y) = 0$, and using the same reasoning as in the proof of Lemma 2.2 we can construct a level curve $\rho' \subset B_R(0) \setminus B_1(0)$ for u with $\rho' \cap \partial B_1(0) = y$. The "pullback" of this curve to the original coordinates yields a curve with the properties stated in this Lemma. Note that the "extension" of ρ may emanate from a point other than x^*; it may also emanate from x^*, but in that case it will extend ρ to the other side of σ. □

Two additional facts that we need are

LEMMA 2.4. *Let u satisfy the boundary value problem (2) with $\gamma = 1$. Let C be any simple closed curve in Ω and suppose u is constant on C. Then u is constant on Ω.*

Sketch of Proof. Let D denote the region enclosed by C and suppose that $u = c$ on C. Define a function $\tilde{u}$ as

$$\tilde{u}(x) = \begin{cases} c, & x \in D \\ u(x), & x \in \Omega \setminus D. \end{cases}$$

It's easy to verify that $\tilde{u} \in H^1(\Omega)$ and that $\tilde{u}$ is constant on each $\sigma \in \Sigma$. However if u is nonconstant on D then we have $Q(\tilde{u}) < Q(u)$, a contradiction (strictly speaking this energy argument should be performed locally, since with the input currents ψ_i, the solution u is not in $H^1(\Omega)$). We conclude that $u \equiv c$ in D, and by unique continuation and the fact that $\Omega \setminus \Sigma$ is connected we must have $u \equiv c$ in Ω. □

REMARK 2.1. *Note that the proof of Lemma 2.3 fails if the values assumed by u on the cracks are specified, rather than the energy minimizing constants, for then equation (5) may not hold. Also in this case the conclusion of Lemma 2.4 is plainly false: We could, for example, have $u \equiv 0$ on the curve C but $u \equiv 1$ on some crack $\sigma \subset D$. In this case the function $\tilde{u}$, constructed in the above proof, is not in the class of functions over which the relevant "minimization" takes place, and we cannot conclude that u is constant on D.*

LEMMA 2.5. *Let u satisfy the boundary value problem (2) with $\gamma = 1$ and flux $\psi = \sum_{i=0}^{2} \beta_i \delta_{P_i}$ where not all $\beta_i = 0$ (but note $\sum_{i=0}^{2} \beta_i = 0$). Then u has no critical points ($\nabla u = 0$) in $\Omega \setminus \Sigma$.*

Proof of Lemma 2.5. We prove this by contradiction. Suppose that $\nabla u(x^*) = 0$ for some $x^* \in \Omega \setminus \Sigma$. Expanding u into a Taylor series in r and using the reasoning of Lemma 2.2 we can find transversal analytic curves ρ and ρ' such that x^* is an interior point for both curves (in particular $\rho \cap \rho' = x^*$).

From Lemma 2.1 we can extend both ends of ρ and ρ' until they either terminate on $\partial\Omega$ or on some crack σ. Let us suppose, for example, that one end of ρ can be extended as a level curve to some crack σ_{k_1}; let us still refer to the extended curve as ρ. By Lemma 2.3 we know that ρ can be extended again as a level curve, either terminating on $\partial\Omega$ or on some crack σ_{k_2}. If the latter occurs, this process can be repeated. It's clear that ρ must eventually terminate on $\partial\Omega$, for ρ cannot intersect the same crack twice or else we find some closed curve C on which u is constant and so by Lemma 2.4 u would be constant in Ω, a contradiction.

We conclude that both ends of ρ and ρ' can be extended as level curves for u to $\partial\Omega$. Let the points at which these level curves intersect $\partial\Omega$ be denoted by x_1, x_2, x_3, and x_4. Note that these points must be distinct or else some subdomain D would be enclosed with ∂D consisting entirely of pieces of ρ and ρ', and so Lemma 2.4 shows that u would be constant on Ω, a contradiction. Also, none of the x_j can coincide with a point P_i, $i = 0, 1, 2$, corresponding to which $\beta_i \neq 0$, since u has a logarithmic singularity at such a point. For simplicity let us suppose all β_i are nonzero.

The set $\partial\Omega \setminus (x_1 \cup x_2 \cup x_3 \cup x_4)$ contains four connected components. Now one of these components, call it S, does not contain any of P_0, P_1, or P_2, and hence $\partial u/\partial n \equiv 0$ on S. It is thus clear that S, together with some portion of ρ and ρ', enclose a region D on whose boundary u is either constant or has zero Neumann data. An argument similar to that in Lemma 2.4 (and again requiring that the constants on the cracks be chosen to "minimize" energy) shows that u must be constant in D, and hence in Ω, a contradiction. □

Proof of Theorem 2.1. We first prove that if $u_i = \tilde{u}_i$ on Γ then $u_i = \tilde{u}_i$ in Ω. Let O be the (possibly empty) open region enclosed by $\Sigma \cup \tilde{\Sigma}$, *i.e.*, those points in $\Omega \setminus (\Sigma \cup \tilde{\Sigma})$ from which it is possible to reach $\partial\Omega$ only by crossing Σ or $\tilde{\Sigma}$. Clearly $\Omega \setminus (O \cup \Sigma \cup \tilde{\Sigma})$ has only one connected component. Since u_i and $\tilde{u}_i$ have the same Cauchy data on Γ it follows by unique continuation that $u_i = \tilde{u}_i$ in $\Omega \setminus (O \cup \Sigma \cup \tilde{\Sigma})$. If O is non-empty then ∂O consists of pieces of Σ and $\tilde{\Sigma}$. On each piece of ∂O, u_i or $\tilde{u}_i$ is constant, and so u_i $(\tilde{u}_i)$ assumes finitely many values on ∂O; indeed u_i $(\tilde{u}_i)$ assumes at most $|\Sigma| + |\tilde{\Sigma}|$ values, where $|\Sigma|$ denotes the number of cracks in Σ. Since u_i $(\tilde{u}_i)$ is continuous it now follows that it is constant on each connected component of ∂O, and so by the maximum principle it is constant on each connected component of O itself. It follows that u_i $(\tilde{u}_i)$ is constant in Ω, a contradiction. We conclude that O is empty. If O is empty we have $u_i = \tilde{u}_i$ on $\Omega \setminus (\Sigma \cup \tilde{\Sigma})$. It follows by continuity that $u_i = \tilde{u}_i$ in all of Ω.

If we assume that $\Sigma \neq \tilde{\Sigma}$ then we can, for example, find some curve ρ contained in $\tilde{\Sigma}$ with $\rho \cap \Sigma = \emptyset$. Since $u_1 = \tilde{u}_1$ and $u_2 = \tilde{u}_2$ in Ω the functions u_1 and u_2 must be constant on ρ (remember, $\tilde{u}_1$ and $\tilde{u}_2$ are by definition constant on ρ). Let x^* be a point in the interior of ρ and $\mathbf{n}$ a consistently oriented unit normal vector field on ρ. Note that $\frac{\partial u_2}{\partial \mathbf{n}}(x^*) \neq 0$, for if not then we would have $\nabla u_2(x^*) = 0$ (since ρ is a level curve for u_2) a contradiction to Lemma 2.5. Now let $u = u_1 - \alpha u_2$ where $\alpha = \frac{\partial u_1}{\partial \mathbf{n}}(x^*)/\frac{\partial u_2}{\partial \mathbf{n}}(x^*)$. Note that u satisfies the boundary value problem (2) with a flux of the form $\psi = \sum_{i=0}^{2} \beta_i \delta_{P_i}$ with not all β_i zero. We find that $\frac{\partial u}{\partial \mathbf{n}}(x^*) = 0$, so that $\nabla u(x^*) = 0$, a contradiction to Lemma 2.5. Thus we conclude that $\Sigma = \tilde{\Sigma}$. □

The precise formulation of the result proved by Kim and Seo in [36] is

THEOREM 2.2. *Let ψ_1, ψ_2 be two nonvanishing piecewise continuous functions on $\partial\Omega$ with $\int_{\partial\Omega} \psi_i \, ds = 0$ and with the property that for each real α the set $\{z \in \partial\Omega : \psi_1(z) - \alpha\psi_2(z) \geq 0\}$ is connected and ψ_1 is not identically equal to $\alpha\psi_2$. Suppose that Σ and $\tilde{\Sigma}$ are collections of C^2 cracks in Ω, and suppose γ is real-analytic. Let u_i, $i = 1, 2$, be the solution to the boundary value problem (2) with flux $\psi = \psi_i$ and let $\tilde{u}_i$ be the corresponding solution with Σ replaced by $\tilde{\Sigma}$. Then $u_i = \tilde{u}_i$ for $i = 1, 2$ on $\partial\Omega$ implies that $\Sigma = \tilde{\Sigma}$.*

Kim and Seo give an example of a suitable choice for ψ_1 and ψ_2. In fact, one can give a rather general class of suitable input fluxes. Let S_1, S_2, and S_3 be a decomposition of $\partial\Omega$ into three disjoint simple arcs. Let g_i, $i = 1, 2, 3$, be non-negative functions with $g_i > 0$ in S_i, $g_i = 0$ in $\partial\Omega \setminus S_i$ and $\int_{\partial\Omega} g_i \, ds = 1$. Then one can easily verify that choosing

$$\psi_1 = g_1 - g_2, \quad \psi_2 = g_1 - g_3 \tag{6}$$

satisfies the conditions in Theorem 2.2.

In [9] Alessandrini and Diaz Valenzuela prove a theorem very similar to Theorem 2.2 but under more general conditions. Specifically, the conductivity γ may be anisotropic and only L^∞, i.e., represented by the 2 by 2 matrix γ, with bounded measurable entries and $\xi^T \gamma(x) \xi \geq \lambda |\xi|^2$ for almost all $x \in \Omega$ and $\xi \in \mathbb{R}^2$. The input fluxes are chosen almost as in Equation (6), except for the fact that it is only required that $g_i \geq 0$ in S_i. This is important in the sense that one may now approximate delta functions by taking the support of the g_i to "narrow" to a single point P_i. A collection of cracks Σ is defined to be a closed set which is a union of finitely many pairwise disjoint closed continua (a connected set containing at least two points) $\sigma_1, \ldots, \sigma_n$ such that each of the sets $\Omega \setminus \sigma_j$, $j = 1$ to n, is connected. Alessandrini and Diaz Valenzuela prove

THEOREM 2.3. *Let Γ be a nonempty simple arc on $\partial\Omega$ and Σ and $\tilde{\Sigma}$ two collections of cracks. Let u_i, $i = 1, 2$ be the solution to (2) with input fluxes ψ_1, ψ_2 chosen in accordance to Equation (6), and $\tilde{u}_i$ the corresponding solutions with $\tilde{\Sigma}$ replacing Σ. Then $u_1 = \tilde{u}_1$ and $u_2 = \tilde{u}_2$ on Γ implies that $\Sigma = \tilde{\Sigma}$.*

They also prove an analogous result for perfectly insulating cracks.

To prove Theorem 2.3 the authors use a quasi-conformal map to reduce the problem of characterizing critical points and the local behavior for solutions to (2) to equivalent problems for harmonic functions. Specifically, if D is a simply connected domain and u a solution to $\nabla \cdot \gamma \nabla u = 0$ in D then one can write $u + it = f \circ \xi$ where t is the associated *stream function*, ξ is a quasi-conformal mapping from D to $B_1(0) \subset \mathbb{R}^2$, and f is an analytic function. The key idea is that the geometric structure of the level lines for u will be the same as that of the harmonic function $\mathrm{Re}(f)$. The authors define critical points for u as those points z for which $\nabla \mathrm{Re}(f)(\xi(z)) = 0$ and show that the function u which satisfies (2) with the given input fluxes can have no critical points in $\Omega \setminus \Sigma$ (a specific example of an analysis characterizing the number of interior critical points for solutions to elliptic boundary value problems in terms of the number of sign changes in the boundary data. See [6] and [7].) The proof that $\Sigma = \tilde{\Sigma}$ if $u_i = \tilde{u}_i$ on Γ, for $i = 1, 2$, is then similar to that for Theorem 2.1.

2.2. Stability in the two dimensional case. Of theoretical and practical interest is the issue of how stably one can determine the shape and location of cracks inside a conducting body by using boundary data, since real data is invariably noisy.

In [8] the authors prove a stability estimate for the identification of a single insulating crack inside a two-dimensional conductive region Ω, which we now outline. A few technical definitions are required before stating the result.

Given a curve $c \subset \mathbb{R}^2$, a point $z \in c$, and $r > 0$ we will say that $c \cap B_r(z)$ is a *Lipschitz graph with norm* M if there is some cartesian coordinate system in which $c \cap B_r(z)$ can be represented as $\{(x, \phi(x)); -r < x < r\}$ where $\|\phi'\|_{L^\infty(-r,r)} \leq M$. Given a curve c with an endpoint z and $r > 0$ we will say that $c \cap B_r(z)$ is a *half Lipschitz graph with norm* M if there is some cartesian coordinate system in which $c \cap B_r(z)$ can be represented as $\{(x, \phi(x)); 0 \leq x < r\}$ where $\|\phi'\|_{L^\infty(-r,r)} \leq M$.

The stability result requires that for some positive constants L, M, and δ the domain Ω and crack σ satisfy the following conditions:

- length$(\partial\Omega) \leq L$;
- For each $z \in \partial\Omega$, $\partial\Omega \cap B_\delta(z)$ is a Lipschitz graph of norm M;
- length$(\sigma) \leq L$;
- dist$(\sigma, \partial\Omega) \geq \delta$.
- The crack σ is a simple curve and if V_1, V_2 are the endpoints of σ then $\sigma \cap B_{\delta/2}(V_i)$ is a half Lipschitz graph of norm M for $i = 1, 2$. Further, for each $z \in \sigma \setminus (B_{\delta/2}(V_1) \cup B_{\delta/2}(V_2))$, $\sigma \cap B_{\delta/2}(z)$ is a Lipschitz graph of norm M.

Let two input fluxes of the form $\psi_i = \eta_0 - \eta_i$, $i = 1, 2$, be applied, with $\eta_j \geq 0$ on $\partial\Omega$, $\int_{\partial\Omega} \eta_j \, ds = 1$, and $\|\eta_j\|_{L^2(\partial\Omega)} \leq M$, $j = 0, 1, 2$. For non-degeneracy suppose the η_j have mutually disjoint supports. The background conductivity γ may be anisotropic, i.e., represented by the 2 by 2 matrix γ, with bounded measurable entries and $\xi^T \gamma \xi \geq \lambda |\xi|^2$ for almost all $x \in \Omega$ and $\xi \in \mathbb{R}^2$. We know from Theorem 2.3 that we can uniquely determine any collection of cracks with these two input fluxes. We quantify the distance between two cracks σ and $\tilde{\sigma}$ using Hausdorff distance,

$$d_{\mathcal{H}}(\sigma, \tilde{\sigma}) = \max\{\sup_{x \in \sigma} \operatorname{dist}(x, \tilde{\sigma}), \sup_{x \in \tilde{\sigma}} \operatorname{dist}(x, \sigma)\}.$$

In [8] Alessandrini and Rondi show

THEOREM 2.4. *Let u_i (resp., $\tilde{u}_i$) be the potential function on Ω with insulating crack σ (resp., $\tilde{\sigma}$) for input flux ψ_i, $i = 1, 2$. Suppose Γ is a simple arc on $\partial\Omega$ with* length$(\Gamma) \geq \delta$. *There exists a positive function ω defined on $(0, \infty)$ such that if*

$$\max_{i=1,2} \|u_i - \tilde{u}_i\|_{L^\infty(\Gamma)} \leq \epsilon$$

then

$$d_{\mathcal{H}}(\sigma, \tilde{\sigma}) \leq \omega(\epsilon) \quad .$$

The function ω satisfies $\omega(\epsilon) \leq K(\ln|\ln \epsilon|)^{-\alpha}$ for $0 < \epsilon < 1/e$ and $\alpha, K > 0$, where α, K depend only the constants L, δ, and M.

If one is willing to make further a priori assumptions about the nature of the crack then the stability estimates can be considerably improved. In [4] the problem of the stability of identifying linear (line segment) cracks which are perfectly conducting is considered. A priori it is assumed that

- Ω is bounded and simply connected in $\mathbb{R}^2$ with length$(\partial\Omega) \leq L$ for some constant L.
- There is some constant $\delta > 0$ such that for all $z \in \partial\Omega$ there exists two circles of radius δ which are tangent to $\partial\Omega$ at z, one circle contained in $\bar{\Omega}$, the other in $\mathbb{R}^2 \setminus \Omega$.
- If $z = z(s)$ parameterizes $\partial\Omega$ then $\|z\|_{C^{2,\alpha}} \leq M$ for some M and $0 < \alpha < 1$.
- length$(\sigma) \geq \delta$ and dist$(\sigma, \partial\Omega) \geq \delta$.

The input fluxes are of the form $\psi_i = \eta_0 - \eta_i$, $i = 1, 2$ where the η_i are non-negative, $\int_{\partial\Omega} \eta_i \, ds = 1$ and supp$(\eta_i) \subset \partial\Omega \cap B_h(P_i)$ with $h < \delta/2$ and P_0, P_1, P_2 are distinct points on $\partial\Omega$ with $|P_i - P_j| \geq \delta$. In this case we have

THEOREM 2.5. *Let u_i (resp., $\tilde{u}_i$) be the potential function on Ω with conductive linear crack σ (resp., $\tilde{\sigma}$) for input flux ψ_i, $i = 1, 2$. If Γ is a simple arc on $\partial\Omega$ with* length$(\Gamma) \geq \delta$ *then there exists constants C and h_0, depending only on the prior constants L, δ and M, such that for $h < h_0$*

$$d_{\mathcal{H}}(\sigma, \tilde{\sigma}) \leq C \sum_{j=1}^{2} \|u_i - \tilde{u}_i\|_{L^2(\gamma)}.$$

The requirement $h < h_0$ quantifies the condition that the input electrodes be sufficiently "concentrated." Moreover, as $h \to 0$ the constant C remains bounded, so the theorem is also valid for delta function current inputs.

2.3. Uniqueness results in the three dimensional case. Some work has been done to establish uniqueness results for two-dimensional cracks inside a three-dimensional conductive body. In [5] Alessandrini and DiBenedetto study uniqueness results for a finite collection of such "hypersurface cracks" inside a three-dimensional object; they consider both perfectly conducting and perfectly insulating cracks.

Let Ω be a bounded region in $\mathbb{R}^3$ with $C^{1,\alpha}$ boundary for some $\alpha \in (1/2, 1)$. A crack σ is a simply connected, Lipschitz, non-self-intersecting surface lying at a positive distance from $\partial\Omega$ and such that $\partial\sigma$ is also Lipschitz. We use Σ to denote a finite collection of such disjoint cracks. The background conductivity of Ω is taken to be identically one.

For the conductive problem the input fluxes are of the form $\delta_P - \delta_Q$ where P and Q are distinct points on the boundary of $\partial\Omega$. The potential u will be harmonic in $\Omega \setminus \Sigma$ with $u = c_k$ on crack σ_k. It is also required that u satisfy the "zero flux" condition

$$\int_{\sigma_k} \left[\frac{\partial u}{\partial \nu}\right] ds = 0$$

for each crack σ_k, where $[\frac{\partial u}{\partial \nu}]$ denotes the jump in $\frac{\partial u}{\partial \nu}$ across σ (this condition determines the constants c_k up to a common additive constant). In this setting Alessandrini and DiBenedetto prove

THEOREM 2.6. *Let $\Gamma \subset \partial\Omega$ be a set open in the relative topology of $\partial\Omega$ and Σ a finite collection of perfectly conducting cracks. Let P_0, P_1 and P_2 be distinct points on $\partial\Omega$ and u_i the potential functions on the domain $\Omega\setminus\Sigma$ with input fluxes $\psi_i = \delta_{P_0} - \delta_{P_i}$, $i = 1, 2$. Let $\tilde{\Sigma}$ denote a second finite collection of perfectly conducting cracks and $\tilde{u}_i$ the corresponding potential functions on $\Omega \setminus \tilde{\Sigma}$. Then $u_i = \tilde{u}_i$ on Γ for $i = 1, 2$ implies that $\Sigma = \tilde{\Sigma}$.*

As noted earlier, unlike the two-dimensional case, in three dimensions there is no simple duality between the perfectly conducting and perfectly insulating cases. However, in [5] Alessandrini and DiBenedetto also prove a uniqueness result for collections of insulating cracks which are planar. We say that a crack σ_i is *planar* if it is an open portion of a two-dimensional plane, simply connected with Lipschitz boundary.

THEOREM 2.7. *Let $\Gamma \subset \partial\Omega$ be a set open in the relative topology of $\partial\Omega$. Let P_1, P_2, Q_1, Q_2 be distinct points on $\partial\Omega$ which are not coplanar and let u_i be the potential functions on the domain $\Omega \setminus \Sigma$ with input fluxes $\psi_i = \delta_{P_i} - \delta_{Q_i}$, $i = 1, 2$, where Σ is a finite collection of insulating planar cracks. Let $\tilde{u}_i$ denote the corresponding potential functions on $\Omega \setminus \tilde{\Sigma}$ for some other finite collection of insulating planar cracks. Then $u_i = \tilde{u}_i$ on Γ for $i = 1, 2$ implies that $\Sigma = \tilde{\Sigma}$.*

The authors also establish a stability result for conducting cracks; see section 2.4 below.

In [34] the author gives a relatively simple proof that given a three dimensional region Ω containing a single two-dimensional crack σ, the full Dirichlet-to-Neumann operator uniquely determines σ for insulating boundary conditions. In fact, the author shows more: let the boundary conditions on the crack be of the form $\frac{\partial u}{\partial \nu}_+ + b_+ u_+ = 0$, $\frac{\partial u}{\partial \nu}_- + b_- u_- = 0$ in which subscripts $\pm$ on u and $\frac{\partial u}{\partial \nu}$ signify the traces on each side of σ. The subscript $+$ signifies the trace on the side of σ into which the normal ν points, and $b_+ \leq 0$, $b_- \geq 0$ are unknown functions on each side of the crack. The author shows that the forward problem is well-posed and that knowledge of the Dirichlet-to-Neumann operator determines not only the crack, but the functions b_+ and b_- as well. Inclusion of the functions b_+ and b_- embodies the case in which one need not have a perfectly conducting or insulating crack, but possibly some other conduction condition. Taking $b_+ = b_- = 0$ one models a perfectly insulating crack; the limit $|b_\pm| \to \infty$ corresponds to a perfectly conducting crack.

In [11] the authors give a simple constructive proof that a single planar two-dimensional crack can be identified, under a certain nondegeneracy condition, by a single input flux and the corresponding voltage measurements. We note that even for a planar crack a single pair of measurements is not always sufficient for identification , *i.e.*, the nondegeneracy condition may exceptionally fail. The ideas in [11] form the basis for a complete reconstruction algorithm [17], which we describe in the section on reconstruction algorithms below.

2.4. Stability in the three dimensional case. In [5] the authors also prove a log-log stability result for perfectly conducting planar cracks in a three-dimensional conductor Ω. It is assumed that $\Omega \subset \mathbb{R}^3$ is bounded and convex with C^2 boundary, that σ lies in some plane π, and that $\partial\sigma$ is C^2. It is also assumed that for some $\delta > 0$,

$$\operatorname{diam}(\sigma) \geq \delta, \quad \operatorname{dist}(\sigma, \partial\Omega) \geq \delta, \quad \operatorname{dist}(P_i, P_j) \geq \delta, \ i \neq j \ ,$$

with input fluxes $\delta_{P_0} - \delta_{P_i}$, $i = 1, 2$ as above, and that the resulting potentials are measured on $\Gamma \subseteq \partial\Omega$, where Γ is open in the relative topology of $\partial\Omega$. It is required that Γ be sufficiently large, in the sense that there exists $y_0 \in \Gamma$ such that $\partial\Omega \cap B_\delta(y_0) \subset \Gamma$, and that $\operatorname{dist}(\{P_0, P_1, P_2\}, \Gamma) \geq \delta$. With these assumptions it is shown that

THEOREM 2.8. *Let u_i and $\tilde{u}_i$, $i = 1, 2$ denote the potential in Ω with perfectly conducting planar cracks σ and $\tilde{\sigma}$, respectively, with fluxes $\delta_{P_0} - \delta_{P_i}$. Then*

$$d_{\mathcal{H}}(\sigma, \tilde{\sigma}) \leq \omega\Big(\max_{i=1,2} \|u_i - \tilde{u}_i\|_{L^\infty(\Gamma)}\Big)$$

where $d_{\mathcal{H}}$ denotes Hausdorff distance and ω is a continuous, non-negative, non-decreasing function satisfying $\omega(s) \leq K(\ln|\ln s|)^{-\alpha}$ for $s \in [0, 1/e)$ and some constants K, $\alpha > 0$ (K and α depend on the a priori data $|\partial\Omega|_{C^2}$, $|\partial\sigma|_{C^2}$, $|\partial\tilde{\sigma}|_{C^2}$, and δ.)

3. Reconstruction algorithms. Much work has been done on the very practical problem of reconstructing the interior conductivity of an object from electrostatic boundary measurements, but as with uniqueness and stability, we should expect that superior results will be obtained by incorporating a priori information about the expected features of the object, in this case, cracks.

The first reconstruction algorithm specifically designed for locating cracks in a two-dimensional conductor was developed by Santosa and Vogelius in [44]. The algorithm assumes that the crack is linear and perfectly conducting; the latter assumption is not restrictive, given the duality between the conducting and insulating problems. As discussed above, in this case we have a Lipschitz stability estimate for the location of the crack. For this algorithm a linear crack σ is specified by giving the cartesian coordinates of one endpoint, the angle of the crack with respect to the horizontal axis, and the length of the crack, a total of four parameters. The algorithm uses current input fluxes of the form $\delta_P - \delta_Q$, where P and Q are distinct points on the boundary of the region Ω; the induced potential is then measured on $\partial\Omega$. One would expect generically that the position of the crack would be overdetermined by this data.

The algorithm, however, distills the boundary data down to just four numbers by integrating the data against specified test functions which depend on the current estimated position of the crack. Specifically, let u^σ denote the solution to $\Delta u^\sigma = 0$ in $\Omega \backslash \sigma$, with u^σ constant on σ, $\frac{\partial u^\sigma}{\partial \nu} = \delta_P - \delta_Q$

on $\partial\Omega$, and $\int_{\partial\Omega} u^\sigma \, ds = 0$. Note that u^σ has logarithmic singularities at P and Q. Let us use u_0 to denote the harmonic function on Ω with this same Neumann data (the response of an "uncracked" domain.) Note that u_0 is in principle known (and indeed, if Ω is a circle we can write u_0 in closed form.) It's not hard to see that the quantity $u^\sigma - u_0$ is smooth on $\partial\Omega$, and we will in fact work with $u^\sigma - u_0$ on $\partial\Omega$ (so an uncracked domain should give a zero response for any input flux).

In order to define the test functions let us take, for the moment, a cartesian coordinate system in which a crack σ lies with one endpoint at the origin, at a zero angle with respect to the horizontal axis, with length L (so that the other endpoint of σ is at coordinates $(L, 0)$.) We denote $z = (x, y)$ (or $z = x + iy$) and define functions

$$w_1(z) = \mathrm{Im}[z], \quad w_2(z) = \mathrm{Im}[z^2],$$

$$w_3(z) = \begin{cases} \mathrm{Re}[(z-L)\sqrt{z(z-L)}], & \mathrm{Re}(z) > L/2, \\ -\mathrm{Re}[(z-L)\sqrt{z(z-L)}], & \mathrm{Re}(z) < L/2 \end{cases}$$

$$w_4(z) = \begin{cases} \mathrm{Re}[\sqrt{z(z-L)}], & \mathrm{Re}(z) > L/2, \\ -\mathrm{Re}[\sqrt{z(z-L)}], & \mathrm{Re}(z) < L/2. \end{cases}$$

The functions w_3 and w_4 extend to $\mathrm{Re}(z) = L/2$ by continuity. Both w_1 and w_2 are smooth everywhere; w_3 has a square root singularity at $z = 0$ and w_4 has a square root singularity at $z = L$. One can check that all w_i are harmonic in $\Omega \setminus \sigma$, and all satisfy

$$\int_{\partial\Omega} \frac{\partial w_i}{\partial \nu} \, ds = 0, \quad \int_\sigma \left[\frac{\partial w_i}{\partial \nu}\right] ds = 0.$$

The algorithm distills the quantity $(u^\sigma - u_0)|_{\partial\Omega}$ into the four numbers

$$F_i(\sigma, w^\sigma) = \int_{\partial\Omega} (u^\sigma - u_0)(z) \frac{\partial w_i^\sigma}{\partial \nu}(z) \, ds_z, \tag{7}$$

for $i = 1$ to 4, where by use of the notation w_i^σ we make explicit the fact that the test functions w_i depend on σ (through the choice of coordinate system). The rationale for the particular choice of test functions is described below.

Let $\bar{\sigma}$ denote the actual location of the linear crack and note that we can determine $F_i(\bar{\sigma}, w^\sigma)$ for any linear crack σ by applying the specified current flux, measuring the potential $u^{\bar{\sigma}}$ on $\partial\Omega$, subtracting u_0 and then computing the F_i using (7). One would expect generically that the four equations $F_i(\sigma, w^\sigma) = F_i(\bar{\sigma}, w^\sigma)$ in four unknowns (the coordinates describing σ) would have a solution at $\sigma = \bar{\sigma}$ which is at least locally unique. The system of equations $F_i(\sigma, w^\sigma) = F_i(\bar{\sigma}, w^\sigma)$ could be solved using any standard root-finding technique, e.g., Newton's method.

However, the algorithm has an additional important feature, a feature which is a major reason for the specific choices for the w_i^σ. In order to introduce this feature, we should first note that for any given crack location σ, certain input fluxes will yield very small values for $u^\sigma - u_0$ on $\partial\Omega$. Suppose, for example, that Ω is the unit ball and consider a perfectly conducting crack σ of length L, vertically oriented in Ω anywhere on the line $x = 0$. In this case it's easy to see that an input flux $\delta_P - \delta_Q$ with $P = (1, 0)$ and $Q = (-1, 0)$ will yield $u^\sigma = u_0$ throughout Ω, since σ lies on an equipotential surface for u_0. In short, such an input flux yields no useful information for recovering σ from boundary data. Note also that σ can be moved up and down the line $x = 0$ and produce no change in the boundary measurements. Moving the points P and Q slightly may yield non-zero values for $u^\sigma - u_0$, but the values will be so small that measurement error will corrupt the data, and we expect reconstructions which use this data will be compromised. Moreover, even relatively large changes in the crack position will likely produce only small changes in the boundary data.

On the other hand, if we place the electrodes at positions $(0, 1)$ and $(0, -1)$ then the crack σ cuts orthogonally across equipotential lines for u_0, and so we expect $u^\sigma - u_0$ to be large and maximally sensitive to changes in the position of σ.

The algorithm in [44] is based on Newton's method for solving $F_i(\sigma, w^\sigma) = F_i(\bar{\sigma}, w^\sigma)$, but also attempts to adaptively change the input flux in a way that maximizes the sensitivity of $u^\sigma - u_0$ to changes in the crack position after each iteration of Newton's method. The manner in which the electrode locations are updated is detailed below.

With the given choices for F_i one can verify that at $\sigma = \bar{\sigma}$ the four by four Jacobian matrix for the system $F_i(\sigma, w^\sigma) - F_i(\bar{\sigma}, w^\sigma) = 0$ is lower triangular, regardless of the input flux used. We should expect that Newton's method (which involves implicitly inverting the Jacobian) will be well-conditioned, or alternatively, that the estimate of the crack location will be most stable with respect to the boundary data, when the diagonal elements of the Jacobian are as large as possible (which tends to improve the conditioning of the associated linear system of equations). Thus at each stage of Newton's method the algorithm adapts the input flux pattern to maximize certain diagonal entries of the lower triangular Jacobian. Indeed, the authors show that at $\sigma = \bar{\sigma}$ the first and second entries of the diagonal are given by m and $2m$ where

$$m = (\xi - w_1)(P) - (\xi - w_1)(Q)$$

where the electrodes are located at points P and Q on $\partial\Omega$ and ξ satisfies $\triangle\xi = 0$ in $\Omega \setminus \bar{\sigma}$ with $\xi = 0$ on $\bar{\sigma}$ and $\frac{\partial\xi}{\partial\nu} = \frac{\partial w_1}{\partial\nu}$ on $\partial\Omega$. Note that in an iterative scheme (an approximation to) the function ξ can be computed from the current estimated crack σ. One can thus maximize m and the associated diagonal entries of the Jacobian by choosing P to maximize the quantity $\xi - w_1$ on $\partial\Omega$ and Q to minimize $\xi - w_1$ on $\partial\Omega$.

The general outline of the algorithm is this. Let $\mathbf{F}(\sigma, w^\sigma) = (F_1(\sigma, w^\sigma), F_2(\sigma, w^\sigma), F_3(\sigma, w^\sigma), F_4(\sigma, w^\sigma))$.

1. Make initial guess $\sigma = \sigma_0$ at the crack location, set $l = 0$.
2. Based on the current estimated crack location σ_l, select the maximally sensitive electrode locations P and Q as defined above.
3. Apply the appropriate currents, measure the resulting potential on $\partial\Omega$, and use Equation (7) and the w_i to compute $\mathbf{F}(\bar{\sigma}, w^{\sigma_l})$.
4. Compute $\mathbf{F}(\sigma_l, w^{\sigma_l})$. If the residual $\|\mathbf{F}(\sigma_l, w^{\sigma_l}) - \mathbf{F}(\bar{\sigma}, w^{\sigma_l})\|$ is sufficiently small (in say, the L^2 norm) then terminate with estimate $\bar{\sigma} = \sigma_l$.
5. Compute the Jacobian and make the appropriate Newton update to compute a new estimated crack location σ_{l+1}. Set $l = l + 1$ and return to step 2.

This algorithm proved quite successful on both computationally generated data and on data collected from an experimental apparatus; see [39].

This algorithm was also adapted to seek out multiple cracks in a conductor in [26]. In this case the algorithm, in seeking out a collection of n cracks, applies n distinct fluxes at each stage of Newton's method. The fluxes are of the form $\delta_P - \delta_{Q_i}$, $i = 1$ to n. The points P and the Q_i are chosen by a procedure similar to the single crack case, in which we seek to maximize diagonal entries on the relevant Jacobian matrix in order to stabilize the estimates of the crack locations. The algorithm was tested on both computationally generated data and experimentally gathered data [28].

One of the difficulties in the multiple crack version of the algorithm is that of determining how many cracks might be present (one must choose this a priori). As originally proposed the algorithm uses an ad hoc procedure for adjusting the number of estimated cracks as the algorithm runs. However, in [10] a modification to the algorithm is proposed, in which the number of cracks is automatically adjusted using a Bayesian statistical approach, which also serves to help regularize the inversion.

Other approaches have also been used for the recovery of cracks in conductors. The so-called "reciprocity gap" principle has been used to recover both linear cracks in two-dimensional conductors and planar cracks in three-dimensional conductors [13, 17].

The basis of the reciprocity gap approach is as follows. Let Ω be a bounded domain in $\mathbb{R}^3$ with unit conductivity and σ a planar insulating crack contained in Ω. Suppose that the plane in which σ lies is described by the Equation $\mathbf{n} \cdot \mathbf{x} = c$, where $\mathbf{n} = (n_1, n_2, n_3)$ denotes a unit normal vector to the plane and $\mathbf{x} = (x_1, x_2, x_3)$. Let u be the electrical potential induced in $\Omega \setminus \sigma$ by input current g, so that $\Delta u = 0$ in $\Omega \setminus \sigma$ with $\frac{\partial u}{\partial \mathbf{n}} = 0$ on σ and $\frac{\partial u}{\partial \nu} = g$ on $\partial\Omega$ (ν an outward unit normal vector field on $\partial\Omega$). We require $\int_{\partial\Omega} g\, ds = 0$, and we can normalize u by $\int_{\partial\Omega} u\, ds = 0$. We use f to denote the measured boundary potential, so $f = u|_{\partial\Omega}$. Let v denote a $C^2(\bar{\Omega})$ harmonic function on Ω and define

(8) $$RG_{[g,f]}(v) = \int_{\partial\Omega} \left(gv - \frac{\partial v}{\partial \nu} f \right) ds,$$

the so-called reciprocity gap functional. Note that given the input current g and response f we can compute $RG_{[g,f]}(v)$ for any given harmonic function v. One can easily verify using the divergence theorem that

(9) $$RG_{[g,f]}(v) = \int_{\sigma} [u] \frac{\partial v}{\partial \mathbf{n}}\, ds$$

where $[u]$ denotes the jump in u in the direction of the normal vector field $-\mathbf{n}$.

In [13] the authors show that if the input flux g is chosen such that $\int_\sigma [u]\, ds \neq 0$, then a normal vector to the plane containing σ is given by $\mathbf{L} = (L_1, L_2, L_3)$ with $L_i = RG_{[g,f]}(v_i)$, where v_i denotes the harmonic function $v_i(\mathbf{x}) = x_i$. Of course then we have $\mathbf{n} = \mathbf{L}/\|\mathbf{L}\|$. Although the condition $\int_\sigma [u]\, ds \neq 0$ is not guaranteed, it is generically expected for a "typical" input flux.

Since we have identified $\mathbf{n}$ we can, by an appropriate change of coordinates, assume that the plane containing σ is of the form $x_3 = c$. We can now obtain c as $c = RG_{[g,f]}(p)/\|\mathbf{L}\|$ where $p(x_1, x_2, x_3) = (x_3^2 - x_2^2)/2$, which is harmonic. Thus if the input flux is such that $\int_\sigma [u]\, ds \neq 0$, then we can identify the plane in which the crack lies. One can perform similar computations to show that one may identify the line on which a linear crack lies in a two-dimensional conductor.

Having determined the plane in which the crack lies, one may attempt to recover the actual shape of the crack. In [17] the authors approach the problem as follows. Let us assume that after appropriate scaling, translation, and rotation the plane containing σ is given by $x_3 = 0$ and that σ is contained in the square $S = \{(x_1, x_2, 0) \in \mathbb{R}^3; -1 < x_1, x_2 < 1\}$. Define harmonic functions

$$\phi^i_{p,q}(x_1, x_2, x_3) = \frac{1}{\pi\sqrt{p^2 + q^2}} \psi^i_{p,q}(x_1, x_2) \sinh\left(\pi x_3 \sqrt{p^2 + q^2}\right)$$

for $i = 1$ to 4 where

$$\psi^1_{p,q}(x_1, x_2) = \cos(p\pi x_1)\cos(q\pi x_2), \quad \psi^2_{p,q}(x_1, x_2) = \cos(p\pi x_1)\sin(q\pi x_2)$$
$$\psi^3_{p,q}(x_1, x_2) = \sin(p\pi x_1)\cos(q\pi x_2), \quad \psi^4_{p,q}(x_1, x_2) = \sin(p\pi x_1)\sin(q\pi x_2)$$

Then

(10) $$RG_{[g,f]}(\phi^i) = \int_S \widetilde{[u]}\psi^i\, ds$$

where $\widetilde{[u]}$ denotes the extension of $[u]$ by zero from σ to S. It's not hard to see then that we can recover the Fourier coefficients of $\widetilde{[u]}$ by computing

$RG_{[g,f]}(\phi^i)$ for all positive integers p and q, with $i = 1$ to 4, and so recover $[\tilde{u}]$. We expect that $[\tilde{u}] \neq 0$ on any open portion of σ and indeed, this is proven in [13], subject again to the condition that $\int_\sigma [u]\, ds \neq 0$. We can thus identify the crack as the support of $[\tilde{u}]$. Analogous formulae hold in the two-dimensional case.

However, any estimate of $[\tilde{u}]$ based on a truncated Fourier series will certainly have as support all of S. In [17] the authors construct an estimate of σ as follows. Let $[\tilde{u}]_n$ denote the estimate of $[\tilde{u}]$ constructed by truncating the Fourier series constructed from Equation (10) at $p, q \leq n$. For $\epsilon > 0$ define

$$\sigma_{n\epsilon} = \{x \in S; |[\tilde{u}]_n(x)| > \epsilon\}.$$

We take $\sigma_{n\epsilon}$ as an approximation to σ. The authors prove convergence results which state that provided a certain "stress intensity factor" does not vanish on $\partial\sigma$, we have

THEOREM 3.1. *Given any $\delta > 0$, there exist two positive constants c and $\tilde{c}$ and some positive real number ϵ_0 such that for $\epsilon \leq \epsilon_0$ and $n > \tilde{c}\epsilon^{-(2+\delta)}$ we have*

$$d_{\mathcal{H}}(\sigma_{n\epsilon}, \sigma) < c\epsilon^2 \quad ,$$

where $d_{\mathcal{H}}$ is Hausdorff distance.

The authors provide a variety of numerical computations to illustrate the effectiveness of the algorithm, in both two and three dimensions. A major limitation of this very direct and elegant approach is that it appears impossible to adapt to the identification of multiple cracks that are not all coplanar.

The reciprocity gap principle has also been applied to the crack identification problem in elastostatics, in which the forward problem is governed by the linear equations of elasticity; see [14, 19].

Another approach for determining cracks has been proposed in [32]. The authors describe an efficient procedure for recovering a piecewise linear crack inside a two-dimensional polygonal domain, in which the "crack" may even self-intersect (enclosing a void) or penetrate the exterior boundary of the region. The approach is to reformulate the conduction problem using a Schwarz-Christoffel transformation, and reduce the crack identification problem to system of non-linear algebraic equations which can be solved efficiently. The authors propose several distinct algorithms, depending on the nature of the crack to be identified.

In addition to the above techniques, more traditional regularized least-squares techniques have been investigated, see e.g., [42] and the references therein. The main novelty in [42] is the design of an efficient boundary integral equation formulation for the forward problem (and for computing derivatives with respect to crack parameters) thus minimizing the cost of the optimization.

Two additional approaches to crack identification have emerged in recent work. In [18] the authors use complex analytic techniques to detect and estimate crack locations from boundary data. Moreover, this technique allows one to estimate the crack location(s) even if one has measurements of the induced potentials on only a portion of the boundary. The algorithm described in [23] also enjoys this advantage.

The basis of the technique in [18] is as follows. Suppose $\gamma \equiv 1$ and that only a single insulating, interior crack σ is present. Let a current g be imposed on $\partial\Omega$. The potential u satisfies $\Delta u = 0$ in $\Omega \setminus \sigma$ with $\frac{\partial u}{\partial \nu} = g$ on $\partial\Omega$ and $\frac{\partial u}{\partial \nu} = 0$ on σ. Suppose we have measurements of u on some segment Γ of $\partial\Omega$. Let v denote the harmonic conjugate to u in $\Omega \setminus \sigma$. The function $F = u + iv$ is then analytic in $\Omega \setminus \sigma$.

Now since $\nabla v = (\nabla u)^{\perp}$, we have $\frac{\partial v}{\partial s} = \frac{\partial u}{\partial \nu} = g$, so that we can compute v on $\partial\Omega$ as

$$v(x) = \int_a^x g(y)\,ds_y$$

for $x \in \partial\Omega$ (with the normalization $v(a) = 0$ for some fixed $a \in \partial\Omega$). We thus know the value of $\mathrm{Im}(F) = v$ on all of $\partial\Omega$ and the value of $\mathrm{Re}(F) = u$ on $\Gamma \subseteq \partial\Omega$, with F analytic on $\Omega \setminus \sigma$. To solve the inverse problem we may then seek a crack σ and bounded analytic function F on $\Omega \setminus \sigma$ such that $\mathrm{Im}(F)$ and $\mathrm{Re}(F)$ take on the known values on $\partial\Omega$ (or Γ) and on Γ, respectively. In practice, of course, the latter two constraints can be achieved only to within some specified tolerance.

The approach taken in [18] consists of finding a meromorphic function ϕ on Ω with at most N poles (N a positive integer which can be adjusted) which minimizes $\sup_\Gamma |\mathrm{Re}(\phi) - u|$ subject to the condition that $\sup_\Gamma |\mathrm{Im}(\phi) - v|$ be smaller than some prescribed tolerance. The authors give a procedure for approximately solving this problem, using ideas from AAK approximation theory (see [1]). The idea is that the poles of ϕ inside Ω somehow approximate the location of σ, at least as N increases. The authors provide some analysis of what kind of convergence can be expected, and give numerical examples. It should be mentioned that for fixed (small) N, related "pole-finding" algorithms have earlier been studied in [41].

Finally, an interesting new approach to crack detection is proposed in [23]. This approach also has the advantage that one need not specify the number of cracks a priori, and this method allows one to reconstruct crack estimates by taking measurements only on some portion of $\partial\Omega$. The method is a modification of ideas developed in [21, 22] and [37].

Let Γ be a subset of $\partial\Omega$ with positive measure and define $L^2_{\#}(\Gamma)$ to be those functions in $L^2(\Gamma)$ with zero integral over Γ. Let Σ be a collection of insulating cracks in Ω. For a function $g \in L^2_{\#}(\Gamma)$ we can solve the problem $\Delta u = 0$ in $\Omega \setminus \Sigma$, with $\frac{\partial u}{\partial \nu} = 0$ on Σ and $\partial\Omega \setminus \Gamma$, and $\frac{\partial u}{\partial \nu} = g$ on Γ. The solution is determined up to an additive constant only, but we can fix a unique solution by requiring $u|_\Gamma \in L^2_\#(\Gamma)$. Define an operator Λ on $L^2_\#(\Gamma)$ by $\Lambda(g) = u|_\Gamma$, the "restricted" Neumann-to-Dirichlet operator. Let Λ_0 be the restricted Neumann-to-Dirichlet operator on $L^2_\#(\Gamma)$ for the case in which no cracks are present. Of course we seek to use the operator Λ (or really, our ability to evaluate this operator for select input currents) to determine Σ.

In [23] the authors provide a factorization $\Lambda - \Lambda_0 = K^*K$, where K maps $L^2_\#(\Gamma)$ to a certain Sobolev space defined on $\Omega \setminus \Sigma$. This can be used to demonstrate that the operator $\Lambda - \Lambda_0$ is symmetric, positive semidefinite, and compact. As such we can find a basis of orthogonal eigenfunctions v_j for $\Lambda - \Lambda_0$ on $L^2_\#(\Gamma)$. Moreover, the operator $(\Lambda - \Lambda_0)^{1/2}$ is well-defined.

Now consider some suitably smooth curve σ_0 in Ω, and let n denote a unit normal field on σ_0. Define

$$v_1(x) = \frac{1}{2\pi} \int_{\sigma_0} \phi(y) \frac{\partial}{\partial n_y} \log\left(\frac{1}{|x-y|}\right) ds_y$$

for $x \in \Omega \setminus \sigma_0$. Here ϕ is a smooth density which is positive except at the endpoints of σ_0, where it vanishes. Let v_0 be a harmonic function on Ω with $\frac{\partial v_0}{\partial \nu} = \frac{\partial v_1}{\partial \nu}$ on $\partial\Omega$; such a function exists since $\int_{\partial\Omega} \frac{\partial v_1}{\partial \nu}\, ds = 0$. For a suitable choice of c the function $v = v_1 - v_0 - c$ (or rather, its trace on Γ) will be in $L^2_\#(\Gamma)$. The authors prove that

THEOREM 3.2. *The trace $v|_\Gamma$ is in the range of $(\Lambda - \Lambda_0)^{1/2}$ if and only if $\sigma_0 \subseteq \Sigma$.*

They also provide a quantitative method, based on knowledge of the eigenfunctions v_j and eigenvalues, for determining whether a function lies (to some given precision) in the range of $(\Lambda - \Lambda_0)^{1/2}$.

This leads to a simple method for reconstructing cracks: divide Ω into a large number of small rectangles; in each rectangle consider a small linear "test" crack σ_0; rotate this test crack through various angles about its center and determine if the corresponding function v lies in the range of $(\Lambda - \Lambda_0)^{1/2}$, to within some tolerance. If so, consider σ_0 part of the collection of cracks Σ we seek to image. The authors also use "test dipoles" (the zero length limit of a short crack) to image Σ. Although computationally intensive, the algorithm yields good results, with no a priori assumptions required concerning the number of cracks. This is an example of an imaging method that reconstructs objects by testing whether certain "signatures" are in the range of a (measured) linear operator. Methods of this type are often associated with the name "linear sampling".

4. Final remarks. The elliptic boundary value problems corresponding to the perfectly insulating and the perfectly conducting cracks we have considered here may be obtained as the limit of problems in which the cracks are approximated by "thin domains" inside which the conductivity goes to zero or infinity at an appropriate rate (relative to the thickness). The crack identification problems we have discussed may thus be seen as important special examples of a more general class of problems involving volumetrically small inhomogeneities. There is a whole body of work associated with such problems. In particular, various authors have developed special purpose algorithms for these identification problems, of a nature similar to some of the algorithms described here. It is well outside the scope of this brief survey to attempt any review of the literature on these subjects. In the absence of such a review the interested reader may, as a point of departure, consult [30] and [24], and the references therein.

REFERENCES

[1] Adamjan V.M., Arov D.Z., and Krein M.G., *Analytic properties of Schmidt pairs for a Hankel operator and the generalized Schur-Takagi problem*, Math. USSR Sbornik, **15**, 1971, pp. 31–73.

[2] Alessandrini G., *Examples of instability in inverse boundary-value problems*, Inverse Probl., **13**(4), 1997, pp. 887–897.

[3] Alessandrini G., *Stable Determination of a crack from boundary measurement*, Proc. Roy. Soc. Edinburgh Sect. A, 123, 1993, pp. 497–516.

[4] Alessandrini A., Beretta E., and Vessella S., *Determining linear cracks by boundary measurements: Lipschitz stability*, Siam J. Math. Anal., **27**(2), 1996, pp. 361–375.

[5] Alessandrini G. and DiBenedetto E., *Determining 2-dimensional crack in 3-dimensional bodies: Uniqueness and stability*, Indiana U Math. J., **46**(1), 1997, pp. 1–82.

[6] Alessandrini G. and Magnanini R., *The index of isolated critical points and solutions of elliptic equations in the plane*, Ann. Scuola. Norm. Sup. Pi. Cl. Sci., **19**(4), 1992, pp. 567–589.

[7] Alessandrini G. and Magnanini R., *Elliptic equations in divergence form, geometric critical points of solutions, and Stekloff eigenfunctions*, Siam J. Math. Anal., **25**, 1994, pp. 1259–1268.

[8] Alessandrini G. and Rondi L., *Stable determination of a crack in a planar inhomogeneous conductor*, Siam J. Math. Anal., **30**(2), 1998, pp. 326–340.

[9] Alessandrini G. and Diaz Valenzuela A., *Unique determination of multiple cracks by two measurements*, Siam J. Cont. Opt., **34**(3), 1996, pp. 913–921.

[10] Andersen K., Brooks S., and Hansen M., *A Bayesian approach to crack detection in electrically conducting media*, Inverse Probl., **17**, 2001, pp. 121–136.

[11] Andrieux S. and Ben Abda A., *Identification de fissures planes par une donneé de bord unique; un procédé direct de localisation et d'identification*, , C.R. Acad. Sci., Paris I, 1992, **315**, pp. 1323–1328.

[12] Andrieux S., Ben Abda A., and Jaoua M., *On the inverse emergent plane crack problem*, Math. Method Appl. Sci., **21**(10), 1998, pp. 895–906.

[13] Andrieux S. and Ben Abda A., *Identification of planar cracks by complete overdetermined data: inversion formulae*, Inverse Probl., **12**, 1996, pp. 553–563.

[14] Andrieux S., Ben Abda A., and Bui H.D., *Reciprocity principle and crack identification*, Inverse Probl., **15**(1), 1999, pp. 59–65.

[15] Aparicio N.D. and Pidcock M.K., *On a class of free boundary problems for the Laplace equation in two dimensions*, Inverse Probl., **14**(1), 1998, pp. 9–18.

[16] Aparicio N.D. and Pidcock M.K., *The boundary inverse problem for the Laplace equation in two dimensions*, Inverse Probl., **12**(5), 1996, pp. 565–577.

[17] Bannour T., Ben Abda A., and Jaoua M., *A semi-explicit algorithm for the reconstruction of 3D planar cracks*, Inverse Probl., **13**(4), 1997, pp. 899–917.

[18] BARATCHART L., LEBLOND J., MANDRÉA F., AND SAFF E.B., *How can the meromorphic approximation help to solve some 2D inverse problems for the Laplacian?*, Inverse Probl., **15**(1), 1999, pp. 79–90.

[19] BEN ABDA A., BEN AMEUR H., AND JAOUA M., *Identification of 2D cracks by elastic boundary measurements*, Inverse Probl., **15**(1), 1999, pp. 67–77.

[20] BEN ABDA A., CHAABANE S., DABAGHI F. EL, AND JAOUA M., *On a non-linear geometrical inverse problem of Signorini type: Identifiablity and stability*, Math. Method Appl. Sci., **21**(15), 1998, pp. 1379–1398.

[21] BRÜHL M. *Explicit characterization of inclusions in electrical impedance tomography*, preprint.

[22] BRÜHL M. AND HANKE M., *Numerical implementation of two non-iterative methods for locating inclusions by impedance tomography*, Inverse Probl., to appear.

[23] BRÜHL M., HANKE M., AND PIDCOCK M., *Crack detection using electrostatic measurements*, M2AN Math. Model. Numer. Anal., **35**, 2001, pp. 595–605.

[24] BRÜHL M., HANKE M., AND VOGELIUS M.S., *A direct impedance tomography algorithm for locating small inhomogeneities*, Numerische Mathematik, to appear.

[25] BRYAN K. AND VOGELIUS M., *A uniqueness result concerning the identification of a collection of cracks from finitely many electrostatic boundary measurements*, Siam J. Math. Anal., **23**(4), 1992, pp. 950–958.

[26] BRYAN K. AND VOGELIUS M., *A computational algorithm to determine crack locations from electrostatic boundary measurements—the case of multiple cracks*, Int. J. Engr. Sci., **32**(4), 1994, pp. 579–603.

[27] BRYAN K. AND VOGELIUS M., *Effective behavior of clusters of microscopic cracks inside a homogeneous conductor*, Asymptotic Anal., **16**(2), 1998, pp. 141–179.

[28] BRYAN K., LIEPA V., AND VOGELIUS M., *Reconstruction of multiple cracks from experimental, electrostatic boundary measurements*, proceedings of the conference on Inverse Problems and Optimal Design in Industry, July 8–10, 1993, Philadelphia PA.

[29] BUI H.D., CONSTANTINESCU A., AND MAIGRE H., *Inverse scattering of a planar crack in 3D acoustics: closed form solution for a bounded body*, CR Acad. Sci. II B, **327**(10), 1999, pp. 971–976.

[30] CAPDEBOSCQ Y., AND VOGELIUS M.S., *A general representation formula for boundary voltage perturbations caused by internal conductivity inhomogeneities of low volume fraction*, M2AN Math. Model. Numer. Anal., to appear.

[31] DIAZ VALENZUELA A., *Unicita' e stabilita per il problema inverse del crack perfettamenta isolante*, thesis, Universita' di Trieste, Trieste, Italy, 1993.

[32] ELCRAT A.R. AND HU C.L., *Determination of surface and interior cracks from electrostatic boundary measurements using Schwarz-Christoffel transformations*, Int J. Engr. Sci., **34**(10), 1996, pp. 1165–1181.

[33] ELCRAT A.R., ISAKOV V., AND NECULOIU O., *On finding a surface crack from boundary measurements*, Inverse Probl., **11**(2), 1995, pp. 343–351.

[34] ELLER M., *Identification of cracks in three-dimensional bodies by many boundary measurements*, Inverse Probl., **12**(4), 1996, pp. 395–408.

[35] FRIEDMAN A. AND VOGELIUS M., *Determining cracks by boundary measurements*, Ind. U Math. J., **38**, 1989, pp. 527–556.

[36] KIM H. AND SEO J., *Unique determination of a collection of a finite number of cracks from two boundary measurements*, Siam J. Math. Anal. **27**, 1996, pp. 1336–1340.

[37] KIRSCH A., *Characterization of the shape of a scattering obstacle using the spectral data of the far field operator*, Inverse Probl., **14**, 1998, pp. 1489–1512.

[38] KOLEHMAINEN V., ARRIDGE S.R., ET AL., *Recovery of region boundaries of piecewise constant coefficients of an elliptic PDE from boundary data*, Inverse Probl., **15**(5), 1999, pp. 1375–1391.

[39] LIEPA V., SANTOSA F., AND VOGELIUS M., *Crack determination from boundary measurements—reconstruction using experimental data*, J. Nondestructive Evaluation, **12**, 1993, pp. 163–174.

[40] MELLINGS S.C. AND ALIABADI M.H., *Three-dimensional flaw identification using inverse analysis*, Int. J. Engr. Sci., **34**(4), 1996, pp. 453–469.

[41] MILLER K., *Stabilized numerical analytic prolongation with poles*, SIAM J. Appl. Math., **18**, 1970, pp. 346–363.

[42] NISHIMURA N. AND KOBAYASHI S., *Determination of cracks having arbitrary shapes with the boundary integral-equation method*, Eng. Anal. Bound. Elem., **15**(2), 1995, pp. 189–195.

[43] RONDI L., *Optimal stability estimates for the determination of defects by electrostatic measurements*, Inverse Probl. **15**(5), 1999 pp. 1193–1212.

[44] SANTOSA F. AND VOGELIUS M., *A computational algorithm to determine cracks from electrostatic boundary measurements*, Int. J. Engr. Sci., **29**, 1991, pp. 917–937.

RIGIDITY THEOREMS IN RIEMANNIAN GEOMETRY

CHRISTOPHER B. CROKE*

1. Introduction. The purpose of this chapter is to survey some recent results and state open questions concerning the rigidity of Riemannian manifolds. The starting point will be the boundary rigidity and conjugacy rigidity problems. These problems are connected to many other problems (Mostow-Margulis type rigidity, isopectral problems, isoperimetric inequalities etc.). We will restrict our attention to those results that have a direct connection to the boundary rigidity problem (see Section 2) or the conjugacy rigidity problem (see Section 4). Even with that restriction the connections are numerous and the author was forced to select the topics covered here in accordance with rather subjective criteria. A few of the topics not covered are mentioned in Section 11 but there are others.

Although this chapter consists mostly of a survey of known results there are a few results that appear here in print for the first time. In particular, many of the applications of the Besson-Courtois-Gallot minimal entropy theorem that occur in Section 6.1 are new. Also the applications of the Pestov-Sharafutdinov theorem in Section 8 (which are due to Kleiner and the author) appear here for the first time. Most of the proofs are omitted and the reader is referred to the original articles. The exceptions are when the results are in print for the first time or when they are short and instructive.

The author would like to thank Bruce Kleiner for feedback on the first draft of this chapter. He would also like to thank Chang-Wan Kim for a careful reading of an earlier draft.

2. The boundary rigidity problem. Let $(M, \partial M, g)$ be a compact connected Riemannian manifold with smooth boundary. Then ∂M inherits a "boundary distance function" d_g, i.e., $d_g(p, q)$ is the "chordal" distance in M between boundary points p and q. Since d_g is (typically) sensitive to the geometry of the interior of M, one may ask to what extent the metric space $(M, \partial M, g)$ determines M up to isometry; this is the *Boundary rigidity problem.* We formalize the problem as follows. If $(M, \partial M, g)$ and $(M_1, \partial M_1, g_1)$ are Riemannian manifolds with boundary distance function d_g and d_{g_1} respectively, we call an isometry $(\partial M, d_g) \to (\partial M_1, d_{g_1})$ a *boundary equivalence.* A compact connected manifold $(M, \partial M, g)$ is *boundary rigid* if whenever $(M_1, \partial M_1, g_1)$ is compact connected then any boundary equivalence $(\partial M, d_g) \to (\partial M_1, d_{g_1})$ extends to an isometry $(M, g) \to (M_1, g_1)$.

*Department of Mathematics, University of Pennsylvania, Philadelphia, PA 19104 (ccroke@math.upenn.edu). Supported by NSF grant DMS 99-71749 and CRDF grant.

All manifolds in this chapter are assumed to be connected, however the boundaries of manifolds with boundary are not always connected. When $(\partial M, d_g)$ and $(\partial M_1, d_{g_1})$ are boundary equivalent we can (and will) identify ∂M_1 with ∂M and say $d_g = d_{g_1}$.

One cannot expect that all compact manifolds with boundary will be boundary rigid. Examples of non boundary rigid manifolds can be easily constructed by choosing a metric so that no minimizing geodesic from boundary point to boundary point passes through a given point x (e.g. make the distance from x to ∂M greater than the diameter of ∂M as a subset of M). In this case a change of metric (or even topology) in a small neighborhood of x will yield a metric g_1 not isometric to g but with $d_g = d_{g_1}$. The round hemisphere also provides an example since any metric which is larger than the standard metric and agrees on the boundary will have the same boundary distance function. More interesting counter examples are provided by surfaces of revolution and their higher dimensional generalizations (see [We] and [Cr-Kl1] which we will discuss below in Section 4). But there are some boundary rigid manifolds. For constant curvature spaces we have:

THEOREM 2.1. *Let $(M^n, \partial M, g)$ be a compact subdomain with smooth boundary ∂M of Euclidean space $\mathbb{R}^n$, hyperbolic space H^n, or the open hemisphere of S^n. Then $(M, \partial M, g)$ is boundary rigid.*

The hemisphere case was proved by Michel [Mi] (also see Section 4) using the solution to the Baschke conjecture for spheres (see [Be-Ka]), the Euclidean space case was proved by Gromov (see Section 5.5B of [Gro] or alternatively [Cr2]). The hyperbolic space case can be proved from the minimal entropy theorem of [B-C-G2] (see Section 6 for a discussion). In fact, the results extend somewhat beyond the subdomain case (e.g. in the flat case one needs only assume that there is an isometric immersion of $(M^n, \partial M, g)$ to $\mathbb{R}^n$ - the reader should consult the original sources for such extensions). The only spaces of nonconstant curvature that are known to be boundary rigid are two dimensional spaces of nonpositive curvature (see Section 5), subdomains of negatively curved symmetric spaces (see Section 6), and some products (see Section 9).

Because of the counterexamples we need to restrict the class of manifolds with boundary that we consider. It is important that one can determine whether $(M, \partial M, g)$ belongs to the class by its boundary distance function d since we will then know that any $(M_1, \partial M_1, g_1)$ which is boundary equivalent to $(M, \partial M, g)$ will also be in that class.

There are two natural (for different reasons) such classes of $(M, \partial M, g)$ we consider, the "simple" case and the "SGM" (strong geodesic minimizing) case.

$(M, \partial M, g)$ will be called simple if the boundary is strictly convex and there is a unique geodesic between any pair of boundary points. It is not hard to see (by looking at geodesics leaving a fixed boundary point) that topologically M is a ball and that this condition is determined by the boundary distance function. The simplicity condition will guarantee that certain problems are elliptic, and also has the advantage that any other metric g_1 with $d_g = d_{g_1}$ will have the same C^∞ jet as g at ∂M. (This was proved by Michel for the C^2 jet. For the C^∞ jet see [L-S-U].)

The SGM condition (which is given in terms of d_g only) is harder to define. A precise definition is given in [Cr1], but loosely speaking it means that all nongrazing geodesic segments are strongly minimizing. By a nongrazing geodesic segment we will mean a segment of a geodesic which lies in the interior of M except possibly for the endpoints. A segment is said to minimize if its length is the distance between the endpoints and to strongly minimize if it is the unique such path. Of course this loose definition seems to rely on more than d_g but the relationship is worked out in [Cr2]. Examples of such $(M, \partial M, g)$ are given by compact subdomains of an open ball, B, in a Riemannian manifold where all geodesics segments in B minimize - in particular all the spaces in Theorem 2.1.

Before going on much longer I should pose the question that is at the heart of the matter:

QUESTION 2.2. *Let g_0 be a Riemannian metric on B, the unit ball in $\mathbb{R}^n$, such that all geodesics minimize distance, and the distance from the origin to any point on the boundary sphere is 1. (You might further assume that the boundary is strictly convex or that the curvature is negative.) Then if g_1 is any other Riemannian metric on B with $d_{g_1} = d_{g_0}$ must there be a point $p \in B$ such that the minimum g_1-distance from p to the boundary is 1.*

The only cases where this question has been answered are cases where g_0 is shown to be isometric to g_1 and p is then the point corresponding to the origin. A positive answer to this question for a large class of g_0 would yield boundary rigidity in a correspondingly large class. The proof of this conclusion is not too hard but we will omit it here.

For a Riemannian manifold with boundary $(M, \partial M, g)$ we will let $U^+\partial M$ represent the open disc bundle of unit vectors at boundary points that point inwards into M. Note that orthogonal projection allows us to identify these vectors with the open unit disc bundle of ∂M (i.e. the vectors tangent to ∂M whose length is less than one). Similarly define as $U^-\partial M$ those that point out. Thus if $(M_1, \partial M, g_1)$ is a Riemannian metric so that $g|_{\partial M} = g_1|_{\partial M}$ (in particular if $d_g = d_{g_1}$) then we can identify the $U^+\partial M$ (and the $U^-\partial M$) from the two metrics.

We will say that $(M, \partial M, g)$ is lens equivalent to $(M_1, \partial M, g_1)$ if for each $V \in U^+\partial M$ the entering g-geodesic, $c_V(t)$ with $c_V'(0) = V$ and the entering g_1-geodesic $c_{1V}(t)$ with $c_{1V}'(0) = V$ exit in the same way in the sense that if t_0 is the first time, $t > 0$ such that $c_V(t) \in \partial M$ then the same is true for c_{1V}, and further $c_V(t_0) = c_{1V}(t_0)$ and $c_V'(t_0) = c_{1V}'(t_0)$. That is if we think of the two spaces acting as lenses with geodesics as light rays, then they bend the light in exactly the same way.

It is easy to see that if $(M, \partial M, g)$ is lens equivalent to $(M_1, \partial M, g_1)$ then $d_g = d_{g_1}$. It is also not hard to see that if $(M, \partial M, g)$ is simple or SGM (and hence so is $(M_1, \partial M, g_1)$) and $d_g = d_{g_1}$ then $(M, \partial M, g)$ is lens equivalent to $(M_1, \partial M, g_1)$. The easiest way to see this is to first consider the metric g and a fixed $p \in \partial M$ and consider the functions $f_p : M \to \mathbb{R}$ defined by $f_p(x) = d(p, x)$ and $h_p : \partial M \to \mathbb{R}$ by $h_p(q) = d(p, q)$. Now consider $q_0 \in \partial M$ such that the minimizing geodesic from p to q_0 does not graze. We see that $\nabla f_p(q_0)$ is the (outward or tangent) unit vector tangent to the geodesic from p to q_0. But since $f_p|_{\partial M} = h_p$ we see that the orthogonal projection of $\nabla f_p(q_0)$ is just $\nabla h_p(q_0)$. Now since h_p (and hence $\nabla h_p(q_0)$) is the same for the metric g_1 we see that the exiting for the g_1-geodesic from p to q_0 agrees with that of g. Reversing the roles of p and q_0 we see that the geodesics have the same initial tangents as well. This will allow us to conclude that the metrics are lens equivalent. Thus the boundary rigidity question is closely related to the corresponding lens rigidity question.

Given $(M, \partial M, g)$ and $(M_1, \partial M, g_1)$ such that $g|_{\partial M} = g_1|_{\partial M}$ then a geodesic conjugacy is a is a C^k diffeomorphism $F : UM \to UM_1$ between the unit tangent bundles such that unit vectors based at boundary points go to corresponding unit vectors (as discussed above), and such that $g^t_{M_1} \circ F = F \circ g^t_M$ wherever either (and hence both) sides are defined. In the above g^t represents the geodesic flow, i.e. for a unit vector, V, $g^t(V) = c'_V(t)$ where c_V is the geodesic with $c'_V(0) = V$. In both the simple and the SGM cases every geodesic in the interior of M is a subset of a minimizing geodesic segment between two boundary points. Thus if $d_g = d_{g_1}$ then we can define (uniquely) such a geodesic conjugacy as follows. For $V \in UM$ let $-t_0$ be the smallest $t \leq 0$ such that $c_V(t) \in \partial M$ and let $W = c'_V(-t_0)$ and hence $g^{t_0}_M(W) = V$. We thus, using again our identification of unit vectors of the two metrics at boundary points, define $F(V) = g^{t_0}_{M_1}(W)$.

The fact that F is well-defined follows from lens equivalence. In the simple case F will be C^∞, while in the SGM case F will be C^0 and C^∞ off a set of measure 0. It follows directly from the definition that $F(-V) = -F(V)$, however one cannot assume that F descends to a map $f : M \to M_1$ (i.e., F may take two vectors based at the same interior point to vectors with different base points). In fact if there is such an f where F takes each unit vector based at $x \in M$ to a unit vector at $f(x)$ then f would have to be an isometry and $F = Df$. To see this simply note that, by looking at the F image of the tangent vector field of geodesic segments, f would have to take each geodesic segment to a geodesic segment of the same length.

It is also not hard to see that such a geodesic conjugacy will preserve the canonical contact form and hence the Liouville volume form. And thus we get:

LEMMA 2.3. *If $(M, \partial M, g)$ is simple or SGM and $(M_1, \partial M_1, g_1)$ is boundary equivalent to $(M, \partial M, g)$ then $Vol(g) = Vol(g_1)$.*

One can also see the above result from Santaló's formula (see Section 3).

Since having conjugate geodesic flows is a stronger condition in general than having the same boundary distance function one might hope to get rigidity for a larger class of metrics than for the simple or SGM cases. However, there are counter-examples based on the Zoll surfaces of revolution [We]. Specifically, examples were created in [Cr-Kl1] of nonisometric metrics on the n-ball for which there is a geodesic conjugacy as above (in particular the metrics and conjugacy agree on the boundary) that preserves the standard contact forms.

In what follows there are many results where some extra assumption (usually a curvature condition) is placed on the manifold $(M_1, \partial M_1, g_1)$. The expression $(M, \partial M, g)$ is *boundary rigid among negatively curved spaces* will mean that for any $(M_1, \partial M_1, g_1)$ of negative curvature then any boundary equivalence $(\partial M, d_g) \to (\partial M_1, d_{g_1})$ extends to an isometry $(M, g) \to (M_1, g_1)$. We will use similar expressions for other curvature type conditions.

3. Inverse kinematic problem of seismology. The history of the boundary rigidity problem goes back to 1905 to papers by Herglotz [Her], and Wiechert and Zoepprtz [Wi-Zo] where they consider the case of radially symmetric metrics on the ball B^3 as a model of the earth. The next advances were due to Romanov [Ro1] who considered the linearized problem about radial metrics. In particular, he found in the radially symmetric case that you could determine a function from its integrals along geodesics.

The next advance concerned isotropic metrics, i.e., $g_1 = \lambda(x) g_0$, where λ is a smooth positive function on the manifold. In the case of simple metrics the major advances were made by Mukhometov. He proved the two dimensional case in 1975 [Mu1, Mu2] (also see [Ar-Mi]) and the higher dimensional case in 1982 [Mu3]. His results also contain nice stability estimates. Others independently proved such results (for example see [Ro2]).

The case when M is SGM was resolved in [Cr2]. The proof is easy enough so we present it here along with a stability estimate. There is a canonical measure $d\gamma$ on the space of geodesic segments (which in our case run from a boundary point to a boundary point) Γ_0 such that for every integrable function f on the unit tangent bundle we have:

$$\int_{\Gamma_0} \int_0^{L(\gamma)} f(\gamma'(t))\, dt\, d\gamma = \int_{UM} f(u) du. \tag{3.1}$$

In our case we can parameterize the geodesic segments by their initial tangents $v \in U^+\partial M$ where $U^+\partial M$ represents the unit tangent vectors to M whose base points are on ∂M and which point "inward" (i.e., $< v, N > \geq 0$ where N is the inward normal). Santaló's formula (which is used in the proof of many of the results we are discussing in this paper) allows us to express $d\gamma$ as $< N, v > dv$ where dv is the measure which locally is the product of the Riemannian measure on the base point with the standard measure on a unit hemisphere. (All of the above have versions in the non SGM case.)

We let M be an n-dimensional manifold with boundary such that there two SGM metrics: g_0 with volume $V_0 = \int_M dx_0$ and $g_1 = \lambda g_0$ of volume $V_1 = \int_M \lambda^n dx_0$. We let Γ_0 be the space of unit speed geodesic segments in M_0 (with endpoints in ∂M) with the usual measure. For $\gamma \in \Gamma_0$ and $i = 1, 2$, we let $L_i(\gamma)$ represent the length of γ in the g_i metric while $\mathcal{L}_i(\gamma)$ will represent the distance in g_i between the endpoints of γ. Of course $L_0(\gamma) = \mathcal{L}_0(\gamma)$ while $L_1(\gamma) \geq \mathcal{L}_1(\gamma)$. Note that $L_1(\gamma) = \int_0^{L_0(\gamma)} \lambda(\gamma(t))\, dt$. Further, by applying equation 3.1 with $f \equiv 1$ we have $\alpha(n-1)^{-1} \int_{\Gamma_0} L_0(\gamma)\, d\gamma = \int_M \; dx = V_0$ where $\alpha(n-1)$ is the volume of the unit $n-1$ sphere. (This gives our alternative proof of Lemma 2.3). In what follows $c_1 = \alpha(n-1)^{-1}$.

By a Hoelder inequality:

$$\int_M \lambda^2 dx_0 \leq \left\{ \int_M \lambda^n dx_0 \right\}^{\frac{2}{n}} \left\{ \int_M dx_0 \right\}^{\frac{n-2}{n}} = V_1^{\frac{2}{n}} V_0^{\frac{n-2}{n}}. \tag{3.2}$$

Using equation 3.1 (twice) we see:

$$\begin{aligned}
\int_M \lambda dx_0 &= c_1 \int_{\gamma\in\Gamma_0} \int \lambda(\gamma(t)) dt d\gamma_0 = c_1 \int_{\gamma\in\Gamma_0} L_1(\gamma) d\gamma_0 \\
&= c_1 \int_{\gamma\in\Gamma_0} L_0(\gamma) d\gamma_0 + c_1 \int_{\gamma\in\Gamma_0} (L_1(\gamma) - L_0(\gamma)) d\gamma_0 \\
&= \int_M dx_0 + c_1 \int_{\gamma\in\Gamma_0} (L_1(\gamma) - L_0(\gamma)) d\gamma_0
\end{aligned}$$

and hence

$$\begin{aligned}
2\int_M (1-\lambda) dx_0 &= 2c_1 \int_{\gamma\in\Gamma_0} (L_0(\gamma) - L_1(\gamma)) d\gamma_0 \\
&\leq 2c_1 \int_{\gamma\in\Gamma_0} \mathcal{L}_0(\gamma) - \mathcal{L}_1(\gamma) d\gamma_0.
\end{aligned} \tag{3.3}$$

Thus since

$$\int_M (1-\lambda)^2 dx_0 = \int_M \lambda^2 dx_0 + 2\int_M (1-\lambda) dx_0 - V_0$$

using 3.2 and 3.3 we arrive at the stability estimate:

$$\int_M (1-\lambda)^2 dx_0 \leq (V_1^{\frac{2}{n}} - V_0^{\frac{2}{n}}) V_0^{\frac{n-2}{n}} + 2c_1 \int_{\gamma\in\Gamma_0} \mathcal{L}_0(\gamma) - \mathcal{L}_1(\gamma)\, d\gamma_0.$$

In particular if $\mathcal{L}_0 = \mathcal{L}_1$ then also $V_0 = V_1$ and we get the boundary rigidity result that $\lambda \equiv 1$. Another consequence is that if $\mathcal{L}_1 \geq \mathcal{L}_0$ (i.e $d_{g_1} \geq d_{g_0}$) then $V_1 \geq V_0$. The difficulty in using this stability estimate is that although V_i is determined by $\mathcal{L}_i$ via $V_i = c_1 \int_{U_i^+\partial M} L_i(\gamma_v) < N, v > dv$ (where γ_v here is the geodesic in g_i), in order to determine how v depends on the endpoints involves the derivatives of $\mathcal{L}_i$.

4. Conjugacy rigidity and related problems. One way of finding manifolds that are boundary rigid is to look at subdomains (simple or SGM) of closed (or more generally complete) Riemannian manifolds without boundary that are geometrically rigid in some way and using a cutting and pasting argument. In particular, let $(M, \partial M, g_0)$ is a compact subset of a complete Riemannian manifold (N, g_0) without boundary. If $(M_1, \partial M, g_1)$ (here M_1 need not a-prioiri be diffeomorphic to M but must have the same boundary ∂M) has $d_g = d_{g_1}$ then we can create a new complete Riemannian manifold (N_1, g_1) by cutting M out of N and replacing it by N_1. If $(M, \partial M, g_0)$ is simple then the new metric g_1 on N_1 will in fact be smooth (as was pointed out in Section 2). However, in the SGM case we only get C^0 and C^∞ on $N - \partial M$, but this is often enough differentiability for our purposes. Now if (N, g_0) has the right type of rigidity we may be able to conclude that g_0 is isometric to g_1 and hence that $(M, \partial M, g_0)$ is boundary rigid. In fact most cases where boundary rigidity has been proved have been proved this way.

For example (Michel [Mi]), let $(M, \partial M, g_0)$ be a subdomain of the standard n-sphere such that $M \cap -M = \emptyset$ (here $-M$ is the image of M under the antipodal map). If $(M_1, \partial M, g_1)$ has $d_g = d_{g_1}$ then we can do the cut and paste to both M and $-M$ to get a new Riemannian manifold such that for every point p all geodesics leaving p minimize to length π, and pass through a point $-p$ at length π. This is a so called Blaschke metric and is isometric to the standard sphere since the Blaschke conjecture for spheres has been solved ([Be-Ka]) and the proof is robust enough to work for metrics with the type of differentiability we get. Thus such $(M, \partial M, g_0)$ are boundary rigid. The same type of argument might be tried for subdomains of the other rank one symmetric spaces of positive curvature, but the corresponding Blaschke conjectures are still open.

DEFINITION 4.1. If M and M' are Riemannian manifolds without boundary a map $F : UM' \to UM$ between the unit tangent bundles is called a C^k conjugacy between the geodesic flows if it is a C^k diffeomorphism and $g_t^M \circ F = F \circ g_t^{M'}$ where $g_t^{M'}$ (resp. g_t^M) is the geodesic flow on M'(resp. M). $\mathcal{F}^k U(M)$ will refer to the C^k conjugacy class of the geodesic flow on M. That is, $M' \in \mathcal{F}^k U(M)$ if there is a C^k conjugacy between the geodesic flows of M and M'. M is called C^k conjugacy rigid if it is isometric to all spaces $M' \in \mathcal{F}^k U(M)$.

The cutting and pasting argument above allows us to conclude:

LEMMA 4.2. *Simple subdomains of a C^∞ conjugacy rigid manifold are boundary rigid.*

Examples of C^∞ conjugacy rigid manifolds are given by RP^n with the standard metric (by the Blaschke conjecture for spheres [Be-Ka]), compact surfaces of nonpositive curvature (see Section 7), compact flat manifolds, compact locally symmetric spaces of negative curvature and some compact locally symmetric metrics built out of such products (see Section 6), and compact manifolds that admit a parallel vector field (see Section 9). This gives new classes of manifolds with boundary that are boundary rigid.

Not all compact manifolds without boundary are C^∞ boundary rigid. In fact for any given smooth manifold M one can put one of the metrics from [Cr-Kl1] on a smooth ball B in M and extend it arbitrarily to the rest of M. One can now deform the metric nontrivially on this ball keeping C^∞ conjugacy. That is, there is a 1 parameter family of metrics g_t such that g_0 is the original metric, g_t agrees with g_0 outside B, the geodesic flow of g_t is C^∞ conjugate to the geodesic flow of g_0 (preserving the contact form), but g_t is not isometric to g_0.

Some properties are always preserved under C^k conjugacies. C^1-conjugacies always preserve the volume [Cr-Kl1]. It is still not known if C^0 conjugacies do (even in the negative curvature setting - but see [Ha2] when one metric is locally symmetric). It is also known that the property "no conjugate points" (see Section 10) is preserved under C^0 conjugacies [Cr-Kl2]

We now give the definitions of some invariants of a compact Riemannian manifold M without boundary and some of their relations. We will be concerned with the extent to which the metric on M is determined by these invariants.

DEFINITION 4.3. The *Spectrum*, Spec(M), is the eigenvalue spectrum of the Laplace Beltrami operator (counted with multiplicity).

The *Length Spectrum*, Lspec(M), is the set of all lengths of closed geodesics of M (multiplicity is not counted). When M has nonpositive curvature (or more generally no conjugate points) then all the closed geodesics in a given homotopy class have the same length.

The *Marked length Spectrum*, MLspec(M), refers to the map which assigns to each conjugacy class in $\pi_1(M)$ the length of the shortest closed geodesic in that free homotopy class. Hence, in the case of nonpositive curvature, Lspec(M) is just the image of MLspec(M).

These invariants are closely related to each other. In the case where M has negative curvature for example we know by [Du-Gu] that Spec(M) determines Lspec(M) (the number of closed geodesics of a given length is not necessarily determined), further $\mathcal{F}^0U(M)$ determines and is determined by MLspec(M) (see [Ha1]), and of course MLspec(M) determines Lspec(M). Whether, in the negative curvature case, MLspec(M) determines $\mathcal{F}^1U(M)$ is an open (and important) question. The 2-dimensional case was solved in [Fe-Or].

The volume of M is determined by Spec(M), but it is not known (and it is an important question) if the volume is determined by Lspec(M),

MLspec(M), or $\mathcal{F}^0U(M)$. On the other hand, as we mentioned above, the volume is always determined by $\mathcal{F}^1U(M)$.

There are by now a number of examples of isospectral, nonisometric manifolds (including 2-dimensional constant negative curvature ones by Vignéras [Vi]). The negatively curved examples thus also have the same length spectrum. On the other hand (see Section 5) surfaces (2-dimensional), M and M', of negative curvature with MLspec(M) = MLspec(M_1) must be isometric.

A major open question in this area is the generalization to higher dimensions:

QUESTION 4.4. *Are compact negatively curved manifolds determined by their geodesic flows or their marked length spectra?*

In negative curvature all of the above rigidity questions "linearize" to the same statement. This becomes: If g_0 is a metric of negative curvature on a compact manifold and g_t a smooth 1-parameter family of metrics with the same length spectrum (hence the same marked length spectrum by continuity) then g_t is isometric to g_0. This question was resolved in [Cr-Sh] (see Section 8). In contrast, without the curvature assumption there are by now many examples of nontrivial isospectral deformations. The first of these was given in [Go-Wi] and the chapter by Ballmann in this book will describe some recent developments. Also see Section 11.

Although there are thus no nontrivial 1-parameter families of negatively curved manifolds:

QUESTION 4.5. *Can there be infinitely many nonisometric compact isospectral manifolds of (pinched?) negative curvature?*

One would approach the question by studying the space of isospectral metrics (in the C^k topology) in two parts; compactness, and "local" uniqueness. Some work has been done on the compactness part (for example [Ch-Ya, B-P-P, An, Br]). By "local" uniqueness we mean uniqueness among metrics in some C^k neighborhood of our given metric. Of course, the inherent infinite dimensionality of the problem makes the local uniqueness problem quite a bit harder than the linearized (i.e., infinitesimal uniqueness) problem. See Section 8 and the chapter by Sharafutdinov for some results of this type.

5. Nonpositive curvature in two dimensions. Other than the conformal case discussed in Section 3 the other case where a large class of manifolds are known to be boundary rigid or conjugacy rigid is in two dimensions when the curvature is non-positive. The initial work was done independently by Otal (see [Ot1] and [Ot2]) and the author (see [Cr3]) under a negative curvature assumption and was extended to the nonpositive case (using the method of Otal) in [C-F-F]. For the case of the flat torus see [Ban]. The best statements in the literature are:

THEOREM 5.1. *Every compact two dimensional Riemannian manifold (without boundary) of nonpositive curvature is C^0 conjugacy rigid.*

And in fact:

THEOREM 5.2. *If g and g_1 are Riemannian metrics on a compact two dimensional manifold (without boundary) having the same marked length spectrum where g has nonpositive curvature and g_1 has no conjugate points then g is isometric to g_1.*

One gets as a corollary that every SGM subdomain of such a surface is boundary rigid. Further:

THEOREM 5.3. *Every SGM two dimensional Riemannian manifold of negative curvature is boundary rigid.*

Underlying all these results is the Gauss Bonnet Theorem which seems to be why the arguments have not extended to higher dimensions. One of the fundamental observations in Otal's method is to notice that if three geodesics pass through a point in M' then the sum of the angles between them is π while if the corresponding geodesics (say via the conjugacy) in the negatively curved space M do not pass through a point then they will form a geodesic triangle where the sum of the angles is $\leq \pi$ by the Gauss-Bonnet Theorem and will be strictly less then π unless the triangle has no area (or in the nonpositive curvature case has integral of curvature $= 0$). By integrating this information appropriately one forces all such triangles to be degenerate forcing the conjugacy to cover a map on the base which must be an isometry by the discussion in Section 2.

The argument in [Cr3] involved the study of Jacobi fields. Jacobi fields are vector fields along a geodesic that arise as the variation fields to variations through geodesics. Since a C^1 conjugacy will give a correspondence between variations of geodesics it also gives a correspondence between Jacobi fields. Besides the Gauss-Bonnet Theorem, [Cr3] used the fact that if $J(t)$ is a Jacobi field along a geodesic of M such that $J(0) = 0$ then the corresponding Jacobi field $\bar{J}$ satisfies $\bar{J}(t_0) = 0$ for some t_0 (it is just the intermediate value theorem). In higher dimensions this no longer clear (though some such Jacobi fields must vanish). However, when $n \geq 3$ if all such $\bar{J}$ vanish somewhere than one can show that M and M' are isometric. So in some sense this vanishing is the crux of the mater.

Cao in [Cao] took up the noncompact finite volume case. He showed that if two orientable, uniform visibility surfaces of finite area and bounded non-positive curvature have the same marked length spectrum, then they must be isometric.

6. Symmetric spaces of negative curvature. In this section we introduce some new manifolds with boundary that are boundary rigid. The main class are the subdomains of symmetric spaces of negative curvature. The fact that these spaces are boundary rigid follow more or less directly from the equality case in the minimal entropy theorem of Besson-Courtois-Gallot (see [B-C-G1] or [B-C-G2]) which we now explain.

A symmetric space M is a Riemannian manifold such that for every point $p \in M$ the geodesic symmetry $I_p : M \to M$ at p is an isometry. I_p is the map (in general only locally defined - but defined globally for symmetric spaces) such that for each geodesic c with $c(0) = p$ we have $I_p(c(t)) = c(-t)$ for all $t \in \mathbb{R}$. The symmetric spaces of negative curvature come in 4 families of metrics on $\mathbb{R}^n$; the constant curvature metrics (all n), the complex hyperbolic spaces ($n = 2m$), the quaternionic hyperbolic spaces (for $n = 4m$), and the hyperbolic Cayley plane (n=16). A locally symmetric space is a Riemannian manifold whose universal cover is a symmetric space. They are also characterized as those metrics whose Riemannan curvature tensor is parallel.

The volume entropy $Ent(g)$ of a Riemannian manifold is the exponential growth rate of the volume, $Vol(B(p,r))$, of the metric ball of radius r in the universal covering space centered at a lift of p. More precisely:

$$Ent(g) = lim_{r\to\infty}\left(\frac{log(Vol(B(p,r)))}{r}\right).$$

It is not hard to see that this is independent of the base point p. Another equivalent definition (which is not hard to see) is that $Ent(g)$ is the infimum of s such that

$$\int_{\tilde{M}} e^{-sd(p,q)}dq < \infty$$

where p is a fixed point in the universal cover, $\tilde{M}$, and the integral is with respect to the Riemannian volume form dq.

When (M,g) is a compact manifold of negative curvature (see [Man]) or more generally has no conjugate points (see [Fr-Ma]) then $Ent(g)$ is just the topological entropy of the geodesic flow.

Let (M, g_0) be a compact manifold with a locally symmetric Riemannian metric g_0 of negative curvature. Although the theorem of Besson, Courtois, and Gallot in [B-C-G2] is more general, for our purposes we need only consider the case of a different metric g_1 on M. (The general version involves maps from (M_1, g_1) to (M_0, g_0) of arbitrary degree.)

THEOREM 6.1. *If (M^n, g_0), $n \geq 3$, is a locally symmetric space of negative curvature and g_1 is another metric on M then*

$$Ent(g_0)Vol(g_0)^{\frac{1}{n}} \leq Ent(g_1)Vol(g_1)^{\frac{1}{n}}.$$

Further, if equality holds then g_0 is isometric to g_1.

A number of important rigidity conjectures (e.g., the Lichnerowitz conjecture and rigidity of negatively curved manifolds with smooth Anosov splitting) were resolved by this theorem in [B-C-G2]. The consequence of this theorem proved in [B-C-G2] relevant to our paper is that if a compact Riemannian manifold (M,g) has a geodesic flow that is C^0 conjugate to the geodesic flow of a locally symmetric space (N, g_0) and if the volumes agree then they are isometric. Thus as pointed out in Section 4 this implies that if the conjugacy is C^1 then the volume assumption can be dropped.

However, even more can be said if both spaces have negative curvature and one is locally symmetric. Hamenstädt [Ha2] has shown that in this case the volumes must be the same and hence the assumption on the volumes can be dropped. Thus one gets:

THEOREM 6.2. *Compact locally symmetric spaces of negative curvature are C^1 conjugacy rigid and C^0 conjugacy rigid among negatively curved spaces.*

Cutting and pasting yields (as was presumably known to the authors of [B-C-G2]).

COROLLARY 6.3. *Compact subdomains of symmetric spaces of negative curvature are boundary rigid.*

Proof. Let $(M, \partial M, g_0) \subset N$ where N is a symmetric space of negative curvature. Let Γ be a cocompact discrete subgroup of the isometry group of N (i.e., N/Γ is a compact locally symmetric space) such that for every $\gamma \in \Gamma$, $\gamma(M) \cap M = \emptyset$. (This can be done since by a result of Malcev [Mal] every finitely generated linear group is residually finite. Thus if D is the diameter of M and we start with any discreet cocompact action by a Γ_0 we can choose a subgroup Γ such that for all $x \in N$ and $\gamma \in \Gamma$ $d(x, \gamma(x)) \geq D$.) This implies that a copy of $(M, \partial M, g_0)$ lies in the compact locally symmetric space N/Γ. The rest of the argument is just the cut and paste construction described in Section 4. The volume is the same since the boundary distance functions are the same by Lemma 2.3. The entropy is the same since the geodesic flows are conjugate. One can also see this directly from the alternative definition of entropy by noting by the triangle inequality that for any p and q in the universal cover N we have $|d_0(p,q) - d(p,q)| \leq 2D$ for D the diameter of M, where d is the distance in the metric induced from the pasted metric on N/Γ. (Notice that the original metric on N will be changed on countably many copies M.) One thing to worry about is that the resulting space will not a-priori be C^∞. However the rigidity proof in [B-C-G2] is robust enough to carry through in this case. □

The ideas in [B-C-G2] were extended in [Co-Fa] to prove rigidity for some products of rank one symmetric spaces of nonpositive curvature.

In [Gro] Gromov introduced the notion of filling volume, $FillVol(N^n, d)$, for a compact[1] manifold N with a metric d (here d is a distance function which is not necessarily Riemannian). To see the actual definition one should see [Gro], but it is shown in [Gro] that when $n \geq 3$

$$FillVol(N^n, d) = inf_g Vol(M^{n+1}, g),$$

where M is any manifold such that $\partial M = N$ (one can even take $M = N \times [0, \infty)$), the infimum is taken over all Riemannian metrics g on M for which the boundary distance function is $\geq d$.

[1]There are also generalizations to the noncompact case (see [Gro]).

REMARK 6.4. In the case where d is the boundary distance function of some Riemannian (M^{n+1}, g_0) one can not only fix the topology of M^{n+1} one can also restrict to metrics g which agree with g_0 when restricted to N.

To see this let g be a Riemannian metric on M whose boundary distance function, d_g satisfies $d_g \geq d$ and consider the metrics on $N \times [0,1]$ defined by

$$g_e = (1-t)g|_N + tg_0|_N + \epsilon^2 dt^2.$$

The volume of g_ϵ goes to 0 as ϵ goes to 0, while $g_\epsilon|_{N\times\{0\}} = g|_N$ and $g_\epsilon|_{N\times\{1\}} = g_0|_N$. Thus we can define new metrics $\bar{g}_\epsilon$ on $M = M \cup N \times [0,1]$ by gluing on a collar with arbitrarily small volume. (The lack of differentiability at the old boundary is not a problem since one can smooth without changing the volumes or distances much.) The only thing to check is that $d_{g_\epsilon} \geq d$. To see this first note that $g|_N \geq g_0|_N$ since $d_g \geq d$. Thus on the collar $g_\epsilon \geq g_0|_N + \epsilon^2 dt^2$. This means that any curve in the collar from (p,s) to (q,t) has length greater than or equal to $d(p,q)$. This together with the fact that $d_g \geq d$ implies that $d_{g_\epsilon} \geq d$.

We will call (M,g) a filling of $(\partial M, d_{g_0})$ if both $d_g \geq d_{g_0}$ and $g|_{\partial M} = g_0|_{\partial M}$.

QUESTION 6.5. *Let (Ω, g_0) be a compact (or simple) subdomain of a simply connected space of nonpositive (or negative) curvature. Is*

$$FillVol(\partial\Omega, d_{g_0}) = Vol(\Omega, g_0)?$$

Further if $(\bar{\Omega}, g)$ is a filling with $Vol(g) = Vol(g_0)$ must $(\bar{\Omega}, g)$ be isometric to (Ω, g_0)?

The rigidity case above would imply the boundary rigidity of (Ω, g_0). In [Gro] Gromov showed that the answer was yes for subdomains of $\mathbb{R}^n$ (in fact for more general flat manifolds) which is how subdomains of $\mathbb{R}^n$ were first proved to be boundary rigid.

Another consequence of Theorem 6.1 (which was also pointed out in [Iv]) is:

COROLLARY 6.6. *The answer to the above question is yes for subdomains of symmetric spaces of negative curvature.*

Proof. We use the cut and paste argument of the previous corollary replacing the metric g_0 in a subdomain of a locally symmetric space with g. There is again more than one way to compare the entropies of g and g_0, but we will look at the alternative definition given above as in the proof of the previous lemma. In this case triangle inequalities (and the condition on the boundary distance functions) yield $d(p,q) \geq d_0(p,q) - 2D$ for some fixed number D. Hence we can conclude that $Ent(g) \leq Ent(g_0)$. Hence by Theorem 6.1 $Vol(g) \geq Vol(g_0)$. The rigidity in Theorem 6.1 yields the rigidity here as in the previous corollary. □

Note that the result in Section 3 says that if we restrict our "fillings" to metrics that are pointwise conformal to g_0 (i.e. the only other metrics we consider are of the form $g_1 = f^2(x)g_0$) then $d_{g_1} \geq d_{g_0}$ implies $Vol(g_1) \geq Vol(g_0)$ with equality if and only if $f \equiv 1$.

The case of surfaces (two dimensions) was taken up in [Iv]. In two dimensions the filling volume question is a little different then in higher dimensions since the topology of the "filling" can affect the minimum of the area. The main result in [Iv] shows that for simple metrics (D, g_0) on the 2-disk D then any metric g_1 on D such that $d_{g_1} \geq d_{g_0}$ we have $Vol(g_1) \geq Vol(g_0)$.

A result in [C-D-S] gives a local answer to the question for simple metrics of negative curvature. In the statement below $C^{3,\alpha}(S^2\tau'_M)$ represents the $C^{3,\alpha}$ topology with respect to a fixed underlying metric on the space of symmetric 2-tensors .

PROPOSITION 6.7. *For any simple metric g_0 of negative curvature there is a neighborhood $W \subset C^{3,\alpha}(S^2\tau'_M)$ of g_0, with any $0 < \alpha < 1$, such that if a metric $g_1 \in W$ induces the same Riemannian metric on the boundary as g_0 and $d_{g_1} \geq d_{g_0}$ then $Vol(g_1) \geq Vol(g_0)$ with equality if and only if g_1 is isometric to g_0.*

In fact the curvature condition above can be substantially relaxed (see the chapter by Sharafutdinov in this book).

By analogy with the above, it is natural to pose the following question (which would resolve the local rigidity question and hence yield finiteness theorems as in Section 4):

QUESTION 6.8. *Let g_0 be a metric of negative curvature on a compact manifold (without boundary). Is there a C^k neighborhood, W, of g_0 such that for all $g_1 \in W \cap \{g_1 | \mathrm{MLspec}(g_1) \geq \mathrm{MLspec}(g_0)\}$ then $Vol(g_1) \geq Vol(g_0)$ with equality holding if and only if g_1 is isometric to g_0?*

In fact one could ask the question for g_1 a more general metric of negative curvature. This will be taken up in a future joint work of the author with Dairbekov where the two dimensional inequality should be resolved along with the $n-$dimensional case when $g_1 = f^2\ g_0$.

7. Nonpositive curvature in higher dimensions. The condition of nonpositive curvature plays a role in many of the rigidity results we have discussed. In this section we will discuss some results where all the manifolds are assumed to have nonpositive curvature.

The first rigidity result of this type is the "higher rank rigidity theorem". The rank, $Rank(M)$ of such a space, M, is the minimum (over all geodesics) of the dimension of the space of parallel Jacobi fields along a geodesic (it is always ≥ 1). $Rank(M) \geq 2$ can be expressed in terms of the existence of flat half planes in the universal cover (e.g. see the theorem below).

The initial proof of the result was due to Ballmann [Ba] and independently by Burns and Spatzier [Bu-Sp]. It was later extended by many others. The version below is proved in [Eb-He] (the result in [Eb-He] is somewhat more general than finite volume):

THEOREM 7.1. *Let M be a finite volume nonpositively curved Riemannian manifold and $\tilde{M} = X_0 \times X_1 \times X_2 \times ... \times X_k$ the de Rham decomposition of the universal cover $\tilde{M}$ (with X_0 the Euclidean part). Then each X_i $(i \geq 1)$ is either an irreducible symmetric space or it contains a geodesic that does not bound a flat half plane in X_i.*

The corresponding result for conjugacy rigidity [C-E-K] is:

THEOREM 7.2. *If (M, g) and (M_1, g_1) are compact manifolds of nonpositive curvature with rank$(M) \geq 2$, and $(M_1, g_1) \in \mathcal{F}^0U((M, g))$ then g_1 is isometric to g. Further, when $n \geq 3$ the isometry above can be chosen to induce the "same" action on the fundamental group as the conjugacy F.*

The easiest nontrivial examples of nonpositively curved manifolds that have rank 1 but whose fundamental group is not hyperbolic (and hence have no metrics of negative curvature) are the so called graph manifolds. Though there have been a number of definitions of "graph manifold" we will use a the rather general one used in [Le]: a graph manifold is a 3 dimensional Haken manifold each of whose components under the minimal decomposition (in the sense of [Ja-Sh, Jo]) is Seifert fibered. The decomposition is by cutting along embedded incompressible tori and Klein bottles. Not all such manifolds admit metrics of nonpositive curvature (see [Le] and [Bu-Ko]) but we will only be concerned with the ones that admit nonpositively curved Riemannian metrics. The following is proved in [Cr1].

THEOREM 7.3. *If (M, g) and (M_1, g_1) are compact nonpositively curved graph manifolds and $(M_1, g_1) \in \mathcal{F}^0U((M, g))$ then g_1 is isometric to g.*

One way to construct some such graph manifolds of nonpositive curvature is to use two building blocks $B_i = \Sigma_i^2 \times S^1$ where Σ_i^2 is a surface of nonpositive curvature with a boundary consisting a closed geodesic (to get smooth results assume that the metric is flat in a neighborhood of the boundary). Notice that the boundary is a totally geodesic flat torus. We can thus glue two such blocks together by reversing the two factors in the boundary tori (assuming the appropriate lengths are compatible) yielding a manifold without boundary of nonpositive curvature. One can easily generalize this to many blocks B_i where the surfaces Σ_i have many geodesics on the boundary.

The topology of graph manifolds forces metrics of nonpositive curvature to have a very restricted geometric form as studied in [Sch] and [Le]. They are the union over totally geodesic boundaries of (one or more) pieces (the "geometric Seifert components") whose universal covers are isometric to a Riemannian product $X \times \mathbb{R}$ where X is a surface of nonpositive curvature with geodesic boundary. They thus become amenable to some techniques developed for the results in Section 9.

8. The linearized and local problem. The *deformation boundary rigidity problem* attempts to answer the question: Can one deform the metric on a simple (or SGM) manifold $(M, \partial M, g)$ keeping the boundary distance function fixed? In other words: If g_t is a 1-parameter family of metrics on M with $g_0 = g$ such that $d_{g_t} = d_g$, must there be diffeomorphisms ϕ_t leaving ∂M pointwise fixed such that $\phi_t^*(g_t) = g$? If this is true then $(M, \partial M, g)$ is called deformation boundary rigid.

Linearizing this problem about g_0 leads to the following question: If T is a symmetric n-tensor such that for every geodesic segment γ from boundary point to boundary point we have

$$0 = I_\gamma(T) \equiv \int_0^{L(\gamma)} T(\gamma'(t), \gamma'(t), ..., \gamma'(t))dt,$$

then $T = d\omega$ for a an $(n-1)$ symmetric tensor ω. In the above the operator d is the symmetrized covariant derivative (see [Cr-Sh] for more details). The metric case corresponds to the question for symmetric 2-tensors (representing in our case $\frac{d}{dt}g_t$) and we call a metric g_0 which satisfies the above for 2-tensors *infinitesimally rigid.* The map $\gamma \to I_\gamma(T)$ is called the ray transform of T. In a separate chapter in this book Sharafutdinov goes into more detail and surveys the known results in this area and in particular discusses how infinitesimal rigidity can yield deformation rigidity via ideas of [Gu-Ka1]. Here we will only mention a few of the results. We will stick to the case of negative curvature and simple metrics (in the "with boundary" case) even though all of the results have been improved to hold for wider classes (e.g. the geodesic flow is Anosov, or subdomains of simple metrics). The reader is referred to the other chapter for these improved results. Our purpose here is to show how one can sometimes use such infinitesimal rigidity results along with symmetry to get global rigidity results. The results of that nature stated here can of course be improved using the improvements covered in the other chapter.

We start with the theorem of Pestov and Sharafutdinov (see [Pe-Sh]);

THEOREM 8.1. *If $(M, \partial M, g)$ is simple and has negative curvature then it is deformation boundary rigid.*

The proof of this theorem uses the Pestov identity which is one of the fundamental tools in the study of the linearized and local versions of these problems.

The analogue of this theorem for manifolds without boundary is [Cr-Sh]:

THEOREM 8.2. *Compact negatively curved manifolds are spectrally rigid.*

It would possibly be better to say "deformation spectrally rigid" but the above terminology was introduced in [Gu-Ka1]. What it says is that if g_t is a smoothly varying family of Riemannian metrics on a compact manifold such that g_0 has negative curvature and for all t the eigenvalue spectrum of g_t is the same as that of g_0, then g_t is isometric to g_0.

Guillemin and Kazhdan had proved this result in two dimensions [Gu-Ka1] and in the n-dimensional case when g_0 satisfied a pointwise pinched negative curvature condition [Gu-Ka2]. Min-Oo [MO] had extended this to the case where g_0 has negative definite curvature operator.

The proof also yields the corresponding rigidity for the length spectrum (i.e. if the set of lengths of closed geodesics are the same for the g_t then they are isometric) and hence the marked length spectrum. Both of the above theorems come from versions that concern symmetric n-tensors (not only 2-tensors) as above. Other consequences (extensions of results in [Gu-Ka2]) which are in some sense global rigidity results can be found in [Cr-Sh]:

THEOREM 8.3. *Let (M, g) be a compact negatively curved Riemannian manifold with simple length spectrum, and $\Delta : C^\infty(M) \to C^\infty(M)$ be the corresponding Laplace-Beltrami operator. If real functions $q_1, q_2 \in C^\infty(M)$ are such that the operators $\Delta + q_1$ and $\Delta + q_2$ have coincident spectra, then $q_1 \equiv q_2$.*

In the above *simple length spectrum* means that no two closed geodesics (traversed once) have lengths that are rational multiples of each other. This is a generic condition.

COROLLARY 8.4. *Let M be a compact negatively curved manifold and $f : M \to \mathbb{R}$ a C^∞ function. If f integrates to zero along every closed geodesic then f must itself be zero. In particular a function is determined by its integral along closed geodesics.*

For the rest of this section I want to discuss some rigidity consequences of Theorem 8.1 that were developed by the author and Kleiner but never published.

PROPOSITION 8.5. *Let $(M, \partial M, g)$ be simple and have negative curvature. If V is a tangent vector field to M defined along ∂M such that for every geodesic segment $\gamma : [0,1] \to M$ with $\gamma(0)$ and $\gamma(1)$ in ∂M, we have*

$$< \gamma'(0), V(\gamma(0)) > = < \gamma'(1), V(\gamma(1)) >$$

then there is an extension of V to a Killing vector field on M.

A Killing field, V, is an infinitesimal isometry in the sense that the local one parameter group of local diffeomorphisms, V^t, that it generates are local isometries. Of course any Killing field when restricted to ∂M must satisfy the above condition by the first variation formula. Killing fields are precisely those fields V which satisfy $L_V g = 0$ or, equivalently, whose dual one form ω satisfies $d\omega = 0$ (with the d from above). If g_1 is a metric with $d_{g_1} = d_g$ then for each g-Killing field V the proposition yields a g_1-Killing field $\bar{V}$ with the same (in the sense of Section 2) restriction to the boundary.

The proposition follows from the version of Theorem 8.1 for one forms:

Proof. Let W be an arbitrary C^∞ extension of V and ω its dual one form. Our condition implies that $d\omega$ integrates to 0 along every geodesic segment. The analogue of Theorem 8.1 gives us a one form $\bar{\omega}$ which is 0 at ∂M and such that $d\omega = d\bar{\omega}$. Hence $d(\omega - \bar{\omega}) = 0$ and the dual of $\omega - \bar{\omega}$ is our required Killing field. □

Let $(M, \partial M, g)$ be a simple space of negative curvature with a Killing field V and let $(M_1, \partial M, g_1)$ also have negative curvature and $d_g = d_{g_1}$. Then the above gives us a Killing field V_1 on M_1 with the same boundary values as V. Now if for some open $U \subset \partial M$ the flow of both V and V_1 are defined for $t \in [0, \epsilon)$ then the neighborhoods $\{V^t(p) \in M | t \in [0, \epsilon)$ and $p \in U\}$ and $\{V_1^t(p) \in M | t \in [0, \epsilon)$ and $p \in U\}$ are isometric via the map f defined by $f(V^t(p)) = V_1^t(p)$. This follows since $Df|_{V^t(p)}$ is the composition of three linear isometries $DV^{-t}|_{V^t(p)}$, the identification of T_pM with T_pM_1, and $DV_1^t|_p$. Note here that for any other Killing field W and its corresponding Killing field W_1 we have $Df(W) = W_1$ by the uniqueness properties of Killing fields (if the C^1 jets of two Killing fields agree at a point then they agree everywhere in a connected set).

It is not hard to extend the argument to see that if each point in M can be written as $V^t(p)$ then g_1 is isometric to g. In fact more is true. On a space $(M, \partial M, g)$ we say that $m \in M$ is "reachable from the boundary via Killing fields" if there is a finite sequence $\{(V_i, t_i) | i = 1..k\}$ of Killing fields V_i and real numbers t_i such that

$$m = V_k^{t_k}\left(V_{k-1}^{t_{k-1}}(...V_1^{t_1}(p)...)\right)$$

for some $p \in \partial M$ where V_i^t represents the local flow generated by V_i for time t. One can extend the above reasoning to get the following rigidity result:

COROLLARY 8.6. *If $(M, \partial M, g)$ is simple, has negative curvature, and every $m \in M$ is reachable from the boundary via Killing fields then it is boundary rigid among negatively curved spaces.*

We will say a space (M, g) (with or without boundary) is a space of revolution if there is a point o (the "origin") such that if I_o is the connected subgroup of the isometry group of M that fixes o then the set of differentials acting on T_oM is the full group $SO(n)$. Another way of saying this is that in normal polar coordinates (r, Θ) the metric has the form $ds^2 = dr^2 + g^2(r)d\Theta^2$ where $d\Theta^2$ is the standard metric on the $n-1$ sphere and $g : \mathbb{R}^+ \to \mathbb{R}^+$ is 0 only at 0.

Let D be a convex subdomain of a simply connected negatively curved space of revolution with $o \notin D$. let $q \in \partial D$ be a closest point to o, and L^{n-2} be any $(n-2)$-dimensional subspace of the parallel translate, L^{n-1}, of $T_q\partial D$ along a minimizing geodesic to o. Then, since $exp_o(L^{n-1}) \cap D = \emptyset$, the Killing field that comes from the rotation leaving L^{n-2} fixed will work in the above to show that D is boundary rigid among negatively curved spaces.

Another quite different way to use Proposition 8.5 on spaces of revolution is the following which generalizes to higher dimensions a two dimensional result proved in [Ar-Mi]:

COROLLARY 8.7. *Let $(M, \partial M, g)$ be a simple manifold of negative curvature which is a space of revolution, then it is boundary rigid among* *negatively curved spaces. Further, any simple manifold of negative curvature $(M, \partial M, g)$ satisfying*

- $g|_{\partial M}$ *is a round metric on the* $n-1$ *sphere*
- $d_g(p,q) = f(\bar{d}(p,q))$ *where* $f : \mathbb{R}^+ \to \mathbb{R}^+$ *is a function and* $\bar{d}$ *is the distance function with respect to the boundary sphere metric*

must in fact be a space of revolution.

We note that all spaces of revolution clearly satisfy the two conditions in Corollary 8.7. Since this corollary does not appear in the literature we provide a proof here.

Proof. We do the second part first. Let $(M, \partial M, g)$ have negative curvature and satisfy the two conditions, and let V be a Killing vector field to ∂M coming from the action of $SO(n-1)$ on ∂M. We can of course think of V as a tangent vector field to M along ∂M. Since for every p and q in ∂M, $\bar{d}(V^t(p), V^t(q)) = \bar{d}(p,q)$ we have $d_g(V^t(p), V^t(q)) = d_g(p,q)$ and $< \gamma'(p), V(p) >=< \gamma'(q), V(q) >$ for the (unique by simplicity) geodesic segment, γ from p to q by the first variation formula. Thus by Proposition 8.5 every such Killing field extends to a Killing field on M. Integrating these fields we see that the action of $SO(n-1)$ on the boundary extends to an isometric action on all of M. For the rest of the argument we show in the next paragraph that there is an origin o.

The normal geodesic leaving any boundary point p must exit at the antipodal point to p and normal to the boundary since the whole geodesic will be fixed by any isometry that leaves p fixed. Let $2r$ be the (constant by the homogeneity of the boundary) length of these normal geodesic segments. The obvious map $H : \partial M \times [0, 2r] \to M$ must have some singular points or else it would be a double cover of M which is diffeomorphic to a ball since it is simple. Such a singular point is a focal point $\gamma(t_0)$, of the boundary along some normal geodesic γ. The focal point has index $n-1$ because of the isometries that fix γ and hence is the only focal point along this geodesic by the curvature assumption. In particular $dH_{(\gamma(0),t_0)}(V) = 0$ for all vectors tangent to $\partial M \times \{t_0\} \subset \partial M \times [0, 2r]$ at $(\gamma(0), t_0)$. The Jacobi fields along γ defining this focal point are the same Jacobi fields along γ that define the boundary focal points to $\gamma(2r)$ along $-\gamma$ since all normal geodesic segments are normal at both endpoints. This means that the distance to the focal locus from $\gamma(2r)$ along $-\gamma$ is $2r - t_0$. The isometry group forces the same picture along every normal geodesic. In particular $t_0 = 2r - t_0$ and so the focal point occurs at distance r from the boundary along every normal geodesic. Thus the focal locus $H(\partial M \times \{r\})$ is a single point o, our origin, since dH is 0 on all tangent vectors to $\partial M \times \{r\}$.

Now to see the first part let $(M_1, \partial M, g_1)$ have negative curvature and $d_{g_1} = d_g$. Then it satisfies the two conditions above so by the part already proven $(M_1, \partial M, g_1)$ is a space of revolution. That means, in particular, that for every 2-plane P in T_oM_1, $exp_o(P)$ is a totally geodesic submanifold and hence has negative curvature. The corresponding surface in M (i.e.

the totally geodesic surface having the same boundary circle) will also have negative curvature and the same boundary distance function. Hence the surfaces are isometric by Section 5. But that means that $g(r) = g_1(r)$ (where $g(r)$ comes from the metric representation $ds^2 = dr^2 + g^2(r)d\Theta^2$) and thus the spaces are isometric. □

9. Products. The only cases of rigidity in higher dimensions without any curvature assumptions involve manifolds that have some sort of product structure. In [Cr-Kl1] it was shown that all compact Riemannian manifolds X are stably C^1-conjugacy rigid in the sense that $X \times S^1$ is C^1 conjugacy rigid. More generally

THEOREM 9.1. *Let M be a compact Riemannian manifold with a nontrivial parallel vector field. Then M is C^1 conjugacy rigid.*

In $X \times S^1$ the parallel vector field points in the S^1 direction. In general, by the DeRham splitting theorem, if M admits a parallel vector field then the universal cover of M splits off a line (though M itself need not be a product). Surprisingly it is still not known if M is C^1 conjugacy rigid if rather than a parallel vector field M admits a parallel line field (hence a double cover admits a parallel vector field). Of course the double covers will be isometric but there is no guarantee that the isometry will commute with the covering maps.

The idea behind the proof of Theorem 9.1 is to study Jacobi fields (or more precisely Jacobi tensors) perpendicular to the "vertical" direction. Using an extremal property of Jacobi fields and the fact that the volume is the same it is shown that the conjugacy must map the parallel vector field in one space to a parallel vector field in the other. This uses the C^1 conjugacy. That reduces the problem to the case where the universal covers of both manifolds are Riemannian products.

Now if two manifolds are nontrivial products [Cr-Kl1] studies C^0 conjugacies between the geodesic flows. The techniques developed here were later applied in [C-E-K] and [Cr1] (also see Section 7). In general one needs to assume an additional assumption that the conjugacy "preserves angles". For a Riemannian product $M_1 \times ... \times M_k$, we define $\alpha = (\alpha_1, ..., \alpha_k) : U(M_1 \times ... \times M_k) \longrightarrow R^k$ by $\alpha_i(v) = |\pi_{i*}(v)|$ where $\pi_i : M_1 \times ... \times M_k \longrightarrow M_i$ is projection onto the ith factor. The most general result is:

PROPOSITION 9.2. *Let $M = M_1 \times M_2$ and $N = N_1 \times N_2$ be Riemannian products such that $dim(M_i) \neq 0$. If there is a uniform conjugacy $F : UM \to UN$ which preserves α then M is isometric to N.*

In the above M and N are not assumed to be compact, but the conjugacy is assumed to be uniformly continuous. In many cases it is not hard to prove that a conjugacy preserves α. A particular consequence is:

PROPOSITION 9.3. *Let $M = M_1 \times M_2$ be a compact Riemannian product such that $dim(M_i) \neq 0$, and let $N = X_1 \times X_2 \times ... \times X_k$ where each X_i is a compact Riemannian manifold of negative curvature (or more* *generally a manifold with Anosov geodesic flow). If there is a C^0 conjugacy $F : UM \to UN$ then M is isometric to N.*

10. No conjugate points. One definition of "no conjugate points" for a Riemannian manifold (M, g) without boundary is that in the universal cover all geodesics are minimizing. The official definition is that no Jacobi field along any geodesic vanishes twice. The easiest examples of manifolds without conjugate points are the nonpositively curved ones. There are other examples (see for example [Gul]), however, there is no example known of a compact manifold that admits a metric with no conjugate points that does not also admit a different metric with nonpositive curvature.

The purpose of this section is to survey the rigidity results related to the no conjugate points condition. It is a natural condition for our setting in that (see [Cr-Kl1]) it is preserved by C^0 conjugacies, i..e, if (M, g) and (M_1, g_1) have C^0 conjugate geodesic flows and (M, g) has no conjugate points then (M_1, g_1) has no conjugate points. On the other hand rigidity results are hard to come by since the condition is not a convexity condition as curvature conditions are, but only a global length minimization condition.

The most celebrated such rigidity result is the so called E. Hopf conjecture solved by Burago and Ivanov [Bu-Iv1] (the two dimensional case had been solved by Hopf [Ho]):

THEOREM 10.1. *A metric on the n-torus without conjugate points is flat.*

One could use this to conclude that any SGM subdomain of a compact flat manifold is boundary rigid. Another way of seeing this had been found in [Cr2]. For $x \in M$ let $Vol(B(\bar{x}, r))$ be the volume of the ball of radius r in the universal cover of M centered at a lift $\bar{x}$ of x. This is well defined since all such balls have the same volume. It was shown in [Cr2] that if one fixes the radius $r > 0$, then the average over x (with respect to the standard volume on M) of $Vol(B(\bar{x}, r))$ is greater than or equal to the volume of the flat ball of radius r with equality if and only if the metric is flat. Also in [Cr2] is a more general result relating the metric entropy to the average volume of balls. To get the boundary rigidity for SGM submanifolds [Cr2] considers the asymptotics as r goes to ∞.

If (M, g_0) is a complete Riemannian manifold satisfying some property $\mathcal{P}$ (such as "no conjugate points"), then we will say that (M, g_0) is rigid under compactly supported perturbations (respecting the property $\mathcal{P}$) if any metric g_1 on M which satisfies property $\mathcal{P}$ and coincides with g_0 outside a compact set $K \subset M$ is isometric to g_0.

Euclidean space was shown to be rigid under compactly supported perturbations with nonpositive (or nonnegative) sectional curvature in [Gr-Wu]. In fact they are rigid under compactly supported perturbations having no conjugate points (see [Gr-Gu] for the two dimensional case and [Cr2] in general).

In [Cr-Kl2] this was extended to:

THEOREM 10.2. *Let X be a 1-connected Riemannian manifold without conjugate points, and let g_0 be the Riemannian product metric on $X \times \mathbb{R}$. Then g_0 is rigid under compactly supported perturbations having no conjugate points.*

In the proof, one first shows that if the metric is changed in a compact region without introducing conjugate points then the boundary distance function of the region does not change. One then proves boundary rigidity for such regions to get the result.

The case where X above is not simply connected was considered in [Cr-Kl3]. This case poses many difficulties not seen in the simply connected case. In particular, so far one cannot handle spaces whose fundamental groups grow too fast. Never-the-less many cases are proven in [Cr-Kl3]. One result is:

THEOREM 10.3. *Let (M, g_0) be a complete flat manifold. Then g_0 is rigid under compactly supported perturbations having no conjugate points.*

The case where M is compact relies on Theorem 10.1 while the non compact case comes from [Cr-Kl3].

11. Some other related topics. There are many related topics that we have not taken up here. We mention a few briefly now.

One might ask why one should consider Riemannian metrics rather than say Finsler metrics (i.e. there is a Banach norm in each tangent space rather than an inner product). The boundary rigidity problem in general is under-determined so one cannot expect uniqueness. Arcostanzo in [Ar] shows among other things how one can explicitly construct many Finsler metrics on the 2-disk with the same boundary distance function as the flat ball. Thus one needs to add extra conditions to have any chance. One natural condition to consider is when the curvature is parallel. Two places where such rigidity results are proved are [Ki-Yi] and [Fo]. Another example where rigidity (in the sense of Theorems 5.1 and 5.2) fails to hold if the metric is not assumed to be Riemannian can be found in [Bo]. Some nice results of filling volume type (as in Section 6) for Finsler metrics can be found in [Iv] and [Bu-Iv2].

Without curvature assumptions there are now many examples of isospectral metrics and isospectral deformations. See for example the chapter in this book by Ballmann. Nilpotent groups and their quotients have served as a rich source of examples of nontrivial isospectral deformations. However there is conjugacy rigidity inside this class (both manifolds assumed to be nilmanifolds) see [G-M-S, Eb, Fa]. A nice survey of this subject can be found in [Go-Go].

Rigidity questions for Lorentz metrics has been studied in [A-D-H]. While [Grog] considers magnetic flows rather than geodesic flows.

REFERENCES

[A-D-H] L. Andersson, M. Dahl, and R. Howard, *Boundary and lens rigidity of Lorentzian surfaces*, Trans. Amer. Math. Soc. **348** (1996), no. 6, 2307–2329.

[An] M. Anderson, *Remarks on the compactness of isospectral sets in low dimensions*, Duke Math. J. **63** (1991), 699–711.

[Ar] M. Arcostanzo, *Des métriques finslériennes sur le disque à partir d'une fonction distance entre les points du bord*, Comm. Math. Helv. **69** (1994), 229–248.

[Ar-Mi] M. Arcostanzo and R. Michel, *Métriques de révolution d'un disque et invariance par rotation de la longueur des géodésiques*, Geom. Dedicata **76** (1999), no. 2, 197–209.

[Ba] W. Ballmann, *Nonpositively curved manifolds of higher rank*, Annals. of Math. (2), **122** (1985), 597–609.

[Ban] V. Bangert, *Geodesic rays, Busemann functions and montone twist maps*, Calc. Var. Part. Diff. Eq. **2** (1994), no. 1, 49–63.

[Be] M. Berger, *Une bourne inférieure pour le volume d'une variété riemannienne en function du rayon d'injectivité*, Ann. Inst. Fourier (Grenoble) **30** (1980), no. 3, 259–265.

[Be-Ka] M. Berger and J.L. Kazdan, *A Sturm-Liouville inequality with applications to an isoperimetric inequality for volume in terms of injectivity radius, and to Wiedersehen manifolds*, in General Inequalities 2 (Proceedings of the second International Conference on General Inequalities, 1978), E.F. Beckenbach (ed.), ISNMA47, Birkhauser Verlag, no. 3, pp. 251–254.

[B-C-G1] G. Besson, G. Courtois, and S. Gallot, *Entropies et rigidités des espaces localement symétriques de courbure strictement négative*, Geom. Funct. Anal. **5** (1995), no. 5, 731–799.

[B-C-G2] G. Besson, G. Courtois, and S. Gallot, *Minimal entropy and Mostow's rigidity theorems*, Ergodic Theory Dynam. Systems **16** (1996), no. 4, 623–649.

[Bo] F. Bonahon, *Surfaces with the same marked length spectrum*, Topology Appl. **50** (1993), no. 1, 55–62.

[B-P-P] R. Brooks, P. Perry, and P. Petersen, *Compactness and finiteness theorems for isospectral manifolds*, J. Reine Angew. Math. **426** (1992), 67–89.

[Br] R. Brooks, *Cheng's theorem in L^p spectral geometry*, Algebra i Analiz **8** (1996), no. 2, 98–102; translation in St. Petersburg Math. J. **8** (1997), no. 2, 255–258.

[Bu-Iv1] D. Burago and S. Ivanov, *Riemannian tori without conjugate points are flat*, Geom. Funct. Anal. **4** (1994), no. 3, 259–269.

[Bu-Iv2] D. Burago and S. Ivanov, *On asymptotic volume of Finsler tori, minimal surfaces in normed spaces, and symplectic filling volume* (preprint).

[Bu-Ka] K. Burns and A. Katok, *Manifolds with non-positive curvature*, Ergod. Th. & Dynam. Syst. **5** (1985), 307–317.

[Bu-Sp] K. Burns and R. Spatzier, *Manifolds of nonpositive curvature and their buildings*, Inst. Hautes Études Sci. Publ. Math. **65** (1987), 35–59.

[Bu-Ko] S.V. Buyalo and V.L. Kobelskii, *Geometrization of graph-manifolds. II. Isometric geometrization* (Russian) Algebra i Analiz **7**, no. 3 (1995), 96–117; translation in St. Petersburg Math. J. **7** (1996), no. 3, 387–404.

[Cao] J. Cao, *Rigidity for non-compact surfaces of finite area and certain Kähler manifolds*, Ergodic Theory Dynam. Systems **15** (1995), no. 3, 475–516.

[Ch-Ya] S.-Y. Chang and P. Yang, *Isospectral conformal metrics on 3-manifolds*, J. Amer. Math. Soc. **3** (1990), no. 3, 117–145.

[Co-Fa] C. Connell and B. Farb, *Minimal entropy rigidity for lattices in products of rank one symmetric spaces* (preprint).

[Cr1] C. CROKE, *Conjugacy rigidity for nonpositively curved graph manifolds* (preprint).

[Cr2] C. CROKE, *Rigidity and the distance between boundary points*, J. Diff. Geom. **33** (1991), 445–464.

[Cr3] C. CROKE, *Rigidity for surfaces of non-positive curvature*, Comm. Math. Helv. **65** (1990), no. 1, 150–169.

[Cr4] C. CROKE, *Volumes of balls in manifolds without conjugate points*, International J. Math. **3** (1992), no. 4, 455–467.

[C-D-S] C. CROKE, N. DAIRBEKOV, AND V. SHARAFUTDINOV, *Local boundary rigidity of a compact Riemannian manifold with curvature bounded above*, Trans. Amer. Math. Soc. **325** (2000), no. 9, 3937–3956.

[C-E-K] C. CROKE, P. EBERLEIN, AND B. KLEINER, *Conjugacy and rigidity for nonpositively curved manifolds of higher rank*, Topology **35** (1996), 273–286.

[C-F-F] C. CROKE, A. FATHI, AND J. FELDMAN, *The marked length-spectrum of a surface of nonpositive curvature*, Topology **31** (1992), no. 4, 847–855.

[Cr-Kl1] C. CROKE AND B. KLEINER, *Conjugacy and Rigidity for Manifolds with a Parallel Vector Field*, J. Diff. Geom. **39** (1994), 659–680.

[Cr-Kl2] C. CROKE AND B. KLEINER, *A rigidity theorem for simply connected manifolds without conjugate points*, Ergod. Th. & Dynam. Syst. **18** (1998), pt. 4, 807–812.

[Cr-Kl3] C. CROKE AND B. KLEINER, *A rigidity theorem for manifolds without conjugate points*, Ergod. Th. & Dynam. Syst. **18** (1998), pt. 4, 813–829.

[Cr-Sh] C. CROKE AND V. SHARAFUTDINOV, *Spectral rigidity of a compact negatively curved manifold*, Topology **37** (1998), 1265–1273.

[Du-Gu] J. DUISTERMAAT AND V. GUILLEMIN, *The spectrum of positive elliptic operators and periodic bicharacteristics*, Invent. Math. **29** (1975), no. 1, 39–79.

[Eb] P. EBERLEIN, *Geometry of 2-step nilpotent groups with a left invariant metric*, Ann. Sci. École Norm. Sup. (4) **27** (1994), no. 5, 611–660.

[Eb-He] P. EBERLEIN AND J. HEBER, *A differential geometric characterization of symmetric spaces of higher rank*, Inst. Hautes Études Sci. Publ. Math. **71** (1990), 33–44.

[Fa] H.-R. FANAI, *Conjugaison géodśique des nilvariétés de rang deux* (French), J. Lie Theory **10** (2000), no. 1, 147–155.

[Fe-Or] J. FELDMAN AND D. ORNSTEIN, *Semi-rigidity of horocycle flows over surfaces of variable negative curvature*, Ergod. Th. & Dynam. Syst. **7** (1987), 49–74.

[Fo] P. FOULON, *locally symmetric Finsler spaces in negative curvature* (preprint).

[Fr-Ma] A. FREIRE AND R. MAÑÉ, *On the entropy of the geodesic flow in manifolds without conjugate points*, Invent. Math. **69** (1982), 375–392.

[Go-Go] C. GORDON AND R. GORNET, *Spectral geometry on nilmanifolds*, Progress in inverse spectral geometry, 23–49, Trends Math., Birkhäuser, Basel, 1997.

[G-M-S] C. GORDON, Y. MAO, AND D. SCHUETH, *Symplectic rigidity of geodesic flows on two-step nilmanifolds*, Ann. Sci. École Norm. Sup. (4), **30** (1997), no. 4, 417–427.

[Go-Wi] C. GORDON AND E. WILSON, *Isospectral deformations of compact solvmanifolds*, J. Diff. Geom. **19** (1984), 241–256.

[Gr-Gu] L. GREEN AND R. GULLIVER, *Planes without conjugate points*, J. Diff. Geom. **22** (1985), 43–47.

[Gr-Wu] R. GREENE AND H.H. WU, *On a new gap phenomenon in Riemannian Geometry*, Proc. Nat. Acad. Sci. U.S.A. **79** (1982), no. 2, part 1, 714–715.

[Grog] S. GROGNET, *Flots magnétiques en courbure négative* (French), Ergodic Theory Dynam. Systems **19** (1999), no. 2, 413–436.

[Gro] M. GROMOV, *Filling Riemannian manifolds*, J. Diff. Geom. **18** (1983), 1–147.

[Gu-Ka1] V. GUILLEMIN AND D. KAZHDAN, *Some inverse spectral results for negatively curved 2-manifolds*, Topology **19** (1980), 301–312.

[Gu-Ka2] V. GUILLEMIN AND D. KAZHDAN, *Some inverse spectral results for negatively curved n-manifolds*, Proc. Sympos. Pure Math. **36**, Amer. Math. Soc., Providence RI, 1980, 153–180.

[Gul] R. GULLIVER, *On the variety of manifolds without conjugate points*, Trans. Amer. Math. Soc. **210** (1975), 185–201.

[Ha1] U. HAMENSTÄDT, *Time-preserving conjugacies of geodesic flows*, Ergod. Th. & Dynam. Sys. **12** (1992), 67–74.

[Ha2] U. HAMENSTÄDT, *Cocycles, symplectic structures and intersection*, Geom. Funct. Anal. **9** (1999), no. 1, 90–140.

[Her] G. HERGLOTZ, *Uber die Elastizitaet der Erde bei Beruecksichtigung ihrer variablen Dichte*, Zeitschr. fur Math. Phys. **52** (1905), 275–299.

[Ho] E. HOPF, *Closed surfaces without conjugate points*, Proc. Nat. Ac. Sc. USA **34** (1948), 47–51.

[Iv] S. IVANOV, *On two-dimensional minimal fillings* (preprint).

[Ja-Sh] W. JACO AND P. SHALEN, *Seifert fibred spaces in 3-manifolds*, Mem. Amer. Math. Soc. **220** (1979).

[Jo] K. JOHANSSON, *Homotopy equivalences of 3-manifolds with boundary*, Springer LNM **761** (1979).

[Ki-Yi] C-W KIM AND J-W YIM, *Rigidity of noncompact Finsler manifolds*, Geom. Dedicata **81** (2000), no. 2, 245–259.

[L-S-U] M. LASSAS, V. SHARAFUTDINOV, AND G. UHLMANN, *Semiglobal rigidity for Riemannian metrics* (preprint).

[Le] B. LEEB, *3-manifolds with(out) metrics of nonpositive curvature*, Invent. Math. **122** (1995), 277–289.

[Man] A. MANNING, *Topological entropy for geodesic flows*, Ann. of Math. (2) **110** (1979), no. 3, 567–573.

[Mal] A. MAL'CEV, *On isomorphic matrix representations of infinite groups* (Russian) Mat. Rec. Math. [Mat. Sbornik] N.S. **8** (1940), no. 50, 405–422.

[Mu1] R.G. MUKHOMETOV, *Inverse kinematic problem of seismic on the plane*, Math. Problems of geophysics, Akad. Nauk. SSSR, Sibirsk.Otdel., Vychisl. Tsentr, Novosibirsk, **6** (1975), no. 2, 243–252 (Russian).

[Mu2] R.G. MUKHOMETOV, *The reconstruction problem of a two-dimensioina; Riemannian metric, and integral geometry*, Soviet Math. Dokl. **18** (1977), no. 1, 27–37.

[Mu3] R.G. MUKHOMETOV, *On a problem of reconstructing Riemannian metrics*, Siberian Math. J. **22** (1982), no. 3, 420–433.

[Mi] R. MICHEL, *Sur la rigidité imposée par la longuer des géodésiques*, Inv. Math. **65** (1981), 71–83.

[MO] M. MIN-OO, *Spectral rigidity for manifolds with negative curvature operator*, Contemp. Math. **51** (1986), 99–103.

[Ot1] J.-P. OTAL, *Le spectre marqué des longueurs des surfaces à courbure négative*, Ann. of Math. **131** (1990), 151–162.

[Ot2] J.-P. OTAL, *Sur les longueurs des géodésiques d'une métrique à courbure négative dans le disque*, Comment. Math. Helv, **65** (1990), no. 2, 334–347.

[Pe-Sh] L. PESTOV AND A. SHARAFUTDINOV, *Integral geometry of tensor fields on a manifold of negative curvature*, Novosibirsk (transl. from Sibirskii Math. Zhurnal **29** (1988), no. 3, 114–130.

[Ro1] V.G. ROMANOV, *Reconstructing a function by means of integrals along a family of curves*, Sibirsk. Mat. Zh. **8** (1967), no. 5, 1206–1208.

[Ro2] V.G. ROMANOV, *Integral geometry on geodesics of an isotropic Riemannian metric*, Soviet Math. Dokl. **19** (1979), no. 4, 847–851.

[Sch] V. SCHROEDER, *Rigidity of nonpositively curved graphmanifolds*, Math. Ann. **274** (1986), 19–26.

[Vi] M. VIGNÉRAS, *Variétés riemanniennes isospectrales et non isométriques*, Ann. of Math. **110** (1980), 21–32.

[We] A. WEINSTEIN, *Fourier integral operators,quantization and the spectra of Riemannian manifolds*, Colloques Internationaux C.N.R.S., no. **273** — Géométrie symplectique et physique mathematique (1975), 289–298.

[Wi-Zo] E. WIECHERT AND K. ZOEPPRTZ, *Uber Erdbebenwellen*, Nachr. Königl. Geselschaft Wiss, Göttingen **4** (1907), 415–549.

THE CASE FOR DIFFERENTIAL GEOMETRY IN THE CONTROL OF SINGLE AND COUPLED PDES: THE STRUCTURAL ACOUSTIC CHAMBER

R. GULLIVER*, I. LASIECKA†, W. LITTMAN*, AND R. TRIGGIANI†

Contents

Part I: The need for differential geometry in the control of PDEs. Its main features . . . 75

0 Informal overview of some control theory concepts and terminology . . . 76

1 Introduction: Aim and scope of the present paper. Qualitative statement of results. Two approaches . . . 83

1.1 Linear single PDEs, or coupled systems of PDEs, with variable coefficients and/or defined on curved surfaces . . . 83

1.2 Two geometric approaches . . . 84

1.3 Summary of main features of differential geometric methods in the control of PDEs . . . 85

1.4 Geometric methods and control of PDEs: A two-way interaction . . . 86

1.5 Systems of coupled PDEs. Nonlinear problems . . . 87

2 A benchmark PDE-control problem advocating a broad use of differential geometry: The structural acoustic problem . . . 88

2.1 Various structural acoustics models with constant coefficients and with flat flexible wall . . . 89

2.2 More realistic models: Variable coefficients and curved walls . . . 91

2.3 Conclusion . . . 92

Part II: Riemannian geometric assumptions. Overview of results 92

3 Basic geometric assumption for the energy method approach in the Riemannian metric . . . 94

3.1 Strictly convex functions and coercive fields in the Riemannian metric for hyperbolic and non-hyperbolic (Petrowski)-problems . . . 94

3.2 Use of assumption (H.1). General statement of Carleman estimates and overview . . . 96

*School of Mathematics, Vincent Hall, University of Minnesota, Minneapolis, MN 55455

†Department of Mathematics, Kerchof Hall, University of Virginia, Charlottesville, VA 22904. Research partially supported by the National Science Foundation under Grant DMS-0104305 and by the Army Research Office under Grant DAAD19-02-1-0179.

4 Survey of geometric results and methods for the construction of strictly convex functions and coercive fields 103
4.1 The need for a geometrical perspective 103
4.1.1 Euclidean Space 104
4.1.2 The Hessian Comparison Theorem 104
4.2 Functions with convex level sets 105
4.3 The Matrix Riccati Equation 106
4.4 Matrix Riccati Comparison Theorem 107
4.5 Nonsmooth distance function 109
4.6 Construction of a strictly convex function in the 2-d case via curvature flows 110
5 A Riemannian geometric approach for exact controllability of second-order hyperbolic equations. A "view from the boundary" which verifies the geometric optics condition . . 111
5.1 Introduction. Counterexamples. Hypotheses. Main exact controllability statement 111
5.2 Chord uniqueness and nondegeneracy 115
5.3 Criteria for chord uniqueness and non-degeneracy . . 116
5.4 Examples 117
5.5 Sketch of proof of Theorem 5.1.1: Verification of the geometric optics conditions 119
Part III: Single PDE equations: Carleman estimates. Exact controllability. Uniform stabilization 120
6 Second-order hyperbolic equations on a Riemannian manifold: Energy methods in the Riemannian metric yielding Carleman and control estimates with lower-order terms . . 120
6.1 Model. Euclidean versus Riemannian domain 120
6.2 Strictly convex functions and coercive fields in the Riemannian metric 121
6.3 A second-order hyperbolic equation with variable coefficients defined on a Euclidean domain 122
6.4 Carleman estimates for problem (6.1.1), (6.1.2) with lower-order terms 123
6.5 Control-theoretic results for (6.1.1)–(6.1.3) 126
6.6 Lower-level energy inequality: stabilization inequality with Dirichlet-boundary feedback in $L_2(\Sigma)$ . . . 131
6.7 Geometric optics and strictly convex functions . . . 133
7 Second-order hyperbolic equations: Energy method in the Riemannian metric yielding Carleman and control estimates with no lower-order terms. Purely Neumann B.C. Global unique continuation 136
8 Schrödinger equations on a Riemannian manifold: Energy methods in the Riemannian metric yielding Carleman/control estimates with lower-order terms 144

9 Schrödinger equations: Energy method yielding Carleman/ control estimates with no lower-order terms. Purely Neumann B.C. Global unique continuation 150
10 Plate equations on the Riemannian manifold: Energy methods in the Riemannian metric yielding Carleman/control estimates with lower-order terms 153
Part IV: A dynamic shell model 159
11 Uniform stabilization of a shallow shell model with nonlinear boundary feedbacks. Critical marriage between Riemannian geometry (continuous observability inequality) and microlocal analysis (boundary trace estimates) 159
11.1 A differential geometric model for a dynamic shallow shell's model in nonlinear, dissipative, feedback form 161
11.2 Main results: Well-posedness and uniform stabilization 163
11.3 Preliminary results 167
11.4 First step of the proof: Local reduction to a Euclidean (flat) coordinate system 169
11.5 Trace regularity for elastic waves (W-component) . . 170
11.6 Trace regularity for normal component w 172
11.7 Stabilizability estimate and completion of the proof of Theorem 11.2.2 172

Part I: The need for differential geometry in the control of PDEs. Its main features.

Foreword. In line with the title of the IMA Summer Program—Geometric Methods in Inverse Problems and PDE Control—the aim of the present article may be summarized as follows: we intend to provide a relatively updated survey (subject to space limitations) of results on *exact boundary controllability* and *uniform boundary stabilization* of certain general classes of single Partial Differential Equations as well as of classes of systems of coupled PDEs (in dimension strictly greater than one), that have become available in recent years *through novel approaches based on differential (Riemannian) geometric methods.*

Thus, this is *not* a survey on *all* relevant methods that have been introduced in the control of PDEs (in dimension greater than one). Rather, this is a review article deliberately restricted and confined to approaches that make explicit and intense use of differential (Riemannian) geometric methods. Accordingly, some cornerstone approaches available in the literature of control of PDEs are in the background rather than on the forefront of this article. They include (but are not necessarily limited to):

(i) The geometric optics approach in an Euclidean domain, which is restricted, however, to *hyperbolic* dynamics. It was initiated by W. Littman [Lit.1] for *hyperbolic* systems controlled on the full boundary, and by J. Ralston [Ra.1] in the study of 'Gaussian beams,' and later carried out to a high level of generality by C. Bardos, J. Lebeau, and J. Rauch [B-L-R.1] for general *second-order hyperbolic equations.* This approach rests on an almost necessary and sufficient geometric optics assumption, which, however, is not readily checkable. It requires C^∞-smoothness of both the Euclidean domain and the coefficients. Later, smoothness was greatly reduced to C^3-domains and C^2-coefficients in Burq [Bu.1], but only for the wave equation with Dirichlet-control, by using H-measures. An adaptation of this method to the pure Schrödinger equation is given in Lebeau [Le.2], under the same geometric optics condition.

(ii) The approach of D. Tataru [Ta.1–3] this time for general evolution equations (to include the single PDEs explicitly treated in the present article), which is based on pseudo-differential calculus (see his review article [Ta.8] on unique continuation on this same volume) to obtain first, as an intermediate step, Carleman-type estimates, under a not-so-easily checkable pseudo-convexity assumption, and next the sought-after control-theoretic inequalities.

The computational Riemannian geometric energy method approach to be treated in the present article shares, in common with the pseudo-differential approach of Tataru, the important feature that it applies to *various* single PDE equations, be they hyperbolic or not, including Petrowski-type PDEs (and other non-hyperbolic PDEs are expected to be included in the feature). As such, the computational Riemannian geometric method serves as a unifying tool applicable also to the study of *systems* of PDEs which couple single PDEs dynamics of *various type*, hyperbolic or not, of interest in applications. A canonical motivating illustration is the structural acoustic chamber of Section 2.

Much of the article—Parts I through III—is focused on general *linear* PDEs, while we refer to Section 1.5 below for information and insight on the corresponding *nonlinear* classes. However, Part IV considers the stabilization problem of the *linear* (coupled system of two hyperbolic-like PDEs known as) shallow shell, subject to suitable *nonlinear* dissipative terms. We quote from the IMA Workshop: "*The IMA workshop goals are to bring together geometers with researchers in inverse problems and control of PDEs to facilitate exchange of ideas and encourage collaboration; to make tools of differential geometry known to those working in inverse problems and control, and to open new areas of research in geometry.*" In this spirit, we open up this paper with an introductory qualitative treatment, written for the benefit of those in areas other than control of PDEs. The expert in control of PDE may skip this introductory part and move directly to the variable coefficient case (Section 1).

0. Informal overview of some control theory concepts and terminology.

Exact boundary controllability and equivalent continuous observability inequality. Qualitatively, by *exact boundary controllability* of an evolution equation we mean the property of steering or transferring any initial condition at time $t = 0$ to any target state at time T (in suitable function spaces) by means of a non-homogeneous boundary function, called *boundary control* (in a suitable function space) acting on (possibly, a portion of) the boundary of the open bounded domain Ω in which the evolution is defined.

The optimal relationship between the function space that defines the class of boundary controls and the function space defining the corresponding solution (state) at $t = T$ is very important in itself, and critical in the area of *optimal control*, which justifies the interest in the property of exact boundary controllability in the first place [L-T.11]. Other questions such as: (i) "minimal" universal time of transfer (independent of the pair of initial and final state); (ii) "minimal" portion of the boundary required by the boundary control to perform any such transfer, are also of interest, though at a far lower order of priority in applications to *optimal control* theory.

Here we shall be concerned only with *time reversible* single or coupled evolution equations (second-order hyperbolic equations, Petrowski-systems such as Schrödinger equations and plate-like equations (hyperbolic or not), system of elasticity, shells, etc.). [An exception is the thermoelastic plate of Section 2.] Accordingly, w.l.o.g. we may take the initial condition to be the *origin*. We then seek to hit any target state (in a suitable function space) at time T (universal) by means of a suitable class of boundary controls. This says, mathematically, that the 'control $\to$ solution map' L_T (which may be either continuous, or else unbounded but closed)

$$(0.1)\qquad \begin{array}{ll} L_T: & \text{from the space of boundary controls be surjective (onto)} \\ \longrightarrow & \text{state space of solutions at } t = T. \end{array}$$

The standard mathematical property (0.1) is re-labelled, in the control theory jargon, as *exact boundary controllability* at T. By a standard result in Functional Analysis [T-L.1, p. 235], the surjectivity property (0.1) is then *equivalent* to the property that the adjoint operator L_T^* is bounded below: there exists a constant $c_T > 0$ such that

$$(0.2)\qquad \|L_T^* z\| \geq c_T \|z\|, \qquad \forall\, z \in \mathcal{D}(L_T^*).$$

A readily accessible step (which may be carried out either at the PDE-level, or at the operator-theoretic level) permits one to recast (0.2)—for each evolution equation of concern, thus for any such L_T—into an *equivalent* PDE estimate: there exists a constant $c_T > 0$ such that

$$(0.3)\qquad \begin{cases} c_T E_w(0) \leq \displaystyle\int_0^T \int_{\Gamma_1} (\text{suitable trace of } w)^2 d\Gamma_1 dt \\ w = \text{solution of } \textit{dual} \text{ problem defined by } L_T^*, \text{ which is} \\ \quad \textit{homogeneous} \text{ on the boundary in the same B.C.} \\ \quad \text{where the boundary control is applied in the original} \\ \quad \text{problem defining } L_T. \end{cases}$$

In (0.3), $E_w(t)$ is the 'energy' (norm) associated with the solution w of the dual, *boundary homogeneous* PDE, whose homogeneous Boundary Condition(s) (B.C.(s)) correspond(s) to the B.C.(s) where the boundary controls are applied in the original problem defining L_T. Moreover, Γ_1 is the 'observed' portion of the boundary $\Gamma = \partial\Omega = \Gamma_1 \cup \Gamma_0$ of the dual problem w.

This equivalently says that Γ_1 is the 'controlled' portion of the boundary of the original problem; that is, Γ_1 is the part of Γ where controls are applied, while on the complementary portion $\Gamma_0 = \Gamma \setminus \Gamma_1$ of the boundary, the boundary conditions are instead *homogeneous*. Finally, T in (0.3) is a (universal) time for boundary observation of the dual problem w, or for boundary controllability of the original problem defining L_T.

We note at the outset that the vanishing trace (restriction) of the boundary homogeneous solution w is 'complementary' to the trace appearing on the RHS of (0.3). For instance, in the case of the wave equation, if the boundary controls are applied to the Dirichlet [respectively, Neumann] B.C. on $(0, T] \times \Gamma_1$, then the dual problem w satisfies homogeneous (*zero*) Dirichlet B.C. [respectively, Neumann B.C.] on $(0, T] \times \Gamma_1$, while the trace of w occurring on the RHS of (0.3) is the Neumann trace [respectively, a Dirichlet trace]. We refer to Part III for specific cases: see (6.5.3) and (6.5.6) below for second-order hyperbolic equations; see (8.14) and (8.17) below for Schrödinger equations, etc.

Inequality (0.3) is traditionally [R.1] called the *Continuous Observability Inequality* (C.O.I.) of the evolution described by the dual boundary homogeneous w-problem (of which (0.2) is the abstract version in terms of L_T).

To recap, we have thus seen that: the exact boundary controllability (surjectivity) property (0.1) is then *equivalent* to the *Continuous Observability Inequality* (0.3). [The equivalence between controllability and observability has been noticed and used extensively in the finite-dimensional control theory, since the work of R. Kalman in the early 1960's.]

The *crux* of the entire problem is, therefore, to *establish the validity* of the C.O.I. (0.3) [possibly with a "minimal" universal time T, with a "minimal" observed/controlled portion of the boundary Γ_1 and, above all] with the optimal relationship between the topology of the trace and the topology of the initial energy. We note that [(0.2) or] (0.3) is an *inverse-type inequality*: it reconstructs the initial energy of the w-evolution equation in terms of information on the boundary (trace, or restriction, of the solution). This establishes a link between this part of control theory and inverse problems. *A-fortiori*, the continuous observability inequality (0.3) *implies a global unique continuation result*: if the boundary homogeneous w-problem has, in addition, equal to zero also the traces on Γ_1 contained on the RHS of (0.3), then $E(0) = 0$ and (when the problem is semigroup well-posed forward in time), then $w \equiv 0$ identically. However, the precise nature of the inequality in (0.3)—where the inverse map from the boundary traces (restrictions) over $[0, T] \times \Gamma_1$ of the boundary homogeneous problem w to the initial energy is continuous—makes the continuous observability inequality much stronger than the corresponding global unique continuation property of the over-determined problem. The latter is a much studied (and already challenging to study) basic property in pure PDE theory [Lit.2], [Hor.1], [Ta.8], [Tay.1]: however, the C.O.I. (0.3) requires much more!

Constructive minimal-norm steering control. One issue of potential interest remains. Suppose that one succeeds in showing the C.O.I. (0.3) for the dual problem w. Then, equivalently, the original problem is exactly controllable on $[0, T]$, by means of boundary controls applied on $(0, T] \times \Gamma_1$. But then: can one construct a boundary control that steers say, the origin [respectively, an original state] at time $t = 0$ to a desired target state [respectively, to rest] at $t = T$? The answer is in the affirmative: once exact boundary controllability is established (by showing the validity of the dual C.O.I. in (0.3)), one may then set up a minimization problem with constraint, to identify (via a Lagrange multiplier, functional analytic argument) the *minimal norm* boundary control; this is the control that, among all controls performing the required transfer, has minimal norm. References [Tr.1, Appendix], [L-T.3, Appendix] produce general formulas of the minimal norm, steering controls related to abstract (linear) equations, which encompass the present case of boundary controls for PDEs.

These formulas are as follows. Henceforth, we call U the abstract control space and $e^{\mathcal{A}t}$ the free dynamics s.c. semigroup on the state space. Then, under the property of exact controllability, we have:

(i) The minimal $L_2(0, T; U)$-norm steering control $u^0(\,\cdot\,; z_0)$ [respectively, $u^0_{z_T}(\,\cdot\,)$], which steers the initial state z_0 at $t = 0$ [respectively, the origin at $t = 0$] to the origin at $t = T$ [respectively, to the target state z_T at $t = T$], along the dynamics described by the operator L_T, is given by

$$u^0(\,\cdot\,; z_0) = -L_T^*(L_T L_T^*)^{-1} e^{\mathcal{A}T} z_0 \quad [\text{resp. } u^0_{z_T}(\,\cdot\,) = L_T^*(L_T L_T^*)^{-1} z_T];$$

(ii) The corresponding minimal control energy is given by

$$\left\{ \int_0^T \|u^0(t; z_0)\|^2 dt \right\}^{\frac{1}{2}} = \|(L_T L_T^*)^{-\frac{1}{2}} e^{\mathcal{A}T} z_0\|$$

[respectively,

$$\left\{ \int_0^T \|u^0_{z_T}(t)\|^2 dt \right\}^{\frac{1}{2}} = \|(L_T L_T^*)^{-\frac{1}{2}} z_T\|. \quad]$$

Uniform boundary stabilization and equivalent inequality (Linear case). One begins with a *conservative* (energy preserving) evolution equation, homogeneous on the boundary. Next, one seeks to identify a suitable dissipative (damping) ("feedback") term on (a suitable portion of) the boundary such that: the new corresponding dissipative problem is (i) well-posed (that is, it defines a semigroup $e^{\mathcal{A}_F t}$, F stands for feedback) and (ii) its energy ("norm") $E(t)$ decays exponentially to zero: there exist constants $M \geq 1$, $\delta > 0$ such that

$$E(t) \leq M e^{-\delta t} E(0), \text{ equivalently } \|e^{\mathcal{A}_F t}\| \leq \overline{M} e^{-\delta t},\ t \geq 0. \tag{0.4}$$

A (necessary and) sufficient condition for achieving the exponential energy decay (0.4) is as follows: there exists a time $T > 0$ sufficiently large, and a corresponding constant $c_T > 0$ such that the following Uniform Stabilization Inequality (U.S.I.) holds true:

$$(0.5)\quad \begin{cases} c_T E_y(T) \le \int_0^T \int_{\Gamma_1} (\text{suitable boundary dissipation of } y)^2 d\Gamma_1 dt; \\ y = \text{solution of boundary dissipative problem.} \end{cases}$$

Indeed, (0.5) would then imply (be equivalent to) the property that the resulting feedback semigroup $e^{A_F t}$ describing the dissipative system satisfies the condition: $\|e^{A_F T}\| < 1$. And this is a well-known characterization for the semigroup $e^{A_F t}$ to be exponentially stable (in the appropriate uniform norm).

We have said that, by construction, the y-problem is dissipative. This means that it satisfies the following *dissipative identity*:

$$(0.6)\quad E_y(T) + \int_0^T \int_{\Gamma_1} (\text{suitable boundary dissipation of } y)^2 d\Gamma_1 dt = E_y(0),$$

which can readily be proved, by and large, by a routine energy method. Here Γ_1 is the portion of the boundary affected by the dissipation, while homogeneous B.C. are in place on $\Gamma \backslash \Gamma_1$. Moreover, the "suitable boundary dissipation of y" appearing in both (0.5) and (0.6) is the same. Thus, in the LHS of the U.S.I. (0.5), we can always replace the energy $E_y(T)$ at $t = T$ with the initial energy $E_y(0)$, modulo a different constant $c_T > 0$. Thus, with the U.S.I. (0.5), with either $E_y(T)$ or else $E_y(0)$ on its LHS, we are back to an *inverse-type* inequality, this time for the *boundary dissipative* problem y (not the *boundary homogeneous* problem w, as in (0.3). This is another link between this part of control theory and inverse problems!

Topological level in exact boundary controllability and in uniform boundary stabilization. It is clear that showing the U.S.I. (0.5) is generally more challenging than showing the C.O.I. (0.3). For one, (0.5) refers to a more complicated boundary dissipative problem y, while (0.3) refers to a simpler boundary homogeneous problem w. In addition, and much more seriously, the energy E_y may very well be at a *lower* Sobolev level than the energy E_w in (0.3). This occurs, e.g., in the case of uniform stabilization of either the (conservative) wave equation or the (conservative) Schrödinger equation with dissipative term in the *Dirichlet* (rather than Neumann) B.C.: for a treatment of wave equations, we refer to the forthcoming Section 6.6). In these cases, the energy level E_y of the dissipative y-problem is one Sobolev unit below than the energy level E_w of the homogeneous case:

(i) $E_y \sim L_2(\Omega) \times H^{-1}(\Omega)$, see [L-T.1], [L-T.12], [L-T-Y.3]; $E_w \sim H^1(\Omega) \times L_2(\Omega)$ for wave equation;

(ii) $E_y \sim L_2(\Omega)$, see [L-T.8]; $E_w \sim H^1(\Omega)$ for Schrödinger equation.

Showing the inverse type inequality at a lower energy level (below the 'physical' energy) is surely much more demanding. It requires a shift of topology. We refer to Remark 6.6.2 for more technical and historical insight.

Relationship between exact boundary controllability and uniform boundary stabilization. There is a *two-way relationship* between uniform boundary stabilization and exact boundary controllability of an evolution equation of the type under consideration.

In one direction, for an originally, say *conservative* problem, a simple, smart idea—due to D. Russell [R.1]—shows that uniform stabilization (for positive times) of the new dissipative problem implies exact controllability of the original conservative evolution equation, on the same function space, constructively, with a steering control which may be constructed by means of two stabilizing feedback controllers. As a result of this argument, the transfer from the initial state to the target state of the original conservative system takes place *continuously in time* with values in the state space: a bonus or desirable property this, that cannot follow from the C.O.I. (e.g., exact controllability in $H^1(\Omega) \times L_2(\Omega)$ of the wave equation with $L_2(0, T; L_2(\Gamma))$-Neumann controls), since the latter is only concerned with what happens at $t = T$ and not with the evolution over $[0, T]$, which is a regularity question. The additional price paid in this approach—where one begins by proving the more demanding U.S.I. (0.5) rather than the less demanding C.O.I. (0.3) is twofold: (i) the original system has to be conservative (a small generalization is possible); (ii) the (universal) time for exact controllability is generally much larger than necessary.

In the other direction, consider a general evolution equation which is exactly controllable on the space of optimal regularity: this means that both the C.O.I. (0.3) (possibly for T large) as well as its *reverse* inequality (for any $T > 0$) hold true (in the same topologies for the initial energy and boundary traces). Then the theory of optimal control with a quadratic cost functional to minimize provides a *different* feedback stabilizer, not expressed as a velocity feedback as discussed heretofore, but expressed by means of a Riccati operator: but this is an altogether different topic, for which we refer to [L-T.11], [L-T.24]. A recent related paper on these questions is [L-T.30].

Thus, this part of control theory for PDEs concerned with exact controllability and uniform stabilization seeks to establish the validity of the equivalent dual inequalities: the C.O.I. (0.3) and/or the U.S.I. (0.5). [There are some exceptions which follow a *direct* approach to exact controllability (not via the dual problem): (i) the use of harmonic or non-harmonic analysis, moment problems, Riesz basis, almost exclusively in one-dimensional problems (or problems reducible to one dimension by spatial symmetry) and thus not in the mainstream interest of the present article; (ii) the approach of W. Littman [Lit.1], [Lit-Ta.1] inspired by the Huyghens' principle. Though these concepts of exact controllability/uniform stabilization are very 'old,' it was only in the early mid-1980s that inequalities such as the U.S.I. (0.5) (or equivalent versions) and the C.O.I. (0.3) were established, at first for the *pure Euclidean wave equation*: with the *Euclidean Laplacian* and *no energy level terms* ($H^1(\Omega) \times L_2(\Omega)$, *that is, gradients in space and time*). Soon thereafter, the successes achieved with the pure Euclidean wave equation, were transferred to establish the corresponding inequalities—C.O.I. (0.3) and U.S.I. (0.5)—also for the pure Euclidean Schrödinger equation, plate equations (of both Euler-Bernoulli type or of Kirchhoff type), etc., with energy method techniques inspired by the pure Euclidean wave equation case.

Inadequacy of, and difficulties with, "classical" energy methods ("multipliers") of the '80s in the case of variable coefficients and energy level terms. By the end of the 80's, one may say that the energy methods used (see the many works cited in the References) (special multipliers), however, worked well *only* for such canonical wave/Schrödinger/plate models *(constant coefficients and no energy level terms), but were inadequate to treat more general models with variable coefficients in the principal part and/or in the energy level terms.* This brings us to the main motivation of the present article: How to establish the C.O.I. (0.3) and the U.S.I. (0.5) for single PDEs, or coupled systems of PDEs, with variable coefficients in both the principal part and in energy level terms and/or defined on curved surfaces? Whatever method is used, this will have to represent a seriously more technical and sophisticated approach over the "classical" one of the '80's, which was successful only for canonical models. One general line of investigation proposed relies on Riemannian geometry: its energy method approach—to be expounded in Section 3 and Part III—may be viewed as a far-reaching generalization of the energy methods of the '80's. This will be made more clear in Remark 6.4.1 below for second-order hyperbolic equations; in Remark 8.1 for Schrödinger equations; and in Remark 10.1 for plate-like equations.

Other approaches to the control of variable coefficient (linear) PDEs include: (i) the geometric optics line of investigation [Lit.1], [B-L-R.1] for second-order hyperbolic equations, with C^∞-domains and with C^∞-coefficients (later reduced to C^3-domains and C^2-coefficients in [Bu.1] via the use of H-measures, but only for the variable coefficient pure wave equation with no energy level terms and Dirichlet control). It is likely that recent improvements in the field of propagation of singularities [Tay.2] will allow to reduce the degree of smoothness required by [Lit.1], [B-L-R.1]; (ii) the pseudo-differential method in [Ta.1–3] for general evolution equations. The Ph.D. thesis [Ta.1] (for general evolution equations) and [K-K.1] (second-order hyperbolic equations with constant principal part and traditional B.C.) were, apparently, the first works to introduce in the control literature Carleman estimates involving *boundary traces*, the object of key interest in control-theoretic inequalities. Prior literature on the vast topic of Carleman estimates (initiated by Carleman around 1937 and much expanded and perfected by Hörmander [Hor.1–2], Taylor [Tay.1]) referred to solutions of PDEs with compact support for purposes of unique continuation, and hence contained no boundary traces. There appears to be a close link [Hor.1], [Ta.8], in both scope and formulation of hypotheses, between the pseudo-differential energy method approach of [Ta.1–3] and the Riemannian energy method of Part III (Sections 6, 8, 10), though the mathematical machineries and technicalities involved are vastly different. The infusion of Riemannian geometric ideas does appear to help, at any rate, to clarify and verify the notions of pseudo-convexity in the pseudo-differential approach. See further Remark 3.2.1 below, and Sections 4 and 5.

1. Introduction: Aim and scope of the present paper. Qualitative statement of results. Two approaches.

1.1. Linear single PDEs, or coupled systems of PDEs, with variable coefficients and/or defined on curved surfaces. Very recently, differential (Riemannian) geometric methods have been introduced in the control of certain broad classes of Partial Differential Equations (PDEs). As explained in the preceding historical summary, the original motivation may be said to have arisen from the need to cope with the following two situations:

(i) The case where the coefficients of the PDE are variable in space in its principal part and, possibly, variable in both space and time in its 'energy level' terms;

(ii) the case where the PDE itself is defined on curved surfaces (manifolds).

Regarding the first need (i), we shall present two approaches, see Section 1.2: one applicable to various single PDEs, be they hyperbolic or not, as well as to coupled systems of various PDEs; and one specifically tuned to hyperbolic problems. As to the second need (ii), it would seem self-explanatory that if an equation is defined on a manifold (as in the case of shells), the natural setting which is called for its analysis should be that of differential geometry. However, this was not the case in the large literature in shell theory.

While a mutually profitable link had been long established, for at least 30 years, between differential geometry and control of nonlinear ordinary differential equations, a comparable relationship between differential geometry and control of PDEs is a new topic, which has already confirmed, in just a few years, several of its original promises. Similarly, a useful interplay between differential geometry and the general theory of PDEs has long been in place and documented [Taylor, Hörmander, Treves]. Yet, the distinguished use of differential geometric methods in the setting which arises in boundary control theory of PDE is novel. Moreover, such an approach is also far from having been exhaustively explored. Nevertheless, to date, a relatively large and useful body of knowledge has emerged over the past few years in the area of control of PDEs, which is based on differential geometric methods. It was originated in [Y.1] for the purely wave equation followed by [L-T-Y.1–2] for the full second-order hyperbolic equation case.

Main aim. Thus, the *main aim* of the present paper is to give an updated and motivated account of this direction of research (within space limits). In the process, we intend to make a strong case illustrating the positive role that differential geometric methods may play in the modern study of control of PDEs. In substantiating their usefulness, we shall see that their impact has been, so far, three-fold:

(1) in facilitating, indeed in contributing to, the very mathematical formulation of PDE models defined on curved surfaces (manifolds) such as shells, see Part IV;
(2) in obtaining certain *a-priori* inverse-type inequalities, such as they are needed in control theory of PDEs, be they hyperbolic or Petrowski-type, with emphasis on variable coefficients and/or on equations defined on manifolds, see Part III.
(3) in the case of second-order hyperbolic equations, in obtaining geometric conditions sufficient to verify the geometric optics condition. See Section 5.

Inverse-type inequalities in control theory: The C.O.I. (0.3) and the U.S.I. (0.5). We shall have to say more below on both points (1), (3). For now, we note that the *a-priori* inequalities mentioned in (2) are those traditionally referred to as "continuous observability inequalities" (equivalent to corresponding "exact controllability" properties) and "uniform stabilization" inequalities, which were noted in (0.3) and (0.5) of Section 0. These will be the object of Part III below for different classes of evolution equations.

A preliminary step: Carleman-type inequalities. The inverse-type C.O.I. and U.S.I. of interest in control theory will be obtained as corollaries of Carleman-type inequalities (one-parameter family of inequalities with exponential weight) of interest in their own sake. For instance, in Sections 7 and 9, Carleman-type inequalities will, *a-fortiori*, imply global uniqueness of over-determined PDE-problems. The key Carleman inequalities are given in Theorem 6.4.2 (with lower-order terms) and in Theorem 7.2 (without lower-order terms) for second-order hyperbolic equations; in Theorem 8.2 (with lower-order terms) and in Theorem 9.1 plus [L-T-Z.2] (without lower-order terms) for Schrödinger equations; and in Theorem 10.2 (with lower-order term) for plate-like equations.

1.2. Two geometric approaches. In this paper we shall review recent Riemannian geometric lines of research for PDEs with variable coefficients as above, or else on manifolds. Two quite different approaches will be presented.

Approach 1. One line of research, which wholly pervades the present article, is 'computational' in flavor. It is *entirely self-contained.* Moreover, it is broad in scope, as it applies to both hyperbolic as well as non-hyperbolic, Petrowski-type classes of PDEs, as well as to coupled systems of such PDEs (e.g., structural acoustic chambers as in Section 2; shells as in Section 11). Its virtue is that, as noted above, it yields general Carleman-type estimates from which observability/stabilization estimates (0.3) or (0.5) may be derived. They are obtained by using energy methods (multipliers) in a corresponding natural Riemannian metric. These multipliers may be viewed as far-reaching extensions of the 'classical multipliers' of the early 80's used in the Euclidean setting for 'canonical' PDEs, with *constant principal* part and *no energy level terms* in the equation, as noted in Section 1. See Remark 6.4.1 for second-order hyperbolic equations; Remark 8.1 for Schrödinger equations, and Remark 10.1 for plate-like equations. In addition, the combination of these Riemannian methods with microlocal sharp trace estimates will be given in Part IV in solving the stabilization problem of a shallow shell with non-linear, dissipative terms in the physically important free B.C.

The Riemannian geometric multipliers to be exhibited in Part III for different classes of PDEs have several advantages: they allow for a readily accessible geometric interpretation of the key assumption (H.1) = (3.1.1) behind this method; they open up the vast reserves of differential geometry for the verification of this assumption (H.1) = (3.1.1) and for constructing classes of nontrivial examples (see Section 4); they reveal themselves as being far-reaching generalizations of the 'classical' multipliers of the '80s, to which they reduce in the case of canonical PDEs with constant coefficient principal part and no energy level terms. See Remark 6.4.1, 8.1, and 10.1 for waves, Schrödinger, and plate equations.

Approach 2. Another line of research is more 'geometric' in flavor. It will be presented in Section 5. It offers a 'view-from-the-boundary approach,' which focuses on geodesics. As such, it is confined to hyperbolic problems, where geodesics are (graph of) the bi-characteristics (carriers of energy). This way, this Riemannian approach merges with the established geometric optics approach for hyperbolic problems [Lit.1], [B-L-R.1]. Indeed, this approach produces specific sufficient conditions, which eventually permit the verification of the geometric optic assumptions and hence the application of the geometric optics results.

A comparison between the two methods on the common ground of applicability—hyperbolic equations—is also given in the examples of Section 5.4.

1.3. Summary of main features of differential geometric methods in the control of PDEs. Over the past 4–5 years, differential (Riemannian) geometric methods have emerged as a powerful new line of research to obtain general inverse-type, *a-priori* inequalities of interest in boundary control theory (continuous observability/stabilization inequalities (0.3), (0.5)) for various classes of PDEs. Their range of applicability now includes: second-order hyperbolic equations; Schrödinger-type equations; various plate-like equations; systems of elasticity; very complicated shell models described more below in Part IV, etc., all with variable coefficients, where the 'classical' energy methods of the early/mid-80's proved *inadequate*, as pointed out repeatedly in Section 0. We refer once more to Remark 6.4.1 for second-order hyperbolic equations; Remark 8.1 for Schrödinger equations; Remark 10.1 for plate-like equations.

In all of these PDEs classes, main features of these differential geometric methods are: (1) they apply to operators with principal part which is allowed to have variable coefficients (in space) with low regularity, C^1; (2) they tolerate energy level terms which are both space- and time-dependent, and only in L_∞ in time and space; (3) they yield rather general and verifiable sufficient conditions, which may serve for the construction of many complicated, variable coefficient examples, as well as for counter-examples (say, in the hyperbolic case, in dimension greater than 2), even when the control acts on the whole boundary; see Part II, Section 5; (4) they provide a good estimate (for some classes, optimal estimate) of the minimal time for observability/exact controllability in the hyperbolic case, and arbitrary short time when there is no finite speed of propagation; (5) they combine well with microlocal analysis methods needed for sharp trace estimates and for shifting topologies, thus producing at the end very general observability/stabilization results, with variable coefficients and with no geometric conditions on the observed (controlled) portion of the boundary; see Section 11, Part IV, on shells; (6) they merge well with techniques for reducing the number of traces in plate-like problems [L-T.14], [L-T-Z.2]; (7) ultimately, and with the same effort, they apply to these classes of PDEs defined on Riemannian manifolds, as is often the case in mathematical physics.

In addition, differential geometric methods have recently provided the intrinsic language for: (i) modeling the motion of dynamic shells far beyond the classical approach (rooted in classical geometry), and (ii) performing observability/stabilization energy methods on their very complicated equations, for which the classical setting based on Christoffel symbols appears to be too complicated or unfeasible. As we shall see in Part IV, a shell is a curved geometric object which can be modeled as a system of two PDEs both of hyperbolic type with strong coupling depending on the curvature: an 'elastic wave-type' equation ('curved system of elasticity') in the in-plane displacement; and a 'curved Kirchhoff plate-like equation' for the vertical displacement.

1.4. Geometric methods and control of PDEs: A two-way interaction. Although the subject of boundary control of PDEs is about a quarter of a century old, and that of Riemannian geometry is much older still, there has been relatively little interaction between the two. It was just over 10 years ago that the role that bi-characteristics play in boundary control of *hyperbolic* PDEs was brought to the forefront [Lit.1], [B-L-R.1]. This then naturally leads one to think about their geometric equivalent (Duff [Du.1, p. 209], Courant-Hilbert [C-H.1]), at least for time-invariant, second-order hyperbolic equations—geodesics, a basic concept in Riemannian geometry. On the other hand, it will be documented below that differential geometric methods play an equally positive role for non-hyperbolic, Petrowski-type PDEs, such as Schrödinger equations, plate-like equations of Euler-Bernoulli type, which have no finite speed of propagation, and where the above critical identification between bi-characteristics and geodesics is no longer available. This aspect is critical when dealing with coupled systems of PDEs of various type, be they hyperbolic or not—more on this later. Each of the two disciplines—control theory of PDEs, and Riemannian geometry—has been pursued in virtually complete independence, or occasionally even ignorance, of the other. It is the belief by the authors that both subjects have much to gain by closer interaction with one another. In this spirit, the authors are pleased to contribute the present paper, in the hope that it will further stimulate a two-way interaction between control of PDEs and differential geometry. On the one hand, we anticipate that the reservoir of as yet untapped Riemannian geometric methods and concepts could be applied productively in boundary control of PDEs. On the other hand, we expect that certain problems which have arisen recently in the boundary control of PDEs will stimulate the disciplines of Riemannian and Lorentzian geometry to undertake new areas of research. One example is the discovery of additional sufficient conditions (over available literature)—to be collected in Section 4 below—which guarantee the existence of a C^2-strictly convex function on a bounded domain Ω of a finite-dimensional Riemannian manifold (M, g), where strict convexity is intended with respect to the Riemannian metric g. Such strictly convex function plays a critical role in the results presented below in Part III for both hyperbolic as well as non-hyperbolic, Petrowski-type PDEs. A new result of this type, stimulated precisely by the interaction with boundary control theory is given in Theorem 4.6.1 of Section 4.6.

1.5. Systems of coupled PDEs. Nonlinear problems. The advantage of both obtaining *explicit estimates*, such as the C.O.I. (0.3) or the U.S.I. (0.5), as well as devising the techniques leading to these, is accrued also in two important cases, beyond the original linear single PDE setting: (i) the case of *systems of coupled* PDEs possibly of different type, to be illustrated in Section 2 below; and (ii) the case of *nonlinear* (semilinear) versions of the single PDE equations (which we do not treat here for lack of space). To elaborate further, more specifically, *global exact controllability* results of nonlinear (semilinear) wave and plate problems are obtained with the help of explicit, accurate continuous observability estimates for the corresponding linear problems [L-T.7]. In this work, global exact boundary controllability results are established for wave equations across all state spaces $H^{\gamma}(\Omega) \times H^{\gamma-1}(\Omega)$, $0 \leq \gamma \leq 1$, $\gamma \neq \frac{1}{2}$, including the most relevant cases at the end points $H^1(\Omega) \times L_2(\Omega)$ $(\gamma = 1)$ and $L_2(\Omega) \times H^{-1}(\Omega)$ $(\gamma = 0)$. Inclusion of the two cases of greatest interest motivated [L-T.7], as previous work [Z.1] had to exclude precisely the two endpoint cases.

Additional instances where explicit continuous observability estimates have a beneficial impact on nonlinear problems may be given. First, in [L-O.1] the proof of global existence of solutions to *quasi-linear* wave equations depends critically on the continuous observability inequality of the linear part. Second, in [Las.5], [Las.6], continuous observability estimates for the linear components are critically used to prove uniform stabilization of the (nonlinear) von Karman equation. Third, Part IV will give a version of this same strategy in the case of a shell: the continuous observability inequality of its linear part [Y.4] will be needed to prove—in the style of [Las.5], [Las.6]—uniform stabilization of a shallow shell [L-T.29] by virtue of nonlinear dissipative terms in the physically relevant and mathematically challenging Free Boundary Conditions. We finally refer to the books of Lagnese [Lag.3] and Komornik [K.1] for further instances on the use of C.O.I. for the stabilization of semi-linear problems.

2. A benchmark PDE-control problem advocating a broad use of differential geometry: The structural acoustic problem.

General considerations. The structural acoustic problem—to be described below in its various forms—provides an excellent carrier to motivate, introduce, justify and advocate the use of differential geometry in boundary control theory for PDEs. Indeed, realistic mathematical models of structural acoustic problems consist of a second-order hyperbolic equation (modeling the acoustic pressure) defined on a bounded domain Ω of $\mathbb{R}^3$, where the restriction of its solution to its elastic wall Γ_0 couples with an elastic equation (modeling its deflection) defined on Γ_0. In turn, the elastic solution couples with the boundary conditions of the second-order hyperbolic equation as expressed on Γ_0. The complementary part Γ_1 of the boundary of Ω is, instead, rigid. The elastic equation defined on Γ_0 may be of various types: (i) a hyperbolic elastic PDE of Kirchhoff type with, or without, strong damping; (ii) a non-hyperbolic elastic PDE of Euler-Bernoulli type with, or without, strong damping, (iii) an elastic PDE of either Kirchhoff type or Euler-Bernoulli type which, in addition, accounts for thermal effects, thus becoming a thermoelastic system on Γ_0, and thus coupling, in turn, an elastic and a heat equation on Γ_0; (iv) a composite (sandwich) plate. Finally, the elastic wall may either be *flat*, in which case the aforementioned equations in (i) to (iv) are elastic or thermoelastic or sandwich *plates*; or else may be *curved* (a manifold), in which case the aforementioned equations in (i) to (iv) are elastic (dynamic) or thermoelastic or composite *shells*. In realistic environments, where properties of the medium vary from point to point, the corresponding PDEs have *space variable* coefficients, even in the flat case. The above qualitative description, therefore,

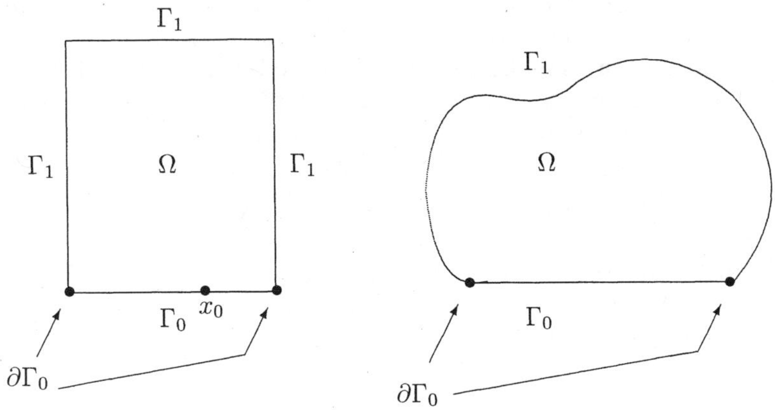

FIG. 1.

catapults up front some of the key features that the present paper intends to address and emphasize. They are: (i) second-order hyperbolic equations with variable coefficients described on a bounded Euclidean domain;(ii) elastic plate-like equations, both hyperbolic-like (Kirchhoff type) and not (Euler-Bernoulli type), with variable coefficients, defined on a Euclidean domain (such as the flat wall Γ_0); (iii) thermoelastic plate-like equations, both with hyperbolic or non-hyperbolic elastic component, and with variable coefficients, defined on a Euclidean domain (such as the flat wall Γ_0); (iv) (dynamic) elastic and thermoelastic shell equations defined on a curved surface (manifold), such as the curved wall Γ_0.

As we shall see in Part IV, an elastic shell equation is a system of two coupled hyperbolic-like PDEs, defined on a curved surface, which comprises: (i) the system of elasticity in the in-plane 2-dimensional displacement of the shell; and (ii) a scalar Kirchhoff type plate-like equation in the normal displacement. In the present paper, to the above list, we shall add one more dynamics: (v) Schrödinger equations with variable coefficients defined on a Euclidean domain (the iteration of two Schrödinger equations yields a plate-like equation).

2.1. Various structural acoustics models with constant coefficients and with flat flexible wall.

The acoustic chamber. Let $\Omega \subset \mathbb{R}^3$ be an open bounded domain ("the acoustic chamber") with boundary $\Gamma = \overline{\Gamma_0 \cup \Gamma_1}$, where Γ_0 and Γ_1 are open, connected, disjoint parts, $\Gamma_0 \cap \Gamma_1 = \emptyset$ in $\mathbb{R}^2$, of positive measure. Throughout this subsection, the sub-boundary Γ_0 is flat and is referred to as the elastic or flexible wall. Instead, Γ_1 is referred to as the rigid or hard wall. The interaction between wave and plate in the models below takes place on Γ_0. We also assume throughout that either Ω is sufficiently smooth (say, Γ is of class C^2), or else Ω is convex. This assumption guarantees that solutions to classical elliptic equations with $L_2(\Omega)$-forcing terms are in $H^2(\Omega)$ [Gri.1], or that the domain of the Laplacian in Ω, with (either Dirichlet or) Neumann B.C., is contained in $H^2(\Omega)$. The acoustic medium in the chamber is described by the wave equation in the variable z with acoustic pressure $\rho_1 z_t$, where ρ_1 is the density of the fluid. Moreover, we let c^2 be the speed of sound. Finally, v denotes the deflection of the plate equation on Γ_0.

Structural acoustic model with elastic flat wall Γ_0: Hyperbolic/hyperbolic interaction. Here the mathematical model is given by the following coupled PDE system:

$$
\begin{cases}
\begin{cases}
z_{tt} = c^2 \Delta z & \text{on } (0,T] \times \Omega \equiv Q; & (2.1.1a)\\
\dfrac{\partial z}{\partial \nu} + d_1 z = 0 & \text{on } (0,T] \times \Gamma_1 \equiv \Sigma_1; & (2.1.1b)\\
\dfrac{\partial z}{\partial \nu} = v_t & \text{on } (0,T] \times \Gamma_0 \equiv \Sigma_0; & (2.1.1c)
\end{cases}\\[2ex]
\begin{cases}
v_{tt} - \gamma \Delta v_{tt} + \Delta^2 v + \rho_1 z_t|_{\Gamma_0} = 0 \quad \text{on } \Sigma_0; & & (2.1.1d)\\
\text{either clamped B.C. on } \partial\Gamma_0, & &\\
v = \dfrac{\partial v}{\partial \tilde{\nu}} \equiv 0 & \text{on } (0,T] \times \partial\Gamma_0; & (2.1.1e_1)\\
\text{or else hinged B.C. on } \partial\Gamma_0, & &\\
v = \Delta v \equiv 0 & \text{on } (0,T] \times \partial\Gamma_0; & (2.1.1e_2)
\end{cases}\\[2ex]
z(0, \cdot\,) = z_0, \ z_t(0, \cdot\,) = z_1 \text{ in } \Omega; \quad v(0, \cdot\,) = v_0, \ v_t(0, \cdot\,) = v_1 \text{ in } \Gamma_0, \qquad (2.1.1f)
\end{cases}
$$

where $\nu(x)$ = unit outward normal vector at $x \in \Gamma$ in (2.1.1b–c) and similarly for $\tilde{\nu}(x)$ at $x \in \partial\Gamma_0$ in (2.2.1e$_1$) and $d_1 > 0$ (constant). Moreover, in the present model $\gamma > 0$ (constant), so that the uncoupled v-equation in (2.1.1d), with $\rho_1 = 0$, is the hyperbolic Kirchhoff equation. Naturally, one may supplement the v-plate equation (2.1.1d) with other B.C.'s, such as the complicated Free B.C. [Lag.1–3]. [L-T.24, Chapter 3].

Structural acoustic model with thermoelastic flat wall Γ_0: Hyperbolic/parabolic coupling ($\gamma = 0$) or hyperbolic/hyperbolic-type coupling ($\gamma > 0$). Here, the mathematical model is given by the following doubly coupled PDE system:

$$\begin{cases} \begin{cases} z_{tt} = c^2 \Delta z & \text{on } (0,T] \times \Omega \equiv Q; & \text{(2.1.2a)} \\ \dfrac{\partial z}{\partial \nu} + d_1 z = 0 & \text{on } (0,T] \times \Gamma_1 \equiv \Sigma_1; & \text{(2.1.2b)} \\ \dfrac{\partial z}{\partial \nu} = v_t & \text{on } (0,T] \times \Gamma_0 \equiv \Sigma_0; & \text{(2.1.2c)} \end{cases} \\ \\ \begin{cases} v_{tt} - \gamma \Delta v_{tt} + \Delta^2 v + \Delta \theta + \rho_1 z_t|_{\Gamma_0} = 0 & \text{on } \Sigma_0; & \text{(2.1.2d)} \\ \theta_t - \Delta \theta - \Delta v_t = 0 & \text{on } \Sigma_0. & \text{(2.1.2e)} \end{cases} \end{cases}$$

Boundary Conditions, such as the clamped or hinged B.C. (2.1.1e) for v and, say, the thermal Dirichlet condition $\theta \equiv 0$ in $(0,T] \times \partial\Gamma_0$ need to be added, along with initial conditions $\{z_0, z_1, v_0, v_1, \theta_0\}$.

For $\gamma = 0$, the $\{v, \theta\}$-dynamics (2.1.2d-e) with $\rho_1 = 0$ generates a s.c. *analytic* semigroup on a natural (energy) state space based on the flat wall Γ_0, under *all* canonical B.C., including the complicated coupled free B.C. for $\{v, \theta\}$ [L-T.18–20], [L-T.24, Appendices to Chapter 3]. Instead, for $\gamma > 0$, the $\{v, \theta\}$-dynamics (2.1.2d-e) with $\rho_1 = 0$ is hyperbolic-dominated, in the technical sense of [L-T.27].

2.2. More realistic models: Variable coefficients and curved walls. More realistic models demand the need of further refining the canonical models of Section 2.1, by introducing additional non-trivial complications. In order of progressively increasing difficulties, they are:

(i) still in the case of flat wall Γ_0, the differential operators, in particular the Laplacian Δ (both on Ω as well as on Γ_0) should be replaced by variable coefficients elliptic operators, to reflect the fact that properties of the medium may vary from point to point. In particular,

(i_1) the wave equation (2.1.1a) should be replaced by its variable coefficient counterpart

$$z_{tt} + \mathcal{A} z = F(z) \text{ in } (0,T] \times \Omega \equiv Q, \tag{2.2.1}$$

where for some positive $c > 0$ and $a_{ij}(x) = a_{ji}(x) \in C^1(\Omega)$,

$$\begin{aligned} \mathcal{A} z &\equiv -\sum_{i,j=1}^{3} \frac{\partial}{\partial x_i} \left(a_{ij}(x) \frac{\partial z}{\partial x_j} \right); \\ \sum_{i,j=1}^{3} a_{ij}(x) \xi_i, \xi_j &\geq c \sum^{2} \xi_i^2, \ x \in \overline{\Omega} \end{aligned} \tag{2.2.2}$$

$$\begin{aligned} F(z) = q_1(t,x) \cdot \nabla z + q_2(t,x) z_t + q_0(t,x) z; \\ |q_1| \in L_\infty(Q); \ q_2 \in L_\infty(Q), \ q_0 \in L_p(Q), \ p = n+1 = 4. \end{aligned} \tag{2.2.3}$$

(i_2) The thermo-plate-equation (2.1.2d) should be replaced by its variable coefficient counterpart [E-L-T.1–4]:

$$\begin{cases} v_{tt} - \gamma \mathcal{A} v_{tt} + \mathcal{A}^2 v + \operatorname{div}(\alpha(x)\nabla\theta) + \rho_1 z_t|_{\Gamma_0} = 0; & (2.2.4a) \\ \theta_t - \mathcal{A}\theta - \operatorname{div}(\alpha(x)\nabla v_t) = 0. & (2.2.4b) \end{cases}$$

(ii) The flat wall Γ_0 should give rise to a *curved* wall Γ_0, so that the elastic or thermoelastic equations in (2.1.1d) and (2.1.2d) respectively, should be replaced by (dynamic) elastic shell and thermoelastic shell systems. A mathematical model of an elastic shell system is given in Section 11 of Part IV below.

2.3. Conclusion. We have seen that a realistic model of a structural acoustic chamber consists of a system of two or more coupled PDEs, all with variable coefficients, defined on different domains, one of which is a curved wall. Accordingly, this model offers a persuasive case of the need of employing a unifying mathematical machinery, equally capable of treating single PDEs of various type, hyperbolic or not, and equally at ease on either a Euclidean domain or a curved surface. A main point of our present paper is that the Riemannian geometric methods presented here represent *an ideal mathematical tool for this purpose.* This is documented in the survey of single PDEs in Part III as well as in the treatment of a dynamic shell in Part IV. It will take us two far afield, however, to review the mathematical literature of a structural acoustic chamber [Las.1,7], [L-T.25], which is presently mostly restricted to constant coefficient models with flat flexible walls as in Section 2.1.

The literature is extensive. It includes many works by the second and fourth author of the present paper, and their coworkers (G. Avalos, C. Lebiedzik, etc.)

Part II: Riemannian geometric assumptions. Overview of results: Carleman estimates ⇒ C.O.I. and U.S.I.

Orientation. Two geometric approaches: (1) general energy methods in a Riemannian metric for hyperbolic or Petrowski type PDEs, and (2) a view from the boundary for second-order hyperbolic equations. (1) A main geometric approach which has been devised in recent years to obtain the sought-after inverse-type inequalities (0.3) and (0.5) of interest in control theory of PDEs is an 'energy method.' As such, it is a computational method, which is carried out in a suitable Riemannian metric. Indeed, we shall present two versions of it, which are progressively more flexible and hence lead to progressively more desirable results. They are also progressively more complicated. They have been so far successfully tested in several of the desirable PDE evolution equations, whether hyperbolic or not. At any rate, in both versions, the 'energy method' relies, principally, on *just one main geometric, fairly general assumption*, which is in common to, and shared by, all specific classes of evolution equations here considered: second-order hyperbolic equations; Schrödinger equations; plate-like equations of either Kirchhoff type (hyperbolic), or else of Euler-Bernoulli type (non-hyperbolic). This assumption postulates the *existence of a strictly convex function on the open bounded set* Ω of interest of the finite-dimensional Riemannian manifold (M, g)—hence roughly an upper or a lower bound on the sectional curvature. More precisely, we can construct a convex function assuming (i) *upper bounds* on the sectional curvature, starting out from a point, or a concave piece of the boundary; or (ii) *lower bounds* on the sectional curvature, starting out from a convex part of the boundary, or from infinity. Once this assumption is satisfied on Ω, then the corresponding control problem of exact controllability and uniform stabilization are generally *always solvable* on Ω, by using control action on a sufficiently large portion of the boundary, *regardless* of the specific class of evolution equation which is defined on Ω, hyperbolic or not. There are additional key features in this energy method: (i) the topology of the obtained estimates is optimal, a critical property, for these results to be useful in *optimal* control theory [L-T.11], [L-T.24]; (ii) the (universal) time in the estimates is often optimal (but not always, e.g., in some classes of hyperbolic problems, where, at any rate, is explicitly estimated, see (3.2.2) below); (iii) the portion of the boundary required (though not necessarily minimal) is, however, explicitly determined by the assumed convex function v as that part of Γ where the explicit vector field $\nabla_g v$ is entering; (iv) the regularity of the coefficients required is essentially optimal, etc. Last but not least, this geometric assumption permits one to construct many non-trivial PDE examples with variable coefficients, and is, in many cases, checkable. See Section 4. These characteristics may compensate for the fact that it may not be the 'most general.' It appears to be fairly general, however. An exception where a further geometrical restriction may be required is the 'purely Neumann problem' for second-order hyperbolic equations, of which we will have to say more in Section 7.

(2) For the class of second-order hyperbolic equations with Dirichlet (or Neumann) boundary control, we shall present an alternative approach, which we label "a view from the boundary." In it, no direct hypothesis on the Riemannian metric in the *interior* of Ω is required, unlike the previously discussed energy method approach. In the "view-from-the-boundary" approach, key emphasis is focused on the geodesics of (M, g), which are the geometric version of the geometric optic notion of bi-characteristics (Duff [Du.1, p. 209], Courant-Hilbert [C-H.1]). For second-order hyperbolic equations, the bi-characteristics are the true carriers of energy of the solutions. The assumptions required by the "view-from-the-boundary" approach will be introduced and discussed in Section 5, along with its main features. Ultimately, they are shown to *be sufficient to guarantee the geometric optics condition* [B-L-R.1], [Lit.1].

3. Basic geometric assumption for the energy method approach in the Riemannian metric.

3.1. Strictly convex functions and coercive fields in the Riemannian metric for hyperbolic and non-hyperbolic (Petrowski)-problems. Regardless of the specific class of evolution equations here considered—second-order hyperbolic equations; Schrödinger equations; plate-like equations of either Kirchhoff type (hyperbolic), or else Euler-Bernoulli type (non-hyperbolic)—the basic assumption that is common to all these different classes in the energy method approach (to be described below) is the following.

Main assumption (H.1). Let (M, g) be a finite-dimensional Riemannian manifold. Let Ω be an open bounded connected subset of M.

We assume that there exists a function $v_0 : \overline{\Omega} \to \mathbb{R}$ *of class* C^2 *which is strictly convex on* $\overline{\Omega}$, *with respect to the Riemannian metric* g. *We then set, henceforth,* $v(x) \equiv v_0(x) - \min_{x \in \overline{\Omega}} v_0(x) \geq 0$.

More specifically, this means the following. Denote by D the Levi-Civita connection in the Riemannian metric g. Then, a quantitative version of this assumption is that there exists a positive constant $\rho > 0$, such that

$$(3.1.1) \quad D^2 v(X, X) \equiv \langle D_X(Dv), X \rangle_g \geq 2\rho |X|_g^2, \quad \forall\, x \in \Omega, \quad \forall\, X \in M_x.$$

In words: for each point $x \in \Omega \subset M$ and each vector field X in the tangent space M_x at x, the Hessian $D^2 v$ (a 2-covariant tensor) is coercive, uniformly in $x \in \Omega$. Since $\overline{\Omega}$ is compact in M, this assumption means that the Hessian $D^2 v$ of v in the Riemannian metric g is positive on $\overline{\Omega}$: $D^2 v(X, X)(x) > 0, \quad \forall\, x \in \overline{\Omega}, \quad \forall\, X \in M_x$.

We finally note that for a scalar function v, we have $Dv = \nabla_g v$, where ∇_g is the gradient in the Riemannian metric g. Assumption (3.1.1) was introduced in [L-T-Y.1–2].

Condition (3.1.1) on the Hessian tensor of v has the following *geometric interpretation*: for any geodesics $\gamma(s)$ parametrized by arc lengths, then the second derivative in s of $v(\gamma(s))$ satisfies: $v(\gamma(s))'' \geq 2\rho$.

The above geometric interpretation of (3.1.1) appears to be related to the notion of pseudo-convexity arising in the framework of pseudo-differential analysis [Hor.1–2]. For instance, for a real non-elliptic symbol p of a partial differential operator P, the notion that "v is strictly convex with respect to the null bicharacteristic flow of P near its critical points," is expressed by Poisson bracket conditions, e.g., [Ta.8, Section 2].

We shall devote all of Section 4 to the issue of verifying assumption (H.1) = (3.1.1).

A first candidate for a suitable v verifying (3.1.1) is $v(s) = d_g^2(x, x_0)$, d_g distance function in the g-metric: this, however, need not be strictly convex on some Ω in general. It becomes strictly convex under some additional properties of the sectional curvature, see Section 4.

Remark 3.1.1. (1) Assumption (H.1) = (3.1.1) will be our *sole* assumption to obtain what we call "Carleman estimates, first version," for any of the classes of PDEs in full generality mentioned above, without any boundary conditions (B.C.) being imposed; see Theorem 3.2.1 below as well as Part III. Such "first version of Carleman estimates" is, in each case, the preponderant part of the proof to obtain the sought-after inverse-type inequalities (0.3), (0.5) (continuous observability/stabilization inequalities) of interest in control theory, modulo—possibly—interior lower-order terms. At any rate, it is for the purpose of establishing the "Carleman estimates, first version," that the *energy method in the Riemannian metric plays its critical, dominant role.* Such 'first version' is a recovery estimate of a time slice of energy, in terms of observations over the entire time interval.

(2) A refinement or improvement of the Carleman estimates, first version, is called here "Carleman estimates, second version," still without imposing B.C.: it is obtained from the first version by a more routine (but, in the case of the Schrödinger equation, non-trivial) energy method argument, which relies on well-posedness of the dynamics and, ultimately, on the Gronwall's inequality. This step may require an additional *structural* assumption, as well as additional regularity assumptions, on the energy level terms (e.g., in the case of a general Schrödinger equation written as in Eqn. (8.1), that the real part of the coefficient of the energy level $H^1(\Omega)$-terms contained in F in (8.3) be the gradient of a scalar function). Some sort of additional structural requirement is, in fact, intrinsic to the equation to obtain 'well-posedness' [Ho.3], [Miz.1].

(3) Finally, by imposing the B.C. on the second version of the Carleman estimates, we ultimately obtain the *sought-after continuous observability/stabilization inequalities* (0.3), (0.5) of control theory, modulo—possibly—interior lower-order terms.

The interior lower-order terms occur in the Carleman estimates—hence in the C.O.I. and U.S.I.—that are obtained by the Riemannian energy methods of Section 6 (second-order hyperbolic equations), Section 8 (Schrödinger equations), and Section 10 (plate-like equations). However, the more technical pointwise energy methods of Section 7 (second-order hyperbolic equations) and Section 9 (Schrödinger equations, presently in the Euclidean case) lead to *pointwise Carleman estimates* with the right sign on the interior lower-order term, which accordingly can be dropped. Hence the final Carleman estimates and corresponding C.O.I. and U.S.I. do not contain lower-order term, a big advantage. In particular, they yield *global uniqueness theorems* for over-determined problem (see Theorems 7.3 and 7.4).

Relaxed version (H.1′) of Main assumption. In cases where the 'energy level' terms are not present in any of the aforementioned classes of equations, it is possible to further relax the Main assumption (H.1) = (3.1.1), with the following slightly more general version. Let (M, g) be a compact, finite dimensional Riemannian manifold. Let Ω be an open bounded, connected subset of M. We assume that there exists a vector field H such that the covariant differential DH of H (a 2-covariant tensor) satisfies

$$DH(X,X) \equiv \langle D_X H, X\rangle_g \geq 2\rho|X|^2, \quad \forall\, x \in \Omega, \quad \forall\, X \in M_x, \tag{3.1.2}$$

for some constant $\rho > 0$, where $D_X H$ is the covariant derivative of H w.r.t. X.

If (H.1) = (3.1.1) is satisfied, with the function v, then the vector field H defined by $H \equiv Dv$ satisfies (H.1′) = (3.1.2). On the other hand, (H.1′) = (3.1.2) may be satisfied with a vector field H, which need not be realized as the covariant differential of any (potential) function v, satisfying (H.1) = (3.1.1) [Y.1]. Assumption (3.1.2) was introduced in [Y.1] and in [L-T-Y.3].

An important case where the energy level terms are not present in the equation—and hence (H.1′) = (3.1.2) would replace (H.1) = (3.1.1)—is the uniform stabilization problem, where the free dynamics is naturally assumed energy-preserving, as explained in Section 0. This is treated in Theorem 6.5.5.

3.2. Use of assumption (H.1). General statement of Carleman estimates and overview. We provide here an overview of the 'energy method' which encompasses in just one main statement the basic Carleman estimates across all aforementioned single PDE classes of evolution equations, though this common statement corresponds to three separate, different proofs, one proof for each single PDE class. In Part III, we shall give more specific details proper to each single class. However, certain common features can be singled out which are shared by all these single PDE equations. This is achieved in the statement of Theorem 3.2.1 below.

Assume (H.1) = (3.1.1). The pseudo-convex function $\phi(x,t)$. Then define the function $\phi : \Omega \times \mathbb{R} \to \mathbb{R}$ by

$$\phi(x,t) = v(x) - c\left(t - \frac{T}{2}\right)^2, \tag{3.2.1}$$

where $v(x) \geq 0$ is the function assumed in (H.1) = (3.1.1). In (3.2.1), the constants c and T are chosen according to the particular evolution equation at hand, as described below.

(1) If the evolution equation is a second-order hyperbolic equation, such as (6.1.1) below, then in (3.2.1) we may choose c and T as follows:

$$0 < c < \rho; \quad T > T_0 = 2\left(\frac{\max_{x\in\bar{\Omega}} v(x)}{\rho}\right)^{\frac{1}{2}}, \tag{3.2.2}$$

ρ as in (3.1.1). The fact that T will have to be 'sufficiently large' is in line with the property that a second-order hyperbolic equation has a finite speed of propagation.

(2) If the evolution equation is either a Schrödinger equation such as (8.1) below, or else a plate-like equation of Euler-Bernoulli type, such as (10.1) below, then we may choose T and c in (3.2.1) as follows:

$$\text{(3.2.3)} \qquad \begin{aligned} &T > 0 \text{ arbitrary}; \;\; c = c_T \text{ sufficiently large,} \\ &\text{such that } \;\; cT^2 > 4 \sup_{x \in \Omega} v(x) + 4\delta, \end{aligned}$$

for some small $\delta > 0$. The fact that T can be arbitrarily small for Schrödinger and Euler-Bernoulli plates is to be expected, since in both cases there is no finite speed of propagation.

PROPERTIES OF $\phi(x,t)$. *The function $\phi(x,t)$ in (3.2.1) has two key properties, which are exploited in the proofs of the Carleman estimate, first version, in Sections 6–10. They are:*

(i) there exists a suitably small constant $\delta > 0$, the one in (3.2.3), such that

$$\text{(3.2.4)} \qquad \phi(x,0) < -\delta \text{ and } \phi(x,T) < -\delta, \text{ uniformly in } x \in \Omega;$$

(ii) there exist t_0 and t_1, with $0 < t_0 < \frac{T}{2} < t_1 < T$, such that

$$\text{(3.2.5)} \qquad \min_{x \in \bar{\Omega}, t \in [t_0, t_1]} \phi(x,t) \geq -\frac{\delta}{2}.$$

Step 1. General statement of Carleman estimates, first version. Denote by $E(t)$ the 'natural (mathematical) energy' associated with the specific evolution equation at hand. This will be made more precise for each evolution equation under consideration, in Part III below, Sections 6–10: see specifically Eqn. (6.1.2) (second-order hyperbolic equations), Eqn. (8.2) (Schrödinger equations); Eqn. (10.2) (plate-like equations). Then, one may give the following general statement (to be made more specific and precise for each evolution under consideration, in Part III below), which is obtained by an energy method approach in the Riemannian metric, an *ad-hoc* method for each specific evolution equation [L-T-Y.1–2], [L-T-Y.4], [T-Y.1].

THEOREM 3.2.1. *For each evolution equation such as: (i) the second-order hyperbolic equation (6.1.1); (ii) the Schrödinger equation (8.1); (iii) the plate-like equation of Euler-Bernoulli type (10.1), denote by $E(t)$ its natural (mathematical) energy [to be given by (6.1.2), (8.2), (10.2), respectively]. Assume the main hypothesis (H.1) = (3.1.1). Let then $\phi(x,t)$ be the function defined in (3.2.1), with the choice of c and T explained above in (3.2.2), (3.2.3). Then: for all values of the parameter $\tau > 0$ sufficiently large, say $\tau \geq$ some $\tau_0 > 0$, the following one-parameter family of Carleman estimates holds true for sufficiently smooth solutions $w(t,x)$:*

$$\text{(3.2.6)} \qquad \begin{aligned} &B_\Sigma(w) + \frac{C_T}{\tau} \int_Q e^{\tau\phi} |f|^2 dQ + lot(w) \\ &\geq \left(p - \frac{c_T}{\tau}\right) e^{-\frac{\tau\delta}{2}} \int_{t_0}^{t_1} E(t)dt - C\tau e^{-\delta\tau}[E(T) + E(0)], \end{aligned}$$

where: (i) $p = \rho - c > 0$ for second-order hyperbolic equations, where ρ is the constant in assumption (H.1) = (3.1.1), and c is the constant in (3.2.1) or

(3.2.2); while $p = \rho > 0$ *for both Schrödinger equations and plate-like equations, where* ρ *is again defined in (3.1.1) [see Eqns. (6.4.2b), (8.6), (10.6), respectively; see also (7.12b) and (9.10) without l.o.t.]; (ii)* $B_\Sigma(w)$ *are boundary terms defined on* Σ *which can be* explicitly *obtained for each class of evolution equations under consideration, see Part III (Eqns. (6.4.3), (7.9), (8.7), (10.7)–(10.9), respectively); (iii) finally,* $lot(w)$ *are interior lower-order terms (below the energy level of* $E(t)$*, topologically). They may occur, as in the approaches of Section 6 (see Eqn. (6.4.2a)), or Section 8 (see Eqn. (8.5)), or Section 10 (see Eqn. (10.5)); or they may* not *occur, as in the more technically demanding approaches of Section 7 or Section 9. We elaborate further on this issue in Remark 3.2.2 below.*

Remark 3.2.1. We point out explicitly that, in effect, there are three different proofs hidden behind the common statement of Theorem 3.2.1: one for each single class of PDEs mentioned there [L-T-Y.1–2], [T-Y.1], [L-T-Y.4]. Though they share a common philosophical strategy, they are technically different: in particular, they employ different "multipliers." The selection of these multipliers, as well as their use, are tuned to each single class of PDE. See Part III below: Remark 6.4.1 (Second-order hyperbolic equations); Remark 8.1 (Schrödinger equation); Remark 10.1 (plate-like equations). It is, however, an advantage of the "energy method" in the Riemannian metric that it leads to final results for each of the various PDE single classes, which can be expressed by a common, shared statement, where the energy $E(t)$ and the boundary terms $B_\Sigma(w)$ depend, however, on the single class, as explicitly given in Part III.

We also note that, as pointed out above, if v is a g-strictly convex function on $\overline{\Omega}$, in the sense of hypothesis (H.1) = (3.1.1), then the function $\phi(x,t)$ in (3.2.1) is pseudo-convex for some constant c. Thus, in principle, the pseudo-differential work of Tataru [Ta.1–3] could be invoked. This is not the case, instead, for coercive vector fields H satisfying the relaxed assumption (H.1′) = (3.1.2), which are *not* the covariant differential of a scalar function (i.e., 'conservative'). Many examples can be given, of course, even in the Euclidean setting, of coercive vector fields that are not conservative. For one example in the Riemannian setting, see [Y.1, Example 3.4]. Thus, the infusion of Riemannian geometric ideas and machinery does help also in the explicit construction, or verifiability, of pseudo-convex functions, as well as in the construction of counter-examples to exact controllability (see Section 5.1: when Ω has a closed geodesic in its interior). On the other hand, at the pseudo-differential level, the 'right' multiplier for a general evaluation equation is given by the derivative in the dual variable of the principal symbol [Hor.1–2], whereby then only one unifying pseudo-differential proof may be given [Ta.1–3] of a statement such as Theorem 3.2.1. The Riemannian geometric proofs seem to us more friendly. □

Step 2. General statement of Carleman estimates, second version. For this step, we need—possibly, but not always—further (mild) restrictions on the *structure* of the 'energy level terms.' These must be such as to guarantee that the following energy inequality holds true: for all $0 \le s \le t \le T$, there exists $C_T > 0$ such that

$$|E(t) - E(s)| \le C_T \left[\int_s^t E(\sigma) d\sigma + G(T) \right], \tag{3.2.7}$$

where

$$G(T) = B_\Sigma(w) + B_{\Sigma,e}(w) + \int_0^T \|f\|^2_{H^\sigma(\Omega)} dt + lot(w). \tag{3.2.8}$$

In (3.2.8), $B_\Sigma(w)$ are the boundary terms occurring in (3.2.6); while $B_{\Sigma,e}(w)$ includes those 'boundary terms' that are generated by the energy method (the subscript 'e' stands for energy), which is employed in seeking to establish inequality (3.2.7).

Comments on the validity of assumption (3.27) for various PDE-classes.

(i) For second-order hyperbolic equations, this energy method consists in multiplying the evolution equation (6.1.1) by w_t and integrating by parts. In this case, (3.2.7) *holds true* with $\sigma = 0$ in (3.2.8), $B_{\Sigma,e}(w) \equiv 0$, and *with no new restriction* imposed on the 'energy level' terms $F(w)$ in (6.1.1).

(ii) For Schrödinger equations, this method consists in multiplying the evolution equation (8.1), e.g., by $\overline{w}_t$ [Tr.2], [T-Y.1], or else by $[i\Delta\overline{w} - i\overline{w}]$ [L-T-Z.2] and integrating by parts. In this case, in order to establish (3.2.7) with $\sigma = 1$ in (3.2.8), with $BT_e(w) \equiv 0$, it is necessary to impose that the energy level terms $F(w)$ in (8.1) be of the type $F(w) = \langle P(t,x), Dw\rangle_g + rw$ with Re $P(t,x) = D\chi(t,x)$ for some real function $\chi(t,x)$ [L-T-Z.2]. This (mild) restriction is intrinsic [Hor.3], [Miz.1] due to well-posedness, or lack thereof.

(iii) For plate-like equations of Euler-Bernoulli type (non-hyperbolic), this method consists in multiplying the evolution equation (10.1) by Δw_t and integrating by parts. This procedure may or may not succeed in establishing inequality (3.2.7) with $\sigma = 1$ in (3.2.8), depending on the structure of the energy level term $F(w)$ in (10.1). In particular:

(iii_1) If $F(w)$ contains, at most, only second-order differential operators on w, and zero-order operators on w_t [that is, if $F(w)$ has terms one unit below the energy level], then inequality (3.2.7) with $\sigma = 1$ in (3.2.8), is satisfied with no further assumptions on $F(w)$.

(iii_2) Energy level terms such as $\frac{\partial}{\partial x} w_t$, $\frac{\partial}{\partial y} w_t$, say on $\Omega = \{(x,y) : c \le y \le d,\ h_1(y) \le x \le h_2(y)\}$ are acceptable for F to satisfy assumption (3.2.7) [L-T-Y.4, Remark 2.1].

Generally, if $F(w)$ is precisely at the energy-level, some further structural restrictions are needed. These are intrinsic, for otherwise examples are known [Hor.3] where inequality (3.2.7) fails to hold true.

THEOREM 3.2.2. *Assume the setting of Theorem 3.2.1. In addition, assume the validity of inequality (3.2.7). Then: for all values of the pa-*

rameter $\tau > 0$ sufficiently large, say $\tau \geq$ some $\tau_0 > 0$, the following one-parameter family of Carleman estimates holds true for sufficiently smooth solutions $w(t,x)$:

$$\begin{aligned} &c_\phi B_\Sigma(w) + C_T(t_1 - t_0)\left\{B_{\Sigma,e}(w) + \int_0^T \|f\|^2_{H^\sigma(\Omega)}dt\right\} \\ &+ lot(w) + \frac{2}{\tau}\int_Q e^{\tau\phi}|f|^2 dQ \;\geq\; k_{\phi,\tau_0}[E(T) + E(0)], \end{aligned} \tag{3.2.9}$$

where $k_{\phi,\tau_0} > 0$, and $B_\Sigma(w)$ [and $B_{\Sigma,e}(w)$] are explicitly *obtained for each class of evolution equations under consideration. The same comments on the occurrence or not of the $lot(w)$ made in point (iii) in the statement of Theorem 3.2.1 apply now, for which again we refer to Remark 3.2.2 below. More details are given in Part III.*

Step 3. Continuous observability/stabilization estimates. (Explicit portion of the boundary involved.) Up to Theorem 3.2.2, no use of B.C. is made. Once Theorem 3.2.2 is available, one obtains the sought-after continuous observability/stabilization inequality (0.3), (0.5) by using the corresponding boundary conditions: that is, homogeneous B.C. in the first case, dissipative B.C. in the second case. Thus, the B.C. are used at the very end of the described procedure. It is in this step that the issue arises of how large should the controlled (or observed) or stabilized portion of the boundary be in the final estimates (0.3) and (0.5). To illustrate, the portion Γ_0 of the boundary where the geometric condition

$$\langle Dv, n\rangle \leq 0 \quad \text{on } \Gamma_0, \qquad Dv = \nabla_g v, \tag{3.2.10}$$

holds can *generally* be taken as uncontrolled (or unobserved) when zero Dirichlet B.C. are here imposed, so that, *generally*, the controlled (observed) portion of the boundary may be taken to be $\Gamma_1 = \Gamma/\Gamma_0$. Here n is the outward normal field to $\partial\Omega$ on M, and D is the Levi-Civita connection of Section 3.1. This result is illustrated by Theorem 6.5.1 (Dirichlet control) and Theorem 6.5.2 (Neumann control) in the case of second-order hyperbolic equations of Section 6; by Theorem 8.3 (Dirichlet control) and Theorem 8.4 (Neuman control) in the case of Schrödinger equations of Section 8; and by Theorem 10.3 in the case of plate-like equations of Section 10. However, again, the situation is geometrically more delicate in dealing with the purely Neumann B.C. say for second-order hyperbolic equations as described in Section 7 below, or Schrödinger equations, as described in Section 9 below. At any rate, this approach yields an *explicit* portion of the boundary which need not be optimal ('minimal') in all cases. However, the control problem admits a solution—in the sense that C.O.I. (0.3) and the U.S.I. (0.5) hold true—as soon as assumption (H.1) = (3.1.1) is satisfied, perhaps at the price of taking more controlled boundary than it is strictly required. See Section 6.7 below, in particular, Example #1 there. We elaborate this important point further in the next paragraph.

Contrast between *constant coefficient* models versus *variable coefficient* models in the solution of the control problems. We focus, for definiteness, on second-order hyperbolic equations. Then, for this class, the exact controllability property is always *achieved* with a sufficiently 'large' controlled portion of the boundary, in particular with control applied over the entire boundary, in the following two cases:

(i) in the case of, say, the *canonical* wave equation in $\Omega \subset \mathbb{R}^n$ (with the Euclidean Laplacian), or even the more general Eqn. (6.3.3) with $\Delta = -\mathcal{A}$, for any dimension $n \geq 1$ [L-T.17],

(ii) the case of the wave equation with *variable* coefficients, this time, however, in dim $\Omega = 1$.

By contrast, the wave equation with variable coefficients in dim $\Omega \geq 2$ may very well be non-controllable (see Section 5.1), even if control is applied to the entire boundary $\partial\Omega$. This occurs whenever the Riemannian metric g generated by the coefficients $a_{ij}(x)$ as described in Section 6.3, Eqns. (6.3.2) possesses a closed geodesic in $\mathbb{R}^n$, $n \geq 2$. Then, taking a bounded domain Ω in $\mathbb{R}^n$ which contains such closed geodesic in its interior, yields a non-controllable system, even if the control is applied to all of $\partial\Omega$ [Ra.1], [Lit.1], [B-L-R.1].

In view of the situation described above, we see that the assumption (H.1) = (3.1.1) on the strictly convex function excludes the danger that the equation—be it hyperbolic or not—be non-controllable. Thus, we consider the "minimal portion of the boundary" issue as subordinated in importance to the issue of being able to solve the control problem in the first place. Thus, in this viewpoint, the sufficient condition (H.1) = (3.1.1) [or its weakened version (H.1′) = (3.1.2), when the energy level terms are not present] is the foundation of this energy method approach in the Riemannian metric.

Geometric conditions. The continuous observability/stabilization inequalities—C.O.I. and U.S.I. in (0.3) and (0.5) of Section 0, respectively—(equivalently, the exact controllability/uniform stabilization results) which we shall present in this paper in Part III, do *not require geometrical conditions on the observed* (equivalently, *controlled*) or stabilized *portion of the boundary.* This feature is in contrast with the great majority of the works on this topic, even those dealing with canonical cases of equations with constant coefficients and no energy level terms. In all these works, geometrical restrictions were imposed that were inherited by the techniques of proof based solely on "classical differential multipliers": see Remark 6.4.1. By contrast, additional tools are needed, such as Lemma 6.5.3, to dispense with unnecessary geometrical conditions—and these tools involve pseudo-differential operators and microlocal analysis: see [L-T.12] for second-order hyperbolic equations, [L-T.13] for plate-like equations, and their critical generalizations and impact in Part IV, dealing with the stabilization of a shallow shell. In particular, Parts III and IV show that differential geometric energy methods combine well with microlocal/pseudo-differential operators techniques and lead to the elimination of such (restrictive) geometrical conditions.

Remark 3.2.2. The energy method approach in the Riemannian metric—to be expounded below in Part III—will have two versions. The first version—which may be viewed as a far-reaching generalization of the classical energy methods (multipliers) of the '80s for canonical models (constant coefficients, no energy level terms) as explained in Section 0—yields ultimately a Carleman estimate, first version, such as (3.2.6), *in integral form*, which contains the lower-order term $lot(w)$. More specifically, in this first approach, it turns out that the lower-order term has the 'wrong' sign in the estimate, and cannot be dropped. The presence of a $lot(w)$ which pollutes the desired final estimates is undesirable, since removal of the $lot(w)$ proceeds by absorbing this term by the appropriate boundary term in the final estimate, via—typically—a compactness/uniqueness argument. This procedure has two disadvantages:

(i) to succeed, it requires appeal to a global unique continuation result under over-determined boundary conditions over the time T of the estimate (identified in (3.2.2), (3.2.3)). This may be a problem for time-dependent $L_\infty(Q)$-energy level terms $F(w)$, as assumed [Hor.2], [Hor.5], [I.1–5], [Ta.4–8]. If the coefficients of $F(w)$ are time independent, one converts the uniqueness requirement for the evolution equation to a uniqueness requirement for the corresponding elliptic problems, for which very general results are available [Hor.1], [Hor.2, p. 14].

(ii) When the procedure succeeds, it loses control of the constant in the estimate, as the aforementioned compactness-uniqueness argument is by contradiction.

Thus, it is desirable to avoid the pollution of $lot(w)$ in the Carleman estimates. This can be done through a more complicated energy method, as we now describe. Indeed, a second version of the energy method approach may be given, which is more general, as its proof includes an additional degree of freedom: this yields a *pointwise* Carleman estimate (at each time t and point x), which this time has the 'right' sign $[\beta\tau^3 + \mathcal{O}(\tau^2)]$, for τ large, with $\beta > 0$, in front of the lower-order term, as least on an explicitly identified set in time and space [L-T-Z.1]. See Section 7 in the hyperbolic case and Section 9 for Schrödinger equations. Thus, in this case, after some further analysis given in Section 7, the lower-order term may be dropped in the final estimate. Thus, one then obtains, ultimately, a Carleman estimate such as (3.2.6) or (3.2.9), *without*, however, the presence of the lower-order term $lot(w)$. See Theorem 7.2. Consequently, one obtains in one shot new (?) global uniqueness results, Theorems 7.3 and 7.4, as well as observability stabilization estimates such as (0.3) and (0.5). The global uniqueness is precisely in the form needed for the control inequalities in the Neumann case: see Theorem 6.5.6(iii).

Through the PDE control theory literature, the sought-after control-theoretic inequalities—C.O.I. and U.S.I.—have always been given first polluted by lower-order terms. This includes [B-L-R.1], [Lit.1], [L-T-Y.1] and all 'classical' multipliers literature [Li.1], [K.1]. Few exceptions were [K-K.1] (second-order equations with constant coefficient principal part, in line with Russian literature [L-R-S.1]) and the very technical paper [Ta.7]. Our Riemannian approach is inspired by [L-R-S.1, Lemma 1, p. 124] and encompasses [L-T-Z.1], [T-Y.2] for Section 7 below (second-order hyperbolic equations) and Section 9 (Schrödinger equations [L-T-Z.2], along with Remark 10.3 for a purely Euler-Bernoulli plate with hinged B.C. (obtained as an iteration of two Schrödinger's problems).

4. Survey of geometric results and methods for the construction of strictly convex functions and coercive fields.

4.1. The need for a geometrical perspective. Many PDE problems of interest, including higher-order problems, may be expressed in terms of a second-order elliptic operator whose principal part is of the form

$$\mathcal{A}u \equiv -\sum_{i,j=1}^{n} \frac{\partial}{\partial x_i}\left(a_{ij}(x)\frac{\partial u}{\partial x_j}\right), \tag{4.1.1}$$

as in (2.2.2). Ellipticity means that

$$\lambda|\xi|^2 \leq \sum_{i,j=1}^{n} a_{ij}(x)\xi_i\xi_j \leq \frac{1}{\lambda}|\xi|^2 \tag{4.1.2}$$

for some $\lambda > 0$. It is of great merit to understand the properties of an operator such as $\mathcal{A}$ in terms which are not based on a specific system of coordinates.

For example, the constant λ in (4.1.2) can often be brought closer to 1 by merely introducing an appropriate change of coordinates (in dimension $n = 2$, $\lambda \equiv 1$ may be achieved in simply-connected Ω by multiplying $\mathcal{A}$ by a positive function and using Morrey's theorem on conformal mapping). The most interesting properties of the operator $\mathcal{A}$ have not been changed by the change of coordinates. This tells us that the constant $\lambda \in (0,1)$ is not an interesting invariant of the operator, even though its existence for any system of coordinates is essential.

What are the invariants of the operator $\mathcal{A}$? Which curves, which real-valued functions, etc., play special roles? Such questions may be addressed by introducing the Riemannian metric (see also Section 6.3)

$$g = \sum_{i,j=1}^{n} g_{ij}(x)\,dx_i\,dx_j,$$

where $(g_{ij}(x)) = (a_{ij}(x))^{-1}$ for each $x \in \Omega$. Then $\mathcal{A}$ becomes the Laplace-Beltrami operator of g, modulo lower-order terms; see (6.3.6). Bicharacteristics $c : [0,T] \to T^*\Omega$ of the wave equation $\frac{\partial^2}{\partial t^2}u + \mathcal{A}u = 0$ are determined by geodesics $\gamma : [0,T] \to \Omega$ of the Riemannian metric g (Duff [Du.1,

p. 209], Courant-Hilbert [C-H.1, p. 565]). The conormal vector becomes the g-normal vector. Pseudoconvex functions become g-convex functions. And the pointwise invariants of $\mathcal{A}$ are the sectional curvatures of the Riemannian metric. For the Riemannian connection ∇, sectional curvature, etc. see e.g. [DoC.1].

4.1.1. Euclidean Space. In Euclidean $\mathbb{R}^n$, a familiar and useful function is $\rho_0(x) = |x|$, the distance from the origin. It is well known, and widely used, that $f(x) \equiv \frac{1}{2}\rho_0(x)^2$ is strictly convex. In fact, the Hessian $D^2 f$ of f is *equal* to the identity matrix at each point, which is a much stronger property than convexity. We would like to extend the usefulness of this function to non-Euclidean spaces, so that equations with coefficients depending on the space variables may be treated, as emphasized in the Introduction.

In an n-dimensional Riemannian manifold M, the distance function ρ from any closed subset Σ_0 of M shares some, but not all, of the special properties of the Euclidean distance ρ_0 from a point. The distance function is defined as

$$\rho(x) \equiv \inf \{L(\sigma) \,|\, \sigma : [0,1] \to M, \sigma(0) \in \Sigma_0, \sigma(1) = x\},$$

where $L(\sigma)$ denotes the length of a C^1 or Lipschitz curve σ. One special property which continues to hold in this more general case is that $|\nabla\rho| \equiv 1$, wherever ρ is smooth. At points where ρ fails to be differentiable (cut points of Σ_0), the distributional inequality $|\nabla\rho| \leq 1$ is valid. However, it might not be true that $\frac{1}{2}\rho(x)^2$ is strictly convex. This property depends on *the curvature of M and on the geometry of Σ_0*. A valuable tool in this regard is the matrix Riccati equation (4.3.1), below. For the case $\Sigma_0 = \{p_0\}$, we have

4.1.2. The Hessian Comparison Theorem. Let M and $\overline{M}$ be two n-dimensional Riemannian manifolds, and choose points $p_0 \in M$, $\overline{p_0} \in \overline{M}$. Choose an orthonormal basis $\{E_1(0), \ldots, E_n(0)\}$ for the tangent space to M at p_0, and similarly an orthonormal basis $\{\overline{E}_1(0), \ldots, \overline{E}_n(0)\}$ for the tangent space to $\overline{M}$ at $\overline{p}_0$. Let γ be the unit-speed geodesic of M with initial conditions $\gamma(0) = p_0$, $\gamma'(0) = E_n(0)$, and similarly let $\overline{\gamma}$ be the geodesic of $\overline{M}$ with initial conditions $\overline{\gamma}(0) = \overline{p}_0$, $\overline{\gamma}'(0) = \overline{E}_n(0)$. Extend $E_i(0)$ as a parallel vector field $E_i(t)$ along γ, and similarly extend $\overline{E}_i(0)$ as a parallel vector field $\overline{E}_i(t)$ along $\overline{\gamma}$, $i = 1, \ldots, n$. Write $\rho(x) = d_M(x, p_0)$ for the Riemannian distance $d_M(x, p_0)$ and $\overline{\rho}(y) = d_{\overline{M}}(y, \overline{p}_0)$.

THEOREM 4.1.1. *[S-Y, p. 4] If for all $0 \leq t \leq T$, for all tangent vectors V to M at $\gamma(t)$, and all tangent vectors $\overline{V}$ to $\overline{M}$ at $\overline{\gamma}(t)$, the sectional curvatures*

$$K_\sigma^M \leq K_{\overline{\sigma}}^{\overline{M}},$$

where σ is spanned by $\gamma'(t)$ and V, and where $\overline{\sigma}$ is similarly spanned by $\overline{\gamma}'(t)$ and $\overline{V}$, then for all $t \in [0,T]$, the Hessian tensors applied to unit tangent *vectors* $V \in T_{\gamma(t)}M$ *orthogonal to* $\gamma'(t)$ *and* $\overline{V} \in T_{\overline{\gamma}(t)}\overline{M}$ *orthogonal to* $\overline{\gamma}'(t)$ *satisfy*

$$D^2\rho(V,V) \equiv \nabla^2_{V,V}\rho \geq \overline{\nabla}^2_{\overline{V},\overline{V}}\overline{\rho} \equiv \overline{D}^2\overline{\rho}(\overline{V},\overline{V}).$$

The proof of the Hessian Comparison Theorem 4.1.1 involves ideas closely connected with H. Rauch's 1951 Comparison Theorem [B-C, p. 250]. We shall indicate a proof based on the matrix Riccati equation below.

COROLLARY 4.1.2. *If* M *has all sectional curvatures* $K^M_\sigma \leq b^2$, *resp.* $\leq -\beta^2$, *for some constant* b *resp.* β, *then for all tangent vectors* V *at points where* ρ *is smooth,*

$$D^2\rho(V,V) \equiv \nabla^2_{V,V}\rho \geq b \cot b\rho \left(\langle V,V\rangle - \langle V,\nabla\rho\rangle^2\right)$$

respectively

$$\beta \coth \beta\rho \left(\langle V,V\rangle - \langle V,\nabla\rho\rangle^2\right).$$

COROLLARY 4.1.3. *If* M *is complete and simply connected, and has nonpositive sectional curvatures, then* $f(x) \equiv \frac{1}{2}\rho(x)^2$ *has Hessian greater than or equal to the identity, and is thus uniformly convex.*

COROLLARY 4.1.4. *If* M *has sectional curvatures* $\leq b^2$, *then for any* $r_0 < \frac{\pi}{2b}$, $f(x) \equiv \log \sec b\rho(x)$ *has Hessian greater than or equal to the identity, and is thus uniformly convex, on the subdomain where* $\rho(x) \leq r_0$ *and* ρ *is smooth.*

4.2. Functions with convex level sets. This section has connections with pseudo-differential literature [Hor.1–2], [I.1–6], [Ta.1–8].

To ensure the existence of a convex function v on a given Riemannian manifold with boundary M, it is roughly sufficient to find a real-valued function f with convex level sets. This fact is well known, but should be kept in mind in the context of establishing boundary controllability. The proof is based on the simple expedient of replacing f with $v(x) = e^{\Lambda f(x)}$ for a sufficiently large constant Λ. Similar, and sometimes sharper, results may be found by composing f with other functions of one real variable. *Convexity of an oriented hypersurface* Σ is measured by *the positivity of its second fundamental* form

$$B_\Sigma(X,Y) \equiv \langle \nabla_X \nu, Y\rangle,$$

where ν is the unit normal vector of Σ, and where X and Y are tangent vectors to Σ. Write Σ_s for the level set $\{x : f(x) = s\}$, oriented in the direction of increasing values of s. A more quantitative statement is as follows [B-G-L.1]:

LEMMA 4.2.1. *Suppose that* $f : M \to I\!R$ *has convex level sets and nonvanishing gradient, and assume that at points of each level surface* Σ_s, *for all unit vectors* X *tangent to* Σ_s, *that*

$$|\nabla f| B_{\Sigma_s}(X,X) \geq c_1 > 0. \tag{4.2.1}$$

Further let $-c_2$ *be a lower bound for* $\nabla^2_{\nu,\nu} f$ *and suppose that* $|\nabla f|^2 \geq c_3 > 0$. *Write* $v(x) \equiv e^{\Lambda f(x)}$.

If Λ *is chosen large enough that* $c_1(c_3\Lambda - c_2) > c_2^2$, *then for all vectors* Y,

$$\nabla^2_{Y,Y} v \geq c\Lambda v |Y|^2,$$

where $c > 0$. *Specifically,* $0 < c < c_1$ *is small enough so that* $(c_3\Lambda - c_2 - c)(c_1 - c) > c_2^2$.

In certain interesting examples, $|\nabla f|$ is zero at an isolated minimum of f, in a neighborhood of which inequality (4.2.1) nonetheless holds uniformly. In such cases the convexity of v at the minimum point may be obtained by a direct argument [B-G-L.1].

A much more general smoothing theory is due to Greene and Wu [G-W.1].

DEFINITION 4.2.1. *A continuous function* $v : M \to \mathbb{R}$ *is uniformly convex if, for some* $c > 0$, *along every geodesic* γ *the function* $v(\gamma(s)) - cs^2/2$ *is convex as a function of arc length* s *along* γ.

Greene and Wu show [G-W.1, p. 214]:

THEOREM 4.2.2. *Let* $v : M \to \mathbb{R}$ *be continuous and uniformly convex in the sense that, for some* $c > 0$, *along every geodesic* γ *the function* $v(\gamma(s)) - cs^2/2$ *is convex as a function of arc length* s *along* γ. *Then* v *is the uniform limit of smooth functions* $v_\varepsilon : M \to \mathbb{R}$ *with* $\nabla^2_{Y,Y} v_\varepsilon > (c - \varepsilon)|Y|^2$ *for all vectors* Y.

As a consequence, they show that: on any *complete, noncompact manifold with positive sectional curvature*, there is *a smooth, uniformly convex function* v_ε. For this situation, they make use of the convexity of the continuous function $v(x) = e^{\Lambda\rho(x)}$, where $\rho(x)$ is the limit of $d_M(x, \sigma(t)) - t$ as $t \to \infty$, and where $\sigma : [0, \infty) \to M$ is a unit-speed geodesic which minimizes length between any two of its points. See [G-W.1, pp. 292–6]. It had been shown by Cheeger and Gromoll in [CG] that the sublevel sets of ρ are totally convex.

4.3. The Matrix Riccati Equation. Consider a smooth real-valued function $\rho : M \to \mathbb{R}$ on a Riemannian manifold M, with the special property that $|\nabla\rho| \equiv 1$. (For the distance function from a set Σ_0, it may be necessary to stay away from any cut points, which form a closed set of measure zero, to achieve smoothness of ρ; cf. Section 5.) Then for various real values s, the level sets Σ_s defined by $\rho = s$ form a smooth one-parameter family of hypersurfaces. Write ν for the vector field $\nabla\rho$, so that $\nu(x)$ is a unit normal vector to the hypersurface Σ_s passing through x. Let $\gamma : \mathbb{R} \to M$ be an orbit of ν, that is, for each $s \in \mathbb{R}$, $d\gamma(s)/ds = \nu(\gamma(s))$. Note that after adding a constant to the independent parameter s, we may assume that $\gamma(s) \in \Sigma_s$ for all s. Moreover, the curve γ has minimum length between any two of its points. In fact, let two points of the curve be $\gamma(s_0)$ and $\gamma(s_1)$. If $\sigma : [0,1] \to M$ is any curve joining $\gamma(s_0) = \sigma(0)$ to $\gamma(s_1) = \sigma(1)$, then

$$\left|\frac{d\sigma(t)}{dt}\right| \geq \left\langle \nu, \frac{d\sigma(t)}{dt} \right\rangle = \frac{\partial \rho(\sigma(s))}{\partial s}.$$

It follows that the length of σ is at least $\rho(\gamma(s_1)) - \rho(\gamma(s_0)) = s_1 - s_0$, which is the length of $\gamma([s_0, s_1])$. In particular, γ is a geodesic. Since there is an orbit of ν passing through each point of M, we conclude that $\nabla_\nu \nu \equiv 0$.

Now the Hessian of ρ may be written $\nabla^2_{V,W}\rho = \langle V, \nabla_W(\nabla\rho)\rangle$ for any tangent vectors V, W. In particular, with $V = W = \nu = \nabla\rho$, we find that $\nabla^2_{\nu,\nu}\rho = 0$; and if $W = \nu$, $\langle V, \nu\rangle = 0$, then $\nabla^2_{V,\nu}\rho = \nabla^2_{\nu,V}\rho = 0$. These properties are exactly as for the Euclidean distance function ρ_0.

The remaining part of the Hessian of ρ can be interpreted as an $(n-1) \times (n-1)$ matrix $\nabla^2_{E_i(s),E_j(s)}\rho = A_{ij}(s) = B_{\Sigma_s}(E_i(s), E_j(s))$. For this purpose, consider a point $x_0 \in \Sigma_{s_0}$ and let $\gamma : I\!R \to M$ be the geodesic passing through $x_0 = \gamma(s_0)$ which is an orbit of ν. Choose an orthonormal basis $E_1(s_0), \ldots, E_{n-1}(s_0)$ for the tangent space to Σ_{s_0} at x_0. As above, we extend this orthonormal basis as parallel vector fields along γ, so that for all s, $E_1(s), \ldots, E_{n-1}(s)$ is an orthonormal basis for the tangent space to Σ_s at $\gamma(s)$.

Note that $A_{ij}(s)$ represents both the second fundamental form of Σ_s and the nontrivial part of the Hessian of ρ. In fact, $A_{ij} = \langle E_i, \nabla_{E_j}(\nabla\rho)\rangle = B_{\Sigma_s}(E_i, E_j) = \nabla^2_{E_i,E_j}\rho = A_{ji}$, $i,j = 1, \ldots, n-1$. We may now compute

$$\frac{\partial}{\partial s} A_{ij}(s) = \nu\left(\langle \nabla_{E_i}\nu, E_j\rangle\right) = \langle \nabla_\nu \nabla_{E_i}\nu, E_j\rangle + \langle \nabla_{E_i}\nu, \nabla_\nu E_j\rangle,$$

where the last term vanishes since E_j is parallel along γ. For the same reason, we have the Lie bracket $[E_i, \nu] = \nabla_{E_i}\nu = A_{ji}E_j$, so that the curvature tensor

$$\begin{aligned} R_{innj} &= R(E_i, \nu, \nu, E_j) := \langle \nabla_\nu \nabla_{E_i}\nu - \nabla_{E_i}\nabla_\nu\nu + \nabla_{[E_i,\nu]}, E_j\rangle \\ &= \frac{\partial}{\partial s}A_{ij}(s) - 0 + A_{ik}(s)A_{kj}(s), \end{aligned}$$

since $\nabla_\nu\nu \equiv 0$. A more familiar form of this Matrix Riccati Equation is

$$\frac{\partial}{\partial s}A_{ij}(s) + A_{ik}(s)A_{kj}(s) = -R_{injn}(\gamma(s)). \tag{4.3.1}$$

The Matrix Riccati Equation has importance in many other contexts, such as optimal control [L-T.11], [L-T.24], [R, p. 215].

4.4. Matrix Riccati Comparison Theorem. Although the Matrix Riccati Equation (4.3.1) is nonlinear, it leads to powerful techniques for comparing functions such as the distance functions from given subsets on two manifolds. Namely:

THEOREM 4.4.1. *Suppose that* Σ_0 *resp.* $\overline{\Sigma}_0$ *are smooth, transversely oriented hypersurfaces of the Riemannian manifolds* M *resp.* $\overline{M}$, *and let* $\rho : M \to \mathbb{R}$ *resp.* $\overline{\rho} : \overline{M} \to \mathbb{R}$ *be the signed distancc to a point of* M *resp.* $\overline{M}$ *from* Σ_0 *resp.* $\overline{\Sigma}_0$. *Assume that* ρ *and* $\overline{\rho}$ *are smooth. Let* γ *resp.* $\overline{\gamma}$ *be unit-speed geodesics meeting* Σ_0 *resp.* $\overline{\Sigma}_0$ *at* $\gamma(0)$ *resp.* $\overline{\gamma}(0)$ *with initial tangent vector* $\gamma'(0) = \nu(\gamma(0))$ *resp.* $\overline{\gamma}'(0) = \overline{\nu}(\overline{\gamma}(0))$, *the unit normal vectors to* Σ_0 *resp.* $\overline{\Sigma}_0$. *Assume that for all* $0 \le s \le s_1$, *for all unit vectors* $W \in T_{\gamma(s)}M$ *orthogonal to* $\gamma'(s)$, *and* $\overline{W} \in T_{\overline{\gamma}(s)}\overline{M}$ *orthogonal to* $\overline{\gamma}'(s)$, *that the sectional curvatures* $R(W, \gamma', W, \gamma') \le \overline{R}(\overline{W}, \overline{\gamma}', \overline{W}, \overline{\gamma}')$. *As initial conditions, assume that for all unit vectors* $V \in T_{\gamma(0)}\Sigma_0$ *and for all unit vectors* $\overline{V} \in T_{\overline{\gamma}(0)}\overline{\Sigma}_0$, *the normal curvatures*

$$B_{\Sigma_0}(V, V) = \nabla^2_{V,V}\rho(\gamma(0)) \ge \overline{\nabla}^2_{\overline{V},\overline{V}}\overline{\rho}(\overline{\gamma}(0)) = B_{\overline{\Sigma}_0}(\overline{V}, \overline{V}).$$

Then for all $0 \le s \le s_1$, *for all unit vectors* $W \in T_{\gamma(s)}M$ *orthogonal to* $\gamma'(s)$, *and* $\overline{W} \in T_{\overline{\gamma}(s)}\overline{M}$ *orthogonal to* $\overline{\gamma}'(s)$, *we have*

$$B_{\Sigma_s}(W, W) = \nabla^2_{W,W}\rho(\gamma(s)) \ge \overline{\nabla}^2_{\overline{W},\overline{W}}\overline{\rho}(\overline{\gamma}(s)) = B_{\overline{\Sigma}_s}(\overline{W}, \overline{W}). \tag{4.4.1}$$

Theorem 4.4.1 has a long statement, but a relatively short proof. First note that the hypotheses of Theorem 4.4.1 allow us to interpose a "scalar" second fundamental form $\widetilde{A}_{ij}(0) = \alpha_0\delta_{ij}$ between $A_{ij}(0)$ and $\overline{A}_{ij}(0)$ and a curvature tensor of "scalar" form $\widetilde{R}_{injn} = \kappa(s)\delta_{ij}$ between R_{injn} and $\overline{R}_{injn}$. Here α_0 is chosen less than or equal to the smallest principal curvature of Σ_0 at $\gamma(0)$, but greater than or equal to the largest principal curvature of $\overline{\Sigma}_0$ at $\overline{\gamma}(0)$. We write the resulting inequality as $(A_{ij}(0)) \ge \left(\widetilde{A}_{ij}(0)\right) \ge \left(\overline{A}_{ij}(0)\right)$, where an inequality between symmetric matrices is to be understood in the sense that the difference of the two sides is positive semi-definite. Also, we choose here $\kappa(s)$ less than or equal to the smallest sectional curvature of $\overline{M}$ in plane sections containing $\overline{\gamma}'(0)$, but greater than or equal to the largest sectional curvature of M in plane sections containing $\gamma'(0)$. We define $\widetilde{A}_{ij}(s) := \alpha(s)\delta_{ij}$, where $\alpha(s)$ is the solution of the scalar Riccati equation

$$\frac{d\alpha(s)}{ds} + \alpha(s)^2 = -\kappa(s), \qquad 0 \le s \le s_1, \tag{4.4.2}$$

with the initial condition $\alpha(0) = \alpha_0$ as already chosen. We shall show that (1) $(A_{ij}(s)) \ge \left(\widetilde{A}_{ij}(s)\right)$ and (2) $\left(\widetilde{A}_{ij}(s))\right) \ge \left(\overline{A}_{ij}(s)\right)$ for all $0 \le s \le s_1$. Thus, it will suffice to prove Theorem 4.4.1 in the two special cases where (1) $(A_{ij}(s))$ and $(R_{injn}(\gamma(s)))$ are in the scalar form, or where (2) $\left(\overline{A}_{ij}(s)\right)$ and $\left(\overline{R}_{injn}(\overline{\gamma}(s))\right)$ are in the scalar form. In both cases, after substituting $\widetilde{A}_{ij}$ for one of A_{ij} or $\overline{A}_{ij}$, we have

$$\sum_k A_{ik}(s)\overline{A}_{kj}(s) = \sum_k \overline{A}_{ik}(s)A_{kj}(s), \tag{4.4.3}$$

that is, for each s, the second fundamental forms $B_{\Sigma_s}(E_i, E_j)$ and $B_{\overline{\Sigma}_s}(\overline{E}_i, \overline{E}_j)$ have a common basis of curvature directions. It remains only to prove inequality 4.4.1 under the additional assumption 4.4.3. In this case, write the difference matrix $D_{ij}(s) = A_{ij}(s) - \overline{A}_{ij}(s)$. Then $\frac{d}{ds}D_{ij}(s) = -A_{ik}A_{kj} + \overline{A}_{ik}\overline{A}_{kj} - R_{injn} + \overline{R}_{injn}$ by the matrix Riccati equation (4.3.1), so that $\left(\frac{d}{ds}D_{ij}(s)\right) \geq \left(-D_{ik}(A_{kj} + \overline{A}_{kj})\right)$ since $(R_{injn}) \leq \left(\overline{R}_{injn}\right)$ and using the commutativity (4.4.3). Also, we have the initial inequality $(D_{ij}(0)) \geq 0$. Let $\lambda(s)$ be the smallest eigenvalue of $(D_{ij}(s))$: $\lambda(s)$ is Lipschitz continuous as a function of s, and $\lambda(0) \geq 0$. Consider any $s_0 \in (0, s_1]$ where $d\lambda/ds$ exists. Assume $E_1(s_0)$ is chosen to be an eigenvector of $(D_{ij}(s_0))$ with eigenvalue $\lambda(s_0)$. Then $D_{11}(s_0) = \lambda(s_0)$ and for all s, $D_{11}(s) \geq \lambda(s)$. Thus

$$\frac{d\lambda}{ds}(s_0) = \frac{dD_{11}}{ds}(s_0) \geq -D_{1k}(A_{k1} + \overline{A}_{k1}) \geq -C\lambda(s_0),$$

where $C := \sup\{B_{\Sigma_s}(V,V) + B_{\overline{\Sigma}_s}(\overline{V},\overline{V}) | 0 \leq s \leq s_1,\ |V| = |\overline{V}| = 1\}$. This implies $\lambda(s) \geq e^{-Cs}\lambda(0) \geq 0$ for all $s \in [0, s_1]$. That is, $(A_{ij}(s)) \geq \left(\overline{A}_{ij}(s)\right)$, which is the conclusion (4.4.1) of Theorem 4.4.1.

As *a corollary, we may prove the Hessian Comparison* Theorem 4.1.1. We first show that in the case where the reference set Σ_0 is a point p_0 of M, all normal curvatures of Σ_s equal $1/s$ plus a term which approaches zero uniformly as $s \to 0$ (proof, based on normal coordinates at p_0, omitted here.) Thus, applying the Matrix Riccati Comparison Theorem 4.4.1 on intervals $[s_0, s_1]$ with s_0 close to 0, we obtain the conclusion of Theorem 4.1.1 within an arbitrarily small tolerance, and hence exactly.

4.5. Nonsmooth distance function. We indicated above that Greene and Wu were able to convert the *distance function from infinity on a complete manifold of positive curvature* into a *smooth function.* But in general, a locally convex distance function may be nonsmooth in a nonconvex way. As an example, we consider the two-dimensional, flat cylinder $S^1 \times I\!R$, with coordinates (θ, s). The distance function ρ from $p_0 = (0,0)$ fails to be smooth along the cut locus $\{\theta = \pi\}$, although everywhere else, $\frac{1}{2}\rho(x)^2$ is strictly convex. In fact, locally near $\{\theta = \pi\}$, $\frac{1}{2}\rho(x)^2$ is the *minimum* of two locally convex functions, whereas the *maximum* of two such functions would be convex, in the sense of Definition 4.2.1 above. In order to obtain a convex function $v(x) = \varphi(\rho(x))$ without discarding the cut locus, we need only to ensure the strict convexity of the level sets Σ_s of ρ and to choose $\varphi : I\!R \to I\!R$ to be strictly *decreasing* as a function of the distance $|\rho(x)|$.

As an interesting case, we consider a result of Burago and Zalgaller ([B-Z.1, p. 259]):

THEOREM 4.5.1. *Suppose Ω is a smooth domain in M which has sectional curvatures bounded below by a constant κ, which lies within a distance R from its boundary, and whose boundary has all normal curva-*

tures greater than a positive constant α_0*. Then there is a convex function* $v : \Omega \to \mathbb{R}$*, provided that either* $\kappa \geq 0$*, or* $\alpha_0 \geq \sqrt{-\kappa} \tanh \sqrt{-\kappa} R$*.*

A proof of this result follows from the Matrix Riccati Comparison Theorem 4.4.1 and Theorem 4.2.2. In fact, let $\alpha(s)$, $-R \leq s \leq 0$, be the solution of the scalar Riccati equation (4.4.2)

$$\frac{d\alpha(s)}{ds} + \alpha(s)^2 = -\kappa,$$

with the "initial" condition $\alpha(0) = \alpha_0$. ($\alpha(s)$ is written as $-Q(f)$ in [B-Z.1].) If $\kappa > 0$, write $\kappa = b^2$; then $\alpha(s) = b \cot b(s - s_0)$, where $s_0 < 0$ is determined by $b \cot b s_0 = -\alpha_0$. If $\kappa < 0$, and we write $\kappa = -\beta^2$, then three cases may occur, depending on the magnitude of α_0: when $\alpha_0 > \beta$, we find $\alpha(s) = \beta \coth \beta(s - s_0)$, where $s_0 < 0$ is determined by $\beta \coth \beta s_0 = -\alpha_0$; when $\alpha_0 = \beta$, we find $\alpha(s) \equiv \beta$; and when $\alpha_0 < \beta$, we find $\alpha(s) = \beta \tanh \beta(s - s_0)$, where $s_0 < 0$ is determined by $\beta \tanh \beta s_0 = -\alpha_0$. In particular, $\alpha(s)$ remains positive as s decreases from 0 to $-R$ under the hypotheses stated.

We apply the Matrix Riccati Comparison Theorem, with M in the role of $\overline{M}$ and the boundary of Ω in the role of $\overline{\Sigma}_0$. The comparison manifold of constant sectional curvature κ plays the role of M, and Σ_0 is a hypersurface of constant normal curvatures α_0. Then the hypersurface Σ_s at constant signed distance s from Σ_0 is totally umbilic with normal curvatures $\alpha(s)$. Note that the parameter domain $-R \leq s \leq 0$ in our case lies on the opposite side of the initial hypersurface at $s = 0$; since the proof follows from a first-order differential inequality, the conclusion will also have the opposite sense from (4.4.1). Thus

$$B_{\Sigma_s}(V, V) \geq \alpha(s)\langle V, V\rangle,$$

$-R \leq s \leq 0$. Thus, under the hypotheses stated, wherever $\rho(x)$ is smooth, $v(x) = e^{\Lambda \rho(x)}$ will be convex for large enough Λ.

Finally, near a general point, since $\rho < 0$ is locally the maximum of a number of locally smooth functions with convex sublevel sets, it has convex sublevel sets. The continuous convex function $v : M \to \mathbb{R}$ may now be approximated by smooth, uniformly convex functions $v_\varepsilon : M \to \mathbb{R}$ according to Theorem 4.2.2.

Acknowledgement. The authors wish to thank Yaroslav Kurylev, Loughborough University, UK, for directing their attention to the results of Burago and Zalgaller.

4.6. Construction of a strictly convex function in the 2-d case via curvature flows. The following result is proved by using, among other things, Lemma 4.2.1 as well as a nonlinear parabolic equation which arises in a quite unrelated geometric problem of curve-shortening flows [Gr.1], [C-Z.1].

THEOREM 4.6.1. *[B-G-L.1] Let $\overline{\Omega}$ be a two-dimensional smooth compact Riemannian surface whose boundary $\partial\Omega$ has positive second fundamental form. Assume there are no closed geodesics in Ω. Then there exists a C^2-strictly convex function v in $\overline{\Omega}$.* □

Remark 4.6.1. It is therefore very important to be able to discern whether a given Ω, with a given Riemannian metric, contains closed geodesics. In the 2-d case, [B-G-L.1, Section 5] describes a computational method which implements the parabolic flow (entering the proof of Theorem 4.6.1), and enables one to either find geodesics homotopic to a component of $\partial\Omega$, or else to conclude that there are no closed geodesics. □

5. A Riemannian geometric approach for exact controllability of second-order hyperbolic equations. A "view from the boundary" which verifies the geometric optics condition.

5.1. Introduction. Counterexamples. Hypotheses. Main exact controllability statement.

Introduction. One of the important advances in the subject of boundary control for (linear) *second-order hyperbolic equations* has been the almost necessary and sufficient criteria for exact controllability and uniform stabilization involving bicharacteristics [Lit.1], [B-L-R.1]. Indeed, the opinion has sometimes been expressed to the effect that this whole subject of exact controllability/uniform stabilization for (linear) second-order hyperbolic equations has been completely resolved. However, one may counter such a hasty opinion with various considerations. To begin with, the conditions involving bicharacteristics are generally difficult to verify, as these are solutions of a system of nonlinear ordinary differential equations (Hamiltonian). Thus, the need arises for the infusion of other ideas into this subject, not only for expressing tests for exact controllability/stabilization in terms of (perhaps more restrictive, but also) more easily recognizable geometric properties, but also for making contact and building links with other (perhaps unsuspected) mathematical areas. The subsequent connection [Ta.1–3] made with pseudo-differential multipliers (symbols) derived from pseudo-convex functions (defined in terms of double Poisson brackets) is one positive case in point. Though providing only, definitely, sufficient conditions in the case of second-order hyperbolic equations, this method of Tataru has a broader scope, as it applies to general non-hyperbolic evolution equations as well. Another similar case in point—which provides the additional benefits of construction and checkability of assumptions, friendlier proofs, with closer and more recognizable links to 'classical methods' of the '80s—is the energy method in the Riemannian metric outlined in Section 3, of which we shall see more definite details in Part III. Here we shall, instead, present a method that singles out an alternative set of *sufficient conditions* for exact controllability of *second-order hyperbolic equations*, even though the resulting proofs are ultimately based on the original 'bicharacteristic' conditions, and hence in technical results in PDE theory [Hor.1–2], [Tay.1]. The focus of this method to be now explained is, for now, on the qualitative relationship between Riemannian geometry and boundary control of second-order hyperbolic PDEs. Thus, the treatment in this section will not attempt to express exact controllability in terms of the optimal choice of Sobolev spaces, or, of portion of the controlled boundary, unlike the energy method of Section 3 (to be detailed in Part III). Nor does the treatment of the present section attempt to find the optimal smoothness of the Riemannian metric and of other coefficients of the hyperbolic equation.

An alternative geometric approach. We shall present another line of investigation—in a sense, even more geometric in flavor—this one valid only for hyperbolic equations. In fact, it will exploit the well-known identification—available exclusively for hyperbolic problems—between the geometric optics notion of bi-characteristics (the carriers of energy) and the Riemannian geometric notion of geodesics with respect to a natural metric defined in terms of the coefficients of the principal part of the elliptic operator in the equation (Duff [Du.1, p. 209], Courant-Hilbert [C-H.1, p. 565]). See Section 6.3 below. This way, Riemannian geometry is merged with geometric optics. Accordingly, in this approach, the focus will be put on the geodesics.

Construction of counterexamples. One advantage is obtained at once, at the level of constructing counterexamples. Riemannian geometry offers plenty of manifolds (M, g) or even $(\mathbb{R}^n, g)$, of dimension at least equal to 2, which possess a *closed geodesic.* It then suffices to take any bounded set Ω, properly containing such closed geodesic in its interior, to have an example (in dimension greater or equal to 2) of a hyperbolic equation as in Section 6.3, where exact controllability fails, even if the control action is applied to the entire boundary $\partial\Omega$. Indeed, since the closed geodesic (bicharacteristic) does not touch the boundary, the necessary condition for exact controllability provided by the geometric optics approach [Ra.1], [Lit.1], [B-L-R.1] permits us to conclude that exact controllability is not possible, even if the boundary is fully controlled. This is an instance of the synergistic advantage that derives from marrying the Riemannian geometric viewpoint with the geometric optics approach. Many specific examples of this phenomenon may be given.

Counterexample #1: See [Y.1, Example 4.1], with $g = (dx\,dx + dy\,dy)/[(1 + x^2 + y^2)^2]$.

Counterexample #2 [G-L.1]: The Frisbee. This example is a large flat n-disk surrounded by a moderate-sized region of positive curvatures and a thin region of negative curvature, resembling the inside surface of the flying toy known *as the Frisbee* (apologies to the Wham-O Corporation).

This is an example with non-constant sectional curvature. We claim that for the Frisbee example, *the Riemannian wave equation defined on it is not exactly controllable from the boundary.* In fact, the sphere $\{r = r_0\}$ is totally geodesic as a submanifold of Ω, and therefore any of the sphere's own great circles will be closed geodesics of Ω. Any one of these closed geodesics suffices to make boundary controllability impossible, as seen above.

Counterexample #3 [G-L.1]: Manifold with a big bulge. Let $\overline{\Omega}$ be a Riemannian manifold-with-boundary, diffeomorphic to the closed ball of $\mathbb{R}^2$, which contains in its interior a smooth subdomain $\overline{D}$ isometric to the closed hemisphere of the unit sphere $S^2 \subset \mathbb{R}^3$. We assume that $\partial\Omega$ has positive second fundamental form. Such a manifold may be constructed as a hypersurface of revolution in $\mathbb{R}^3$, with a generating curve which begins orthogonally to the axis of revolution with a quarter-circle of radius 1, and ends at a moment when it is traveling away from the axis of rotation. Alternatively, such an example may be constructed by "pasting in" an isometric copy of the closed hemisphere in the interior of a reference manifold whose boundary has positive second fundamental form.

One may show that there exist closed geodesics in the boundary of D. Hence, again, the *Riemannian wave equation cannot be exactly controllable on such manifold* Ω with a big bulge, even if its entire boundary $\partial\Omega$ is controlled.

The manifold $\overline{\Omega}$ also *does not allow any convex functions*, as follows from [G-L.1, Proposition 5.3] using a domain slightly larger than D.

A view from the boundary. The computational energy method in the Riemannian metric described in Section 3 (and to be further analyzed in Part II) makes its main assumption (H.1) = (3.1.1) in the interior of Ω. By contrast, the presently described alternative line of research takes a "view from the boundary" as its distinguishing feature.

Hypotheses. This approach is based on three assumptions: (i) that the boundary $\partial\Omega$ has positive second fundamental form: this analytical hypothesis is slightly stronger than the geometric assumption that $\partial\Omega$ be strictly convex (in the Riemannian metric); (ii) that 'chords' be unique; (iii) that chords be nondegenerate.

A definition and analysis of these assumptions will be given in Section 5.2–5.4 below.

Statement of exact controllability results of the Riemannian wave. Consider a compact, n-dimensional Riemannian manifold-with-boundary $\overline{\Omega}$. We assume that $\partial\Omega$ is smooth and nonempty, and that the metric of $\overline{\Omega}$ is smooth, i.e., C^∞. We are interested in the boundary control of the following natural hyperbolic partial differential equation (Riemannian wave equation) on $\Omega \times [0,T]$:

$$\frac{\partial^2 u}{\partial t^2} = \Delta_g u = \sum_{i,j=1}^{n} \frac{1}{\gamma}\frac{\partial}{\partial x_i}\left(\gamma g^{ij}(x)\,\frac{\partial u}{\partial x_j}\right), \tag{5.1.1}$$

for all $(x,t) \in \Omega \times [0,T]$, where $(x_1,\ldots,x_n)$ are arbitrary local coordinates, $g^{ij}(x)$ are the entries of the inverse matrix to the coefficients $g_{ij}(x)$ of the Riemannian metric, and γ is the Riemannian volume integrand: $\gamma(x) = \sqrt{\det(g_{ij}(x))}$. We consider the problem of the control in time T of equation (5.1.1) from the *entire* boundary $\partial\Omega$. More precisely, we consider the boundary conditions

$$u(x,t) = U(x,t) \quad \text{for all} \quad (x,t) \quad \text{on} \quad \partial\Omega \times [0,T], \tag{5.1.2}$$

where $U \in H^{\frac{1}{2}}(\partial\Omega \times [0,T])$ is the control, i.e., a function which may be chosen as needed. The controllability question is whether, given any initial conditions

$$u(x,0) = u_0(x), \qquad \frac{\partial u}{\partial t}(x,0) = u_1(x), \tag{5.1.3}$$

with finite energy, there is a choice of controls $U \in H^{\frac{1}{2}}(\partial\Omega \times [0,T])$ such that the solution of (5.1.1) with initial conditions (5.1.3) and boundary conditions (5.1.2) vanishes identically on $\Omega \times [T,\infty)$. Equivalently, we ask whether for some choice of controls U the terminal Cauchy values vanish: $u(x,T) = 0$, $\frac{\partial u}{\partial t}(x,T) = 0$ for all x in Ω.

In this section we wish to describe two main results (Theorems 5.1.1, 5.1.2). We shall refer to [G-L.1] for complete proofs.

We define a chord to be a curve in $\overline{\Omega}$ of shortest length between two boundary points.

THEOREM 5.1.1. *[G-L.1] Suppose that any two boundary points of the manifold $\overline{\Omega}$ are connected by a unique chord, which is nondegenerate. Assume that $\partial\Omega$ has positive second fundamental form. Then the hyperbolic equation (5.1.1) is controllable from $\partial\Omega$ by means of boundary conditions (5.1.2) in any time $T > T_0 = diam_{\overline{\Omega}}(\partial\Omega)$.* □

Here, the *diameter* of the boundary of Ω is the maximum distance between any two of its points, with respect to the distance measured in $\overline{\Omega}$, that is: the length of the longest chord of $\overline{\Omega}$. Our convention for the sign of the second fundamental form is such that if Ω is a ball of radius r in $\mathbb{R}^n$, with the Euclidean metric, then $\partial\Omega$ has a positive second fundamental form $B = r^{-1}ds^2$.

The "uniqueness" of a chord $\gamma : [0,a] \to \overline{\Omega}$ is understood *modulo* reparameterizations $s \mapsto \gamma(As + B)$, $(A, B \in \mathbb{R})$ of the independent variable s.

Remark 5.1.1. The hypothesis of Theorem 5.1.1 that $\partial\Omega$ has positive second fundamental form at each point $p \in \partial\Omega$ may be computed most easily by making a linear change of coordinates so that the coordinate hyperplane $x_n =$ const., which passes through p is tangent to $\partial\Omega$ there, so that the nth coordinate vector is the inward unit normal to $\partial\Omega$ at p. Then, in local coordinates, the condition that $\partial\Omega$ has a positive second fundamental form amounts to having a *symmetric* matrix $\{B_{ij}\}$, with explicit entries B_{ij} in terms of the Christoffel symbols [G-L.1, p. 150], be *positive definite* at p. The matrix $\{B_{ij}\}$ represents the *second fundamental* form of $\partial\Omega$ in these coordinates. Equivalently, if $\bar{\Omega}$ is extended to be a smooth subdomain of a Riemannian manifold M, one requires that any geodesic of M which is tangent to $\partial\Omega$ at p remains outside of $\bar{\Omega}$ to second order at p. This property has been called "pseudo-convexity" in the PDE literature.

We shall refer to the infimal value T_0 as the "optimal time of control," even though T_0 itself may not be a control time. More generally, we shall consider the hyperbolic equation with additional lower-order terms:

$$\frac{\partial^2 u}{\partial t^2} = \Delta_g u + \sum_{i=1}^{n} V^i(x,t)\, \frac{\partial u}{\partial x_i} + a(x,t)u, \tag{5.1.4}$$

where $(V^1, \ldots, V^n)$ are the components, in any local system of coordinates $x = (x_1, \ldots, x_n)$ for $\overline{\Omega}$, of a vector field V on $\overline{\Omega}$, and $a : \overline{\Omega} \to \mathbb{R}$. The first-order term $\sum_{i=1}^{n} V^i(x,t)\frac{\partial u}{\partial x_i}$ is invariant under change of coordinates. We shall assume throughout that $V(x,t)$ and $a(x,t)$ describe real-analytic mappings from t to the space of smooth vector fields and smooth functions, resp., on $\overline{\Omega}$.

THEOREM 5.1.2. *Theorem 5.1.1 continues to be valid if Equation (5.1.1) is replaced by Equation (5.1.4).*

As in the proof of Theorem 5.1.1 (sketched in Section 5.5) will show this result de facto provides sufficient conditions implying the geometric optics condition [Lit.1], [B-L-R.1], which originally was given in $\mathbb{R}^n$. The proof of [L-T.1] continues to hold true in the Riemannian setting of this present section. The conditions of Theorem 5.1.1 are far from necessary, and indeed, examples are given in [G-L.1] and below where controllability holds without the condition of the uniqueness of the chords being satisfied.

5.2. Chord uniqueness and nondegeneracy. The relevant definitions are as follows:

DEFINITION 1. *The* chord *is a length-minimizing geodesics of* $\overline{\Omega}$ *joining two given points of its boundary* $\partial\Omega$.

The existence of a chord joining any two points of $\partial\Omega$ is well-known, although in general it may have non-geodesic segments lying in $\partial\Omega$ (see, e.g., [DoC.1, pp. 147–148]). If, however, we assume that $\partial\Omega$ has positive second fundamental form as in (i), then for any length-minimizing curve $\sigma : [s_0, s_1] \to \overline{\Omega}$ between two points of $\overline{\Omega}$, we have $\sigma((s_0, s_1)) \subset \Omega$ (see [G-L.1, Corollary 3.3 or Lemma 4.1]). In particular, every chord of $\overline{\Omega}$ is a geodesic in this case.

As restatement of assumption (ii) above is that any two points of $\partial\Omega$ are connected by at most one (and hence exactly one) chord.

DEFINITION 2. *Two points* $\gamma(s_1)$ *and* $\gamma(s_2)$ *of a geodesic* γ *are called* conjugate points *if* $s_1 \neq s_2$ *and there exists a nontrivial Jacobi field* J *along* γ *with* $J(s_1) = 0$ *and* $J(s_2) = 0$. *[A* Jacobi field *is a solution to the linearized geodesic equation, or* Jacobi equation.*] Although a chord, since it has minimum length, may in general have its endpoints conjugate to each other, no two interior points may be conjugate, by Jacobi's theorem. The case where a chord* $\gamma_0 : [0, a_0] \to \overline{\Omega}$ *has conjugate endpoints is therefore somewhat special, and we call such a chord* degenerate.

DEFINITION 3. *We say that a chord* $\sigma : [s_0, s_1] \to \overline{\Omega}$ *is* non-degenerate *if there are no conjugate points to* $\sigma(s_0)$ *along* $\sigma((s_0, s_1]))$.

Another way of viewing these hypotheses of 'chord uniqueness' and 'non-degeneracy' is as follows. Consider any two boundary points p and q. Suppose that, among all the light rays leaving q simultaneously and propagating in Ω (without reflection), only one ray reaches p first. This is equivalent to the uniqueness of the chords. Non-degeneracy of chords is equivalent to the statement that in the situation just described, when a ray leaving q turns out to be a chord from p to q, the boundary point p depends on a diffeomorphic ray in the initial direction of the ray at q.

5.3. Criteria for chord uniqueness and non-degeneracy. In this subsection, we collect several useful criteria which provide sufficient conditions for the hypotheses of chord uniqueness and non-degeneracy, required by the main exact controllability result under the present approach (see Theorem 5.1.1).

The first criterion is especially appropriate in a situation where observations about $\overline{\Omega}$ can only be made from its boundary:

PROPOSITION 5.3.1. *[G-L.1] Assume that $\partial\Omega$ has a positive second fundamental form. For fixed $q \in \partial\Omega$, let $\phi : \overline{\Omega} \to [0, \infty)$ be given by $\phi(x) \equiv d^2_{\overline{\Omega}}(x, q)$.*

(1) If, for each $q \in \partial\Omega$, $\nabla\phi$ is continuous along $\partial\Omega$, then any two points of $\partial\Omega$ are connected by a unique chord.

(2) If, moreover, for each $q \in \partial\Omega$, $\nabla\phi$ is continuously differentiable along $\partial\Omega$, then the chords are non-degenerate. □

Remark 5.3.1. Somewhat surprisingly, it is not true that the continuity of $\nabla\phi$ implies disconjugacy of the chords of $\overline{\Omega}$. That is, the stronger hypothesis of part (2) of Proposition 5.3.1 is required to imply the stronger conclusion. See Example 5.5 in [G-L.1].

The second, rather different, criterion concludes that chords are unique, assuming a condition which implies their nondegeneracy (compare Proposition 5.3.3 below):

PROPOSITION 5.3.2. *[G-L.1] Consider a Riemannian manifold-with-boundary $\overline{\Omega}$, whose boundary has positive second fundamental form. Suppose that for all $q \in \partial\Omega$, each geodesic starting from q leaves $\overline{\Omega}$ strictly before any conjugate point along the geodesic. Then $\overline{\Omega}$ has unique chords (which are nondegenerate).*

(The geodesic γ given by $\gamma(t) = \exp_p(tv) : [0, b] \to M$ has a conjugate point at $t_0 \in (0, b)$ if and only if $D\exp_p$ is singular at $t_0 v$. Thus, on S^2, the north pole is conjugate to the south pole along any meridian.)

The next criterion refers to a general class of examples, with properties analogous to a manifold with a bulge of moderate size.

PROPOSITION 5.3.3. *[G-L.1] Let $\overline{\Omega}$ be a smooth, compact subdomain of a Riemannian manifold M, whose sectional curvatures at $x \in M$ are bounded above by $f(r(x))$, where $r(x) = d(x, x_0)$, for some $x_0 \in \Omega$. Assume that $\exp_{x_0}$, the exponential map of M at x_0, is defined and injective on the closed ball $\overline{B}_R(0) \subset T_{x_0}M$, and that $r(x) \leq R$ on $\overline{\Omega}$. We assume that $f : [0, R] \to \mathbb{R}$ satisfies: (1) f is monotone decreasing; (2) the solution u_1 of the ODE $u'' + fu = 0$ with initial conditions $u_1(0) = 1$, $u_1'(0) = 0$ remains positive on $[0, R]$; and (3) the solution u_2 of the same ODE with* *initial conditions* $u_2(0) = 0$, $u_2'(0) = 1$ *has positive first derivative on* $[0, R]$. *Suppose also that* $\partial\Omega$ *has positive second fundamental form. Then* $\overline{\Omega}$ *has unique chords, which are non-degenerate.*

COROLLARY 5.3.4. *Suppose that for some radius* R, *the Riemannian manifold-with-boundary* $\overline{\Omega}$ *satisfies the hypotheses of Proposition 5.3.3. Then the wave equation (5.1.1) on* $\overline{\Omega}$ *is controllable from* $\partial\Omega$ *in any time* $T > T_0 = 2R$.

In Section 5.1, we have noted that the Frisbee provides an example where the Riemannian wave equation is *not* exactly controllable from the boundary. The following criterion may be used [G-L.1, Proposition 5.4] to claim that the Frisbee example does *not* have unique chords.

PROPOSITION 5.3.5. *[G-L.1] Assume that* $\partial\Omega$ *has positive second fundamental form. Suppose there is a subdomain* $\overline{D} \subset \Omega$ *such that* ∂D *has negative second fundamental form. If the dimension* $n > 2$, *assume further that* $\overline{\Omega}$ *and* $\overline{D}$ *are rotationally symmetric. Then* $\overline{\Omega}$ *does not have unique chords.*

5.4. Examples.

Example 5.4.1: Manifold with a big bulge (negative example). We have already seen in Section 5.1 that the Riemannian wave equation is not exactly controllable on a manifold with a big bulge, even when the control acts on the entire boundary. The reason is that this manifold has closed geodesics in its interior. Examining this example from the viewpoint of the sufficient conditions listed at the beginning of Section 5, we may claim that $\overline{\Omega}$ will not have unique chords. This is proved in [G-L.1].

Example 5.4.2: Manifold with a bulge of moderate size (a gentle hill) (positive example). Let $\overline{\Omega}$ be a spherical cap, of intrinsic radius $R < \frac{\pi}{2}$, in the unit n-sphere S^n. Then $\partial\Omega$ has positive second fundamental form. Also, $\overline{\Omega}$ has unique chords, which are non-degenerate and have lengths $\leq 2R$. Thus, all three assumptions at the beginning of Section 5.1 hold true. Accordingly, Theorem 5.1.1 may be applied to show that the spherical Riemannian wave equation may be controlled from the boundary in any time $T > T_0 = 2R$. Note that the requirement $R < \frac{\pi}{2}$ is sharp, since the normal curvatures of $\partial\Omega$ equal cot R, which becomes negative for $R > \frac{\pi}{2}$.

One may give a strictly convex function as well in this example. An apparently optimal choice for the spherical cap example here considered would be $v(x) = -\cos r(x)$, where $r(x)$ is the distance from x to the center x_0.

Example 5.4.3: The Frisbee (negative example). We have already seen in Section 5.1 that the Riemannian wave equation defined on a Frisbee is not exactly controllable, even when the control acts on the entire boundary. The reason is that this manifold has closed geodesics in its interior. With reference to the sufficient conditions listed at the beginning of Section 5, we may add that chords will not be unique [G-L]. There is no strictly convex function $v : \overline{\Omega} \to \mathbb{R}$ which has positive normal derivative on $\partial\Omega$, either.

Example 5.4.4: The salt shaker or a high mountain (convex function but non-unique chords). In this rotationally symmetric example, positive sectional curvature $\equiv 1$ is concentrated in a ball $B_{R_0}(x_0)$ near the center of $\overline{\Omega}$, while the sectional curvature $K_1(r)$ which affects Jacobi fields along a cross-section becomes identically zero outside that ball. This allocation of curvatures is opposite to Example 5.3.3. Moreover, we shall construct the metric so that there is a convex function $v : \overline{\Omega} \to \mathbb{R}$, although there will be a conjugate points along the cross-sections, and chords will *not* be unique. Control is possible in a finite time, but the optimal time of control may be much less than $\text{diam}_{\overline{\Omega}}(\partial\Omega)$. The example is a truncated cone, topped off at the smaller end with a spherical cap. This resembles a design for salt shakers which are commonly found in American roadside diners, for example.

It is shown in [G-L] that chords of $\overline{\Omega}$ are not unique. Thus a sufficient condition of the present "view from the boundary" approach fails.

On the other hand $\overline{\Omega}$ *does* support a convex function v with $\frac{\partial v}{\partial \nu} > 0$ on $\partial\Omega$. For example, the function v may be constructed in the rotationally-symmetric form $v = \psi(r)$, with $\psi(0) = 0$ and $\frac{d\psi}{dr} = u(r)$. We may apply the results of [L-T-Y.1–2], see Section 6 below, to obtain boundary control in any time greater than $T_1 = 2\,\frac{\max|\nabla v|}{c_0}$, where c_0 is a positive lower bound on convexity of v, as in the second paragraph of subsection 5.2 above. With the convex function v just constructed, we find $c_0 = \cos R_0$ and $\max|\nabla v| = u(R_1)$, so that $T_1 = 2(L + \tan R_0)$. Here we have written $L = R_1 - R_0$ for convenience.

Remark 5.4.1. [Y.1] has also used Greene-Wu's Theorem as a starting point to arrive at sufficient conditions yielding that the function $v(x) = [d_g[x, x)]^2$, d_g being the g-distance function, be strictly convex. We recall from Section 4.2 that Greene-Wu's Theorem is as follows: *If $\{M, g\}$ is a smooth, noncompact complete Riemannian manifold of everywhere positive sectional curvature, then there exists a smooth strictly convex function v on $\{M, g\}$.*

When $M = \mathbb{R}^n$, it is possible to construct a metric g (hence coefficients a_{ij} of a second-order elliptic operator $\mathcal{A}$, see Section 6.3 below), so that by Greene-Wu's Theorem a strictly convex function v exists on all of $\{\mathbb{R}^n, g\}$. In this case, *any* sufficiently smooth bounded domain $\tilde{\Omega}$ in $\mathbb{R}^n$ can be taken for the corresponding PDE, see Section 6.3.

Alternatively, $\{\mathbb{R}^n, g\}$ with non-positive sectional curvature yields $v(x) = [d_g(x^0, x)]^2$ to be globally strictly convex. Again, *any* sufficiently bounded domain $\tilde{\Omega}$ in $\mathbb{R}^n$ can be taken for the corresponding PDE. Additional specific non-trivial examples of globally strictly convex functions in $\{\mathbb{R}^n, g\}$ can be found in [Y.1], [L-T.1–2], [T-Y.1]. □

5.5. Sketch of proof of Theorem 5.1.1: Verification of the geometric optics conditions. As the reader will verify immediately, Theorems 5.1.1 and 5.1.2 follow from Propositions 5.5.1, 5.5.2, and 5.5.3 below. For the propositions below, we let $\overline{\Omega}$ be extended to become a subset of an open n-dimensional Riemannian manifold M.

PROPOSITION 5.5.1. *Let M be a Riemannian manifold. Then the bicharacteristics of equations (5.1) and (5.1_t) are the graphs of $M \times \mathbb{R}$ of geodesics of M, with unit-speed parameter identified with time $\in \mathbb{R}$. See [Du.1], p. 209.*

PROPOSITION 5.5.2. *If every bicharacteristic in $\overline{\Omega} \in (0, T)$ enters or leaves $\overline{\Omega} \times (0, T)$ across the lateral boundary $\partial\Omega \times (0, T)$, then boundary control is available in any time $\geq T$. conversely, if there is a single bicharacteristic in $\Omega \times (0, T)$ that enters $\overline{\Omega} \times [0, T]$ through the open bottom $\Omega \times \{0\}$ and leaves through $\Omega \times \{T\}$, without hitting the lateral boundary, then boundary control in time T is not possible.*

Proof. See for example, [Lit.1] where the proof which is given for bounded domains in $\mathbb{R}^n$ carries over without difficulties to manifolds (this is because it involves only interior propagation of singularity results). The assumption of real analytic coefficients is easily removed for time independent coefficients. The proof in [B.1] given for $\mathbb{R}^n$ holds with optimal Sobolev spaces. The proof of [Lit.1] also works if the lower-order terms (i.e., not in the principal part) are real-analytic mappings from t to the space of C^∞ vector fields or C^∞ functions on $\overline{\Omega}$. The proof should then be supplemented by uniqueness Theorem 2 of [Lit.2]. Recent results in propagation of singularities by M. Taylor (see [Tay.2]) further indicate that the required smoothness in x of the coefficients can be reduced to $C^{2,a}$.

The converse follows from the propagation of singularities (see [Ra.1]). Namely, initial data can be constructed which is not C^∞ near a point inside Ω such that this singularity is propagated along the bicharacteristic, without being affected by boundary values, to form a singularity at time T inside Ω. Thus the solution cannot have terminal Cauchy conditions $u(\,\cdot\,, T) \equiv 0$, $\frac{\partial u}{\partial t} \equiv 0$. □

Remark 5.5.1. In particular, if Ω contains a closed geodesic, then boundary control is impossible in any finite time. This was the basis for the construction of counterexamples in Section 5.1.

Remark 5.5.2. It will be observed that if the hypothesis of the first part of Proposition 5.5.2 holds for a Riemannian manifold $\overline{\Omega}$, then it also holds for any compact subdomain $\overline{\Omega} \subset \Omega$.

Remark 5.5.3. Consider any compact subdomain $\overline{\Omega}_1 \subset \Omega$, and any solution u of (5.1.4) having finite energy. It follows from Remark 5.5.2 above and from Tataru's trace theorem [Ta.5] that the trace of the conormal derivative of u will be in $L^2(\partial\Omega_1 \times (0, T))$. This gives us boundary control for either Neumann or Robin controls in optimal Sobolev spaces.

For example, under the hypotheses of Proposition 5.5.2, we have boundary control for either Neumann or Robin controls in optimal Sobolev spaces for $\Omega \times (0, T)$.

PROPOSITION 5.5.3. *Assume that the boundary $\partial\Omega$ of the compact Riemannian manifold-with-boundary $\overline{\Omega}$ has a positive second fundamental form. Suppose that any two points of $\partial\Omega$ are connected by a unique chord, which is nondegenerate. Then any interior geodesic segment $\gamma : (b-\epsilon, b+\epsilon) \to \Omega$ may be extended to a geodesic $\gamma : [s_0, s_1] \to \overline{\Omega}$ which is a chord, that is, which realizes the minimum length between two distinct points $\gamma(s_0)$ and $\gamma(s_1)$ in $\partial\Omega$.*

The proof uses an "open and closed" connectedness argument in the unit tangent bundle in $\overline{\Omega}$.

Part III: Single PDE equations: Carleman estimates. Exact controllability. Uniform stabilization. The control-theoretic expansion of the energy method in the Riemannian metric outlined in Section 3 begins with the present Part III. In it, we shall examine, in more details and more precisely than in Section 3, each of the single classes of evolution equations, individually, and present the relevant results. *Both assumptions and techniques of proof are rooted in differential geometry. The proofs are essentially self-contained*, which is a further advantage of the approach. It will be clear that we have, basically, two choices of presentation. We may:

(i) either begin with a PDE evolution equation with variable coefficients as defined on a Euclidean open bounded domain of $\mathbb{R}^n$, and then transform it in a corresponding version on a suitable Riemannian manifold $(\mathbb{R}^n, g)$ (this approach is outlined in Section 6.3);

(ii) or else, we may at the outset consider a Riemannian evolution equation on a Riemannian manifold (M, g) involving the corresponding Laplace-Beltrami operator, and then recover the variable coefficient case defined on a Euclidean domain as a special case.

Which form of presentation to choose is largely a subjective question of taste. Mathematically, it makes little difference in the proof, though the second approach is technically more general. Historically, as pointed out in Section 0, it was the variable coefficient case on a Euclidean domain that motivated and generated the infusion of Riemannian methods. Papers [L-T-Y.1–3], [Y.1–2], [T-Y.1] follow this route: from the Euclidean domain to the Riemannian manifold. On the other hand, papers [L-T-Y.4], [T-Y.2] follow the reverse route, by working directly on a Riemannian manifold. Here, in our presentation, we shall opt for the second approach. However, in Section 6.3, we sketch the foundation of the first approach.

6. Second-order hyperbolic equations on a Riemannian manifold: Energy methods in the Riemannian metric yielding Carleman and control estimates with lower-order terms.

6.1. Model. Euclidean versus Riemannian domain. Throughout this section, M is a compact, finite-dimensional Riemannian manifold with metric $g(\cdot\,,\,\cdot) = \langle\cdot\,,\,\cdot\rangle$ and norm $|X| = \{g(X,X)\}^{\frac{1}{2}}$, and Ω is an *open, bounded* set of M with smooth boundary $\Gamma = \Gamma_0 \cup \Gamma_1$. We let n

denote the outward unit normal field along the boundary Γ. Further, we denote by Δ_g the Laplace (Laplace-Beltrami) operator on the manifold M, and by D the Levi-Civita connection on M.

Model. In this section, we study the following Riemannian wave equation with 'energy level terms,' on Ω:

$$\begin{aligned} &w_{tt} = \Delta_g w + F(w) + f \quad \text{in } (0,T] \times \Omega \equiv Q; \\ &f \in L_2(Q) = L_2(0,T; L_2(\Omega)). \end{aligned} \tag{6.1.1}$$

For this equation, the '*energy level*' is $H^1(\Omega) \times L_2(\Omega)$ for the pair $\{w, w_t\}$. We define:

$$E(t) = \int_\Omega [|Dw(t)|^2 + w_t^2(t)]d\Omega. \tag{6.1.2}$$

Accordingly, we make the following assumption:
(A.1) the 'energy level' differential term

$$F(w) = \langle P(t,x), Dw\rangle + p_1(t,x)w_t + p_0(t,x)w, \tag{6.1.3a}$$

where $P(t, \cdot\,)$ is a vector field on M for $t > 0$ and p_0 and p_1 are functions on Q, satisfies the following estimate: there exists a constant $C_T > 0$, such that

$$|F(w)|^2 \leq C_T\{|Dw|^2 + w_t^2 + w^2\}, \quad \forall\, x,t \in Q \text{ a.e.} \tag{6.1.3b}$$

where $Dw = \nabla_g w$ (the gradient of w in the Riemannian metric for the scalar function w). So Dw is a vector field, $Dw \in \mathcal{X}(M)$ = the set of all vector fields on M. Two vertical bars $|\cdot|$ may denote the norm in the tensor space T_x or $L_2(\quad)$ [He.1]. Furthermore, we may assume throughout that the forcing term f in (6.1.1) satisfies $f \in L_2(Q) \equiv L_2(0,T; L_2(\Omega))$, where $dQ = d\Omega\, dt$, and $d\Omega$ are the volume element of the manifold M in its Riemannian metric g.

Remark 6.1.1. Property (6.1.3b) is fulfilled if $P \in L_\infty(0,T,\Lambda)$, $p_0, p_1 \in L_\infty(Q)$. In effect, we could relax the regularity assumption on the lower-order coefficient p_0 and just require that $p_0 \in L_p(Q)$ for $p = \dim \Omega + 1$, by using a Sobolev embedding theorem. □

6.2. Strictly convex functions and coercive fields in the Riemannian metric. We elaborate, with some overlapping and repetition, on Section 3.1, as it applies to the specific problem (6.1.1), (6.1.2).

Main assumption: Our main assumption is hypothesis (H.1) = (3.1.1) in Section 3.1: *there exists a strictly convex function* $v : \overline{\Omega} \to \mathbb{R}$ *of class* C^2, in the Riemannian metric g, which we take non-negative after translation, without loss of generality: $v(x) \geq 0$, $\forall\, x \in \Omega$.

Under the common setting of Section 3.2, two related (or closely related) yet different energy methods in the Riemannian metric have become available. We shall treat them separately in Sections 6.4–6.5 and in Section 7, respectively, after pointing out explicitly how the variable coefficient, second-order hyperbolic equation defined on a Euclidean open bounded domain is contained in the abstract treatment of Sections 6.1 and 6.2.

6.3. A second-order hyperbolic equation with variable coefficients defined on a Euclidean domain. In this section $\tilde{\Omega}$ is an open bounded domain in $\mathbb{R}^n$, with boundary $\partial\tilde{\Omega} = \tilde{\Gamma}$ of class, say, C^2. Let $x = [x_1, \dots, x_n]$, and let, as in (2.2.2):

$$\mathcal{A}w = -\sum_{i,j=1}^{n} \frac{\partial}{\partial x_i}\left(a_{ij}(x)\,\frac{\partial w}{\partial x_j}\right),$$
$$\sum_{i,j=1}^{n} a_{ij}(x)\xi_i\xi_j \geq a\sum_{i=1}^{n}\xi_i^2, \quad x \in \tilde{\Omega}, \tag{6.3.1}$$

be a second-order differential operator, with real coefficients $a_{ij} = a_{ji}$ of class C^1, see Remark 6.3.1, satisfying the uniform ellipticity condition for some positive constant $a > 0$. Thus, we can extend $a_{ij}(x)$ smoothly to all of $\mathbb{R}^n$ so that the matrices

$$A(x) = (a_{ij}(x)); \qquad G(x) \equiv [A(x)]^{-1} = (g_{ij}(x)),$$
$$i, j = 1, \dots, n, \ x \in \mathbb{R}^n \tag{6.3.2}$$

are positive definite on any $x \in \mathbb{R}^n$. The second-order hyperbolic equation defined on $\tilde{\Omega}$ is

$$w_{tt} + \mathcal{A}w = F(w) \qquad \text{in } (0, T] \times \tilde{\Omega}, \tag{6.3.3}$$

with $\mathcal{A}$ defined by (6.3.1) and $F(w)$ the first-order differential operator defined by (2.2.3).

Riemannian metric. Let $\mathbb{R}^n$ have the usual topology and $x = [x_1, x_2, \dots, x_n]$. For each $x \in \mathbb{R}^n$, define the inner product and the norm on the tangent space $\mathbb{R}^n_x \equiv \mathbb{R}^n$ via (6.3.2) by

$$g(X, Y) = \langle X, Y\rangle_g \equiv \sum_{i,j=1}^{n} g_{ij}(x)\alpha_i\beta_j, \quad \forall\, X, Y \in \mathbb{R}^n_x; \tag{6.3.4}$$

$$|X|_g = \{\langle X, X\rangle_g\}^{\frac{1}{2}}; \quad X = \sum_{i=1}^{n}\alpha_i\,\frac{\partial}{\partial x_i}, \quad Y = \sum_{i=1}^{n}\beta_i\,\frac{\partial}{\partial x_i} \in \mathbb{R}^n_x. \tag{6.3.5}$$

Then $(\mathbb{R}^n, g)$ is a Riemannian manifold with Riemannian metric g. One may prove that [Y.2], [L-T-Y.4]

$$\mathcal{A}w = -\Delta_g w + Dw, \tag{6.3.6}$$

Δ_g = corresponding Laplace-Beltrami operators; that is, under the change of metric, from the original Euclidean metric to g, we have that the second-order elliptic operator (6.3.1) becomes the Laplace-Beltrami operator on

$(\mathbb{R}^n, g)$, modulo a first-order term. Thus Eqn. (6.3.3) is turned into (6.1.1) on $(\mathbb{R}^n, g)$, where (6.1.3b) is satisfied. This shows the benefit accrued by the change of metric: The energy method is conducted with respect to Δ_g.

Remark 6.3.1. Let the coefficients a_{ij} in (6.3.1) be of class C^1, as assumed. Then the entries g_{ij} in (6.3.4) are of class C^1 as well. Thus, the connection coefficients (Christoffel symbols) Γ^{ℓ}_{ik}, see [DoC.1] are of class C^0. The geodesic-solutions to a corresponding second-order nonlinear ordinary differential equation [DoC.1] are then of class C^2. Thus, the square of the distance function $d_g^2(x, x_0)$ is in C^2. Typically, but by no means always, the required strictly convex function is taken to be $d_g^2(x, x_0)$, under suitable assumptions on the sectional curvature. See Section 4. We also notice that in our case, where the manifolds are complete, the geodesics exist globally. □

Remark 6.3.2. We may actually start with Eqn. (6.3.1) on a Riemannian manifold $\{M, g_1\}$. After a change similar to the one in (6.3.2), we obtain (6.1.1) on $\{M, g\}$. □

6.4. Carleman estimates for problem (6.1.1), (6.1.2) with lower-order terms.

THEOREM 6.4.1. *[L-T-Y.1–2] (Carleman estimate, first version) With reference to Eqn. (6.1.1), assume (A.1) = (6.1.3) on the energy level term $F(w)$; (H.1) = (3.1.1) on the strictly convex function v; and $f \in L_2(Q)$ as in (6.1.1). Let w be a solution of (6.1.1) in the following class*

$$\begin{cases} w \in H^{1,1}(Q) \equiv L_2(0,T; H^1(\Omega)) \cap H^1(0,T; L_2(\Omega)); & (6.4.1a) \\ w_t, \dfrac{\partial w}{\partial n} \in L_2(0,T; L_2(\Gamma)). & (6.4.1b) \end{cases}$$

Let $\phi(x,t)$ be the function defined by (3.2.1), so that $T > T_0$, with T_0 defined by (3.2.2). Let $E(t)$ be the 'energy' (6.1.2). Then, for all values $\tau > 0$, the following one-parameter family of estimates holds true:

$$B_\Sigma(w) + \frac{C_T}{\tau}\int_Q e^{\tau\phi} f^2 dQ + TC_{T,\tau}\|w\|^2_{C([0,T];L_2(\Omega))}$$

$$\geq \left(\rho - c - \frac{C_T}{\tau}\right)\int_0^T\int_\Omega e^{\tau\phi}[|Dw|^2 + w_t^2]dQ - C(1+\tau)e^{-\delta\tau}[E(T)+E(0)] \tag{6.4.2a}$$

$$\geq \left(\rho - c - \frac{C_T}{\tau}\right)e^{-\frac{\tau\delta}{2}}\int_{t_0}^{t_1} E(t)dt - C(1+\tau)e^{-\delta\tau}[E(T)+E(0)], \tag{6.4.2b}$$

where $0 < c < \rho$ as in (3.2.2) and $\delta > 0$ in (3.2.4). Moreover, the boundary terms $B_\Sigma(w)$ are given explicitly by

$$\begin{aligned} B_\Sigma(w) = &\int_\Sigma e^{\tau\phi}\frac{\partial w}{\partial n}[Dv(w) - \phi_t w_t]d\Sigma + \int_\Sigma \frac{\partial w}{\partial n} w\left[\frac{1}{2}\mu - (1+c)e^{\tau\phi}\right]d\Sigma \\ &+ \frac{1}{2}\int_\Sigma e^{\tau\phi}[w_t^2 - |Dw|^2]\langle Dv, n\rangle d\Sigma, \end{aligned} \tag{6.4.3}$$

where $\mu(x,t)$ is a suitable function depending on ϕ, see [L-T.17], [L-T-Y.2]. (Eqn. (6.4.2a) is obtained by use also of property (3.2.4) of ϕ. To obtain (6.4.2b) from (6.4.2a), one further uses property (3.2.5) of ϕ.) □

A related result is in [Ta.7] in the pseudo-differential language. Our Theorem 6.4.1 has a more friendly proof and the estimate is more explicit.

Remark 6.4.1. (Idea of proof) [L-T-Y.1–2] The proof of Theorem 6.4.1 is by energy method in the Riemannian metric: it uses the following main multipliers, where $Dv = D\phi$, recalling (3.2.1):

$$(6.4.4) \qquad e^{\tau\phi(x,t)}[\langle Dv, Dw\rangle - \phi_t w_t]; \qquad \left[\operatorname{div}(e^{\tau\phi}Dv) - \frac{d}{dt}\left(e^{\tau\phi}\phi_t\right)\right] w.$$

That is, one multiplies (6.1.1) by the multipliers in (6.4.4) and integrates by parts. They represent a far-reaching generalization of the energy methods (multipliers) used since the 'classical' period of early/mid-80's:

(1) Under the following two conditions, that (i) Eqn. (6.1.1) is defined on a Euclidean domain Ω (thus $\Delta_g = \Delta =$ the usual Laplacian, $g_{ij} = \delta_{ij}$), and moreover, (ii) $F(w) \equiv 0$ (no energy level terms!), then we recover the 'classical' main multipliers of the early/mid-80's: $h \cdot \nabla w$, w div h, where $h(x)$ is a smooth coercive vector field on Ω, not necessarily conservative (i.e., not necessarily $h = \nabla v$, for some scalar function $v(x)$) [L-L-T.1], [L-T.1], [L-T.3], [Tr.1], [Ho.1] These are special cases of (6.4.4) when $\tau = 0$ (that is, no free parameter is used) and the metric is Euclidean.

Prior to energy methods efforts in establishing Continuous Observability Inequalities (0.3), the multiplier $h \cdot \nabla w$, $h|_\Gamma = \nu$, had been critically used in [L-L-T.1] to obtain the *trace regularity inequality*; this is the *reverse* inequality (for all $T > 0$) of the C.O.I. (6.5.3) below [which is the present specialization of the C.O.I. (0.3)]. By duality, the *reverse* inequality yields the optimal regularity (6.5.8) of second-order hyperbolic equations with Dirichlet control. This work was instrumental in spurring further research in C.O.I. In fact, the proof of [L-L-T.1], leading to a key energy identity, serves also as a main first component of the proof of the C.O.I. (0.3). The second part of the proof of the C.O.I. (0.3) uses the second main multiplier w div h to take care of the difference: "kinetic energy – potential energy."

Control-theoretic inequalities—C.O.I. (0.3) and U.S.I. (0.5)—versus reverse trace regularity inequalities. The following considerations are of paramount importance in understanding the topic at hand. The proof of the *reverse trace regularity* works *equally well* for variable coefficients (even all depending on time and space!) with the appropriate regularity requirement [L-L-T.1], as noted in (6.5.8). By stark contrast, the presence of variable coefficients in both principal part and/or energy level terms introduces most serious additional difficulties over the constant coefficient, canonical case in showing the relevant control-theoretic inequalities: for instance the C.O.I. (6.5.3) (Dirichlet case), or (6.5.6) (Neumann case); or the U.S.I. (6.5.16) (Neumann case) or (6.6.10) (Dirichlet case). These considerations apply to all present classes of evolution equations, not only second-order hyperbolic equations.

(2) Generalizing case (1) in one direction, we let now $\Delta_g = \Delta$ still but $F(w)$ be a first-order operator satisfying the Euclidean version of (6.1.3b): In this case, the exact Euclidean versions of the multipliers in (6.4.4) were used in [L-T.17], following the general approach, based on pseudo-differential calculus, of [Hor.1–2] (solutions with compact support) and [Ta.1–3] (solutions with traces).

(3) Generalizing case (1) in another direction, we return to (6.1.1) defined on a manifold this time with $F(w) \equiv 0$. Then the multiplier $\langle H, Dw\rangle_g$, H coercive vector field as in assumption (H.1′) = (3.1.2), was used in [Y.1] in the context of Section 6.3, Eqn. (6.3.3) with $F(w) \equiv 0$, a critical assumption for the method of [Y.1] to work.

(4) Finally, the combination of case (2) and case (3) leads to the multipliers in (6.4.4) for the full Eqn. (6.1.1), first used in [L-T-Y.1–2]. The proof in this case is a parallel development of that in [L-T.17]; computations in the Euclidean metric performed in [L-T.17] are replaced by counterpart computations in the Riemannian metric g, using the Levi-Civita connection D.

More historical details are given in the introduction of [L-T-Y.2]. □

By far, the bulk of the proof of the control-theoretic inequalities of the forthcoming Section 6.5 rests with the Carleman estimates, first version, of Theorem 6.4.1. It is particularly in achieving this result that the Riemannian geometric energy method, based on the multipliers in (6.4.4), plays a critical role. The improvement from the first to the second version of the Carleman estimate, as in Theorem 6.4.2 below, is more routine for second-order hyperbolic equations, see below.

THEOREM 6.4.2. *(Carleman estimate, second version) Assume the hypotheses and the notation of Theorem 6.4.1. Then, for all $\tau > 0$ sufficiently large, there exists a constant $k_{\phi,\tau} > 0$ (recall $c < \rho$ from (3.2.2)) such that the following one-parameter family of estimates holds true:*

$$\overline{B}_\Sigma(w) + \frac{C_T}{\tau}\int_Q e^{\tau\phi} f^2 dQ + C_{T,\tau}\|w\|^2_{L_2(0,T;L_2(\Omega))}$$

$$\geq e^{-\frac{\tau\delta}{2}}\left\{\left(\rho - c - \frac{C_T}{\tau}\right)\frac{e^{-C_T T}}{2}(t_1 - t_0) - C(1+\tau)e^{-\frac{\tau\delta}{2}}\right\}[E(T)+E(0)] \tag{6.4.5a}$$

$$\geq k_{\phi,\tau}[E(T)+E(0)], \tag{6.4.5b}$$

where, recalling $B_\Sigma(w)$ from (6.4.3), the boundary terms $\overline{B}_\Sigma(w)$ are given by

$$\overline{B}_\Sigma(w) = B_\Sigma(w) + const_{\phi,\tau}\int_\Sigma \left|\frac{\partial w}{\partial n}\, w_t\right| d\Sigma. \tag{6.4.6a}$$

a 'brutal' estimate of which is

$$|\overline{B}_{\Sigma}(w)| \le C_{\phi} \int_{\Sigma} [|Dw|^2 + w_t^2 + w^2] d\Sigma. \tag{6.4.6b}$$

Assume, further, that $\Gamma = \overline{\Gamma_0 \cup \Gamma_1} \equiv \partial\Omega$, $\Gamma_0 \cap \Gamma_1 = \emptyset$ *and that the solution* w *of Eqn. (6.1.1) satisfies*

$$w|_{\Sigma_0} \equiv 0, \ \Sigma_0 = (0,T] \times \Gamma_0, \ \textit{and that } \langle Dv, n\rangle \le 0 \textit{ on } \Gamma_0, \tag{6.4.7}$$

where $n(x)$ *is the unit outward normal field at* $x \in \Gamma$. *Then: estimate (6.4.5) holds true for all* $\tau > 0$ *sufficiently large with the boundary terms* $\overline{B}_{\Sigma}(w)$ *replaced by the boundary terms* $\overline{B}_{\Sigma_1}(w)$, *that is, evaluated only on* $\Sigma_1 = (0,T] \times \Gamma_1$, $\Gamma = \Gamma_0 \cup \Gamma_1$, *with* Γ_0, Γ_1 *open disjoint subsets of* Γ; *while* $\overline{B}_{\Sigma_0}(w) \le 0$. □

Again, a related result is in [Ta.7] in the pseudo-differential language.

Idea of proof. [L-T.17], [L-T-Y.1–2] By multiplying Eqn. (6.1.1) by w_t we arrive at inequality (3.2.7), from which, using Gronwall inequality, we obtain the inequality

$$|E(t) - E(s)| \le \Lambda(T) e^{C_T(t-s)}, \qquad 0 \le s \le t \le T; \tag{6.4.9}$$

$$\Lambda(T) = \int_0^T \int_{\Omega} f^2 dQ + 2 \int_0^T \int_{\Gamma} \left| \frac{\partial w}{\partial n} w_t \right| d\Sigma + C_T \|w\|^2_{L_2(Q)}, \tag{6.4.10}$$

where $\Lambda(T)$ is the present specialization of $G(T)$ in (3.2.7). ¿From here we obtain [L-T.17], [L-T-Y.1–2]

$$E(t) \ge \frac{E(0) + E(T)}{2} e^{-C_T T} - \Lambda(T), \qquad 0 \le t \le T, \tag{6.4.11}$$

which substituted in the right side of (6.4.2b) yields readily (6.4.5), with τ sufficiently large.

6.5. Control-theoretic results for (6.1.1)–(6.1.3). We preliminarily let $\partial\Omega \equiv \Gamma = \overline{\Gamma_0 \cup \Gamma_1}$, $\Gamma_0 \cap \Gamma_1 = \emptyset$, Γ_0, Γ_1 open in Γ, where Γ_0 is defined by (6.4.7), that is:

$$\Gamma_0 = \{x \in \Gamma : \ \langle Dv, n\rangle \le 0\}, \tag{6.5.1}$$

with $n(x)$ the unit outward normal field at $x \in \Gamma$, where v is the strictly convex function of assumption (H.1) = (3.1.1), in the Riemannian metric g.

Continuous observability inequalities.

THEOREM 6.5.1. *(Dirichlet case) Assume the setting of Section 6.1: hypothesis (H.1) = (3.1.1) (thus* $T > T_0$ *with* T_0 *given by (3.2.2)) and (A.1) = (6.1.3) on* F. *Let* $f \equiv 0$. *Assume further that (6.1.1) possesses the following unique continuation property: if* ψ *is a solution of (6.1.1) in the class (6.4.1) which satisfies the over-determined B.C., with* Γ_0 *as in (6.5.1):*

$$\psi|_{\Sigma} \equiv 0 \quad and \quad \left.\frac{\partial \psi}{\partial n}\right|_{\Sigma_1} \equiv 0, \qquad \Gamma_1 = \Gamma \setminus \Gamma_0, \tag{6.5.2}$$

then, in fact, $\psi \equiv 0$ *in* Q*. Finally, let* w *be a solution of (6.1.1) in the class (6.4.1) which satisfies the homogeneous Dirichlet B.C.:* $w|_{\Sigma} \equiv 0$*. Then, the following continuous observability inequality holds true for* $T > T_0$*: there exists* $C_T > 0$ *such that*

$$\int_0^T \int_{\Gamma_1} \left(\frac{\partial w}{\partial n}\right)^2 d\Sigma_1 \geq C_T \|\{w_0, w_1\}\|^2_{H_0^1(\Omega)\times L_2(\Omega)}. \tag{6.5.3}$$

[This is the C.O.I. (0.3), as specialized to the present Dirichlet case.]

THEOREM 6.5.2. *(Neumann case) Assume the setting of Section 6.1; hypothesis (H.1) = (3.1.1) (thus* $T > T_0$ *with* T_0 *given by (3.2.2)) and (A.1) = (6.1.3) on* F*. Let* $f \equiv 0$*. Assume further that (6.1.1) possesses the following unique continuation property: if* ψ *is a solution of (6.1.1) in the class (6.4.1) which satisfies the over-determined B.C., with* Γ_0 *as in (6.5.1):*

$$\left.\frac{\partial \psi}{\partial n}\right|_{\Sigma} \equiv 0 \quad and \quad \psi|_{\Sigma_1} \equiv 0, \quad \Gamma_1 = \Gamma \setminus \Gamma_0, \tag{6.5.4}$$

then, in fact, $\psi \equiv 0$ *in* Q*. Finally, let* w *be a solution of (6.1.1) in the class (6.4.1) which satisfies the following B.C.:*

$$w|_{\Sigma_0} \equiv 0 \quad and \quad \left.\frac{\partial w}{\partial n}\right|_{\Sigma_1} \equiv 0, \quad \Gamma_0 \neq \emptyset. \tag{6.5.5}$$

Then, the following continuous observability inequality holds true: for $T > T_0$*, there exists* $C_T > 0$ *such that*

$$\int_0^T \int_{\Gamma_1} w_t^2 d\Sigma_1 \geq C_T \|\{w_0, w_1\}\|^2_{H^1_{\Gamma_0}(\Omega)\times L_2(\Omega)}. \tag{6.5.6}$$

[This is the C.O.I. (0.3), as specialized to the present Neumann case (6.5.5).]

Remark 6.5.1. We note the following sequence. One first proves the Carleman estimates (6.4.2b) and (6.4.5b)—hence, the C.O.I. (6.5.3) (Dirichlet) and (6.5.6) (Neumann)—for $H^{2,2}(Q)$-solutions w, as required by the computations involved. The subsequence passage from $H^{2,2}(Q)$-solutions to $H^{1,1}(Q)$-solutions is readily accomplished in the Dirichlet case, by virtue of a limit process which critically uses the "reverse inequality of (6.5.3)," which is available in this case [L-L-T.1] as discussed above. However, this passage from $H^{2,2}(Q)$- to $H^{1,1}(Q)$-solutions is far more delicate in the Neumann case, where a comparable "reverse inequality of (6.5.6)" is false (in dim $\Omega \geq 2$) [L-T.6]. The literature on continuous observability/exact controllability of the mid-80's/early 90's seems to have overlooked this point, and unduly jumps from $H^{2,2}(Q)$-solutions to $H^{1,1}(Q)$-solutions in the Neumann C.O.I. (6.5.6) as well. A proper justification of this passage can be given, however, and can be found in [L-T-Z.1, Section 8]. See also [Ta.7]. The case of U.S.I. (6.5.16) below in the Neumann case is instead more regular: here a reverse inequality to (6.5.16) (even over the infinite time interval, not only on $[0,T]$) is again available at the outset, by the very property of dissipativity of the feedback problem (6.5.13) with $F(w) \equiv 0$. Hence, a limit process, as in the case of C.O.I. for the Dirichlet case, will justify the passage from $H^{2,2}(Q)$- to $H^{1,1}(Q)$-solutions [Lag.3]. □

LEMMA 6.5.3. *(i) Let w be a solution of Eqn. (6.1.1) in the class (6.4.1). Given $\epsilon > 0$, $\epsilon_0 > 0$ arbitrarily small, given $T > 0$, there exists a constant $C_{\epsilon,\epsilon_0,T} > 0$ such that*

$$\int_\epsilon^{T-\epsilon}\int_\Gamma \left(\frac{\partial w}{\partial s}\right)^2 d\Gamma\, dt \leq C_{\epsilon,\epsilon_0,T}\left\{\int_0^T\int_\Gamma \left[\left(\frac{\partial w}{\partial n}\right)^2 + w_t^2\right] d\Gamma\, dt + \|w\|^2_{L_2(0,T;H^{\frac{1}{2}+\epsilon_0}(\Omega))}\right\},$$

where $\frac{\partial w}{\partial s} = \langle Dw, s\rangle$ = tangential gradient.

(ii) Assume further that the solution w satisfies hypothesis (6.4.7). Then the above estimate holds true with $\int_\Gamma$ replaced by $\int_{\Gamma_1}$. □

Exact controllability. Dirichlet case. Consider the following mixed (controlled) problem:

$$\begin{cases} y_{tt} = \Delta_g y + F(y) & \text{in } Q = (0,T] \times \Omega; \quad (6.5.7a) \\ y(0,\,\cdot\,) = y_0,\ y_t(0,\,\cdot\,) = y_1 & \text{in } \Omega; \quad (6.5.7b) \\ y|_{\Sigma_0} \equiv 0,\ y|_{\Sigma_1} \equiv u & \text{in } \Sigma_i,\ i = 0,\ i = 1, \quad (6.5.7c) \end{cases}$$

with F satisfying (H.1) = (3.1.1). By [L-L-T.1] (which was given in the Euclidean setting of Section 6.3 but works also in the present setting where Ω is a bounded set of a Riemannian manifold M), we have the following *regularity result* for problem (6.5.7): the map

$$(6.5.8) \quad y_0 = 0,\ y_1 = 0,\ u \in L_2(\Sigma_1) \to \{y, y_t\} \in C([0,T]; L_2(\Omega) \times H^{-1}(\Omega))$$

is continuous.

DEFINITION 6.5.1. *Making Section 0 more precise, we call problem (6.5.7) "exactly controllable" over the state space $L_2(\Omega) \times H^{-1}(\Omega)$ (of regularity) in the interval $[0,T]$, with respect to the class of controls $L_2(0,T; L_2(\Gamma_1)) \equiv L_2(\Sigma_1)$, in case the map L_T, $L_T u = \{y(T), y_t(T)\}$, see (6.5.8), is* surjective *(onto) $L_2(\Omega) \times H^{-1}(\Omega)$.*

A well-known result [T-L.1, p. 235] then gives the following equivalence: the above map L_T is surjective $L_2(\Sigma_1)$ onto $L_2(\Omega) \times H^{-1}(\Omega)$ if and only if its adjoint L_T^* is bounded below (abstract observability inequality):

$$\|L_T^* z\|_{L_2(0,T;L_2(\Gamma_1))} \geq C_T \|z\|_{L_2(\Omega)\times H^{-1}(\Omega)}, \tag{6.5.9}$$

which is the present specialization of (0.3). One can readily prove (by PDEs methods or by operator methods [L-T.24, Section 10.5]) that, then, inequality (6.5.9) is equivalent to inequality (6.5.3) for the w-problem (6.1.1) satisfying the homogeneous B.C. $w|_\Sigma \equiv 0$, as guaranteed by Theorem 6.5.1. Thus, by duality on Theorem 6.5.1, we obtain

THEOREM 6.5.3. *Assume (H.1) = (3.1.1) and (A.1) = (6.1.3) on F and the unique continuation property of Theorem 6.5.1. Let $T > T_0$, with T_0 as in (3.2.2). Let Γ_0 be as in (6.5.1). Then the mixed problem (6.5.7) is exactly controllable in the sense of Definition 6.5.1.* □

Exact controllability. Neumann case. We next consider the following mixed (controlled) problem, consisting of Eqn. (6.5.7a–b) and

$$y|_{\Sigma_0} \equiv 0, \quad \left.\frac{\partial y}{\partial n}\right|_{\Sigma_1} = u \ \text{ in } \Sigma_1, \ \Gamma_0 \neq \emptyset. \tag{6.5.10}$$

The case $\Gamma_0 = \phi$ is considered in Section 7.

DEFINITION 6.5.2. *Making Section 0 more precise, we call problem (6.5.7a–b), (6.5.10) "exactly controllable" over the (energy) space $H^1_{\Gamma_0}(\Omega)\times L_2(\Omega)$ in the interval $[0,T]$, with respect to the class of controls $L_2(0,T;$ $L_2(\Gamma_1)) \equiv L_2(\Sigma_1)$, in case the corresponding map related to this problem*

$$\begin{gathered} y_0 = 0, \qquad y_1 = 0, \\ u \in \mathcal{D}(\mathcal{L}_T) \subset L_2(\Sigma_1) \to \mathcal{L}_T u = \{y(T,\cdot), y_t(T,\cdot)\} \in H^1_{\Gamma_0}(\Omega)\times L_2(\Omega) \end{gathered} \tag{6.5.11}$$

is surjective (onto).

In the present Neumann case, the map $\mathcal{L}_T$ is *not* continuous when dim $\Omega \geq 2$ [L-T.6], [Ta.5], unlike the Dirichlet case [L-L-T.1]. However, $\mathcal{L}_T$ is closed.

The well-known general result [T-L.1, p. 235] then gives the following equivalence: the map $\mathcal{L}_T$ in (6.5.11) is surjective if and only if its adjoint $\mathcal{L}_T^*$ is bounded below (abstract observability inequality):

$$\|\mathcal{L}_T^* z\|_{L_2(0,T;L_2(\Gamma_1))} \geq C_T \|z\|_{H^1_{\Gamma_0}(\Omega)\times L_2(\Omega)} \tag{6.5.12}$$

for $z \in \mathcal{D}(\mathcal{L}_T^*)$, which is the present specialization of (0.3). One can readily prove (by PDE methods or by operator methods [L-T.3]) that then, inequality (6.5.12) is equivalent to inequality (6.5.6) for the w-problem (6.1.1) satisfying the homogeneous B.C. (6.5.5), as guaranteed by Theorem 6.5.2. Thus, by duality on Theorem 6.5.2, we obtain

THEOREM 6.5.4. *Assume (H.1) = (3.1.1) and (A.1) = (6.1.3) on F and the unique continuation property of Theorem 6.5.2. Let $T > T_0$,* *with T_0 as in (3.2.2), and let Γ_0 be as in (6.5.1). Then the mixed problem (6.5.7a–b), (6.5.10) is exactly controllable in the sense of Definition 6.5.2.*

Uniform stabilization: Neumann case. Consider the following closed loop problem with explicit dissipative feedback in the Neumann B.C.

$$\begin{cases} w_{tt} = \Delta_g w + F(w) & \text{in } (0,T] \times \Omega; & (6.5.13a)\\ w(0,\ \cdot\ ) = w_0,\ w_t(0,\ \cdot\ ) = w_1 & \text{in } \Omega; & (6.5.13b)\\ w|_{\Sigma_0} \equiv 0,\ \left.\dfrac{\partial w}{\partial n}\right|_{\Sigma_1} = -w_t & \text{in } \Sigma_i,\ i = 0,1. & (6.5.13c)\end{cases}$$

THEOREM 6.5.5. *[L-T-Y.2, p. 32] With reference to the closed loop problem (6.5.13), we have:*

(i) when $\Gamma_0 \neq \emptyset$, under assumption (A.1) = (6.1.3) for F, problem (6.5.13) generates a s.c. semigroup $\{w_0, w_1\} \in Y \to \{w(t), w_t(t)\} \in C([0,T];Y)$, $Y \equiv H^1_{\Gamma_0}(\Omega) \times L_2(\Omega)$;

(ii) when $\Gamma_0 = \emptyset$, the same result, under (A.1) = (6.1.3) for F, holds true, with $Y \equiv H^1_{\Gamma_0}(\Omega) \times L_2(\Omega)$ replaced now by its proper subspace

$$Y_0 \equiv \left\{[u_1, u_2] \in Y : \int_\Gamma u_1 d\Gamma + \int_\Omega u_2 d\Omega = 0\right\} \tag{6.5.14}$$

topologized (via Green's theorem) by

$$\|\{u_1, u_2\}\|^2_{Y_0} \equiv \int_\Omega [|Du_1|^2 + u_2^2] d\Omega, \tag{6.5.15}$$

which is a norm on Y_0 (but only a semi-norm on Y) [L-T-Y.2].

(iii) Under the additional assumption (H.1) = (3.1.1) and with Γ_0 defined by (6.5.1), and the unique continuation property in Theorem 6.5.2, the following inequality holds: for all T sufficiently large, there exists a positive constant $k_{\phi,T} > 0$ such that

$$\int_0^T \int_{\Gamma_1} \left(\frac{\partial w}{\partial n}\right)^2 d\Sigma_1 \geq k_{\phi,T} \begin{cases} E(T) \\ E(0) \end{cases} \tag{6.5.16}$$

[For $F(w) \equiv 0$, Eqn. (6.5.16) is the U.S.I. (0.5) as specialized to the present case].

Remark 6.5.2. When $F \equiv 0$ in (6.5.13a)—in which case we may well assume the weaker hypothesis (H.1′) = (3.1.2) instead of (H.1) = (3.1.1) as explained there—the U.S.I. (6.5.16) implies (is equivalent to) *uniform stabilization* of problem (6.5.13): there exist constants $M \geq 1$, $\omega > 0$ such that

$$E(t) \leq M e^{-\omega t} E(0),\ t \geq 0, \tag{6.5.17}$$

where $\sqrt{E(\ \cdot\ )}$ is the $Y \equiv H^1_{\Gamma_0}(\Omega) \times L_2(\Omega)$-norm in case (i) and the Y_0-norm in (6.5.14) in case (ii) for the solution $\{w, w_t\}$ of (6.5.13). □

6.6. Lower-level energy inequality: stabilization inequality with Dirichlet-boundary feedback in $L_2(\Sigma)$. In this section we treat an important situation—the stabilization problem of a wave equation (6.3.3) in its conservative form $F(w) \equiv 0$, with variable coefficient principal part and dissipative feedback in the Dirichlet-boundary condition (rather than Neumann-boundary condition as in Theorem 6.5.6). There is an important difference between the two cases: the Dirichlet—versus the Neumann—boundary feedback. The energy level in the latter is $H^1(\Omega) \times L_2(\Omega)$, see Theorem 6.5.6. By contrast, the energy level of the former case has to be $L_2(\Omega) \times H^{-1}(\Omega)$—the space of optimal regularity, with $L_2(\Sigma)$-forcing term in the Dirichlet B.C. as in (6.5.8) [L-L-T.1]. Thus, the Dirichlet-boundary feedback case requires a shift of topology downward: from $H^1(\Omega) \times L_2(\Omega)$ to $L_2(\Omega) \times H^{-1}(\Omega)$, for the final estimates. This is not a trivial problem. It can best be accomplished by a pseudo-differential change of variable that shifts upward $L_2(\Omega) \times H^{-1}(\Omega)$ to the level $H^1(\Omega) \times L_2(\Omega)$ already achieved and which is natural for energy level computations.

Dissipative model with Dirichlet boundary feedback in $L_2(\Sigma)$. In the notation of this Section 6, where Ω is a bounded set in the Riemannian manifold (M, g), we consider the following (closed loop) Riemannian wave equation

$$\begin{cases} w_{tt} = \Delta_g w + f & \text{in } Q = (0,\infty) \times \Omega; \quad (6.6.1a) \\ w(0, \cdot\,) = w_0, \; w_t(0, \cdot\,) = w_1 & \text{in } \Omega; \quad (6.6.1b) \\ w|_{\Sigma_0} \equiv 0, \; w|_{\Sigma_1} = \frac{\partial(\mathcal{A}_0^{-1} w_t)}{\partial n} & \text{in } \Sigma_i = (0,\infty) \times \Gamma_i, \; i = 0, 1, \quad (6.6.1c) \end{cases}$$

with dissipative feedback in the Dirichelt B.C., where

$$\begin{cases} \mathcal{A}_0 w = \Delta_g w, \; \mathcal{D}(\mathcal{A}_0) = H^2(\Omega) \cap H_0^1(\Omega); & (6.6.2a) \\ \mathcal{D}(\mathcal{A}_0^{\frac{1}{2}}) = H_0^1(\Omega), \; [\mathcal{D}(\mathcal{A}_0^{\frac{1}{2}})]' = H^{-1}(\Omega) \text{ (equivalent norms).} & (6.6.2b) \end{cases}$$

Here Γ_0 is defined with reference to the coercive field H of assumption (H.1$'$) = (3.1.2), that is by

$$\Gamma_0 = \{x \in \Gamma : \langle H, n \rangle \leq 0\} \tag{6.6.3}$$

in the Riemannian metric g, where n is a normal field to Γ [compare with (6.4.7)].

Remark 6.6.1. The counterpart of problem (6.6.1) in the Euclidean case is obtained as in Section 6.3 with Δ_g being replaced now by the operator (6.3.1).

Next, we introduce the lower-level energy $\mathcal{E}_w(t)$ of problem (6.1.1) defined by

$$\mathcal{E}_w(t) = \|w(t)\|^2_{L_2(\Omega)} + \|\mathcal{A}_0^{-\frac{1}{2}} w_t(t)\|^2_{L_2(\Omega)} = \|\{w(t), w_t(t)\}\|^2_Z; \tag{6.6.4}$$

$$Z \equiv L_2(\Omega)\times[\mathcal{D}(\mathcal{A}_0^{\frac{1}{2}})]' \equiv L_2(\Omega)\times H^{-1}(\Omega) \text{ (equivalent norms)}. \tag{6.6.5}$$

We begin with a well-posedness result.

THEOREM 6.6.1. *[L-T.1], [L-T.12] (i) Problem (6.6.1) defines a s.c. contraction semigroup* $e^{A_D t}$ *on the space* Z *defined by (6.6.5);*

(ii) the energy $\mathcal{E}_w(t)$ *satisfies the dissipative identity (a specialization of (0.6)),*

$$\mathcal{E}_w(t) + 2\int_s^t \left\langle w|_{\Gamma_1}, \frac{\partial \mathcal{A}_0^{-1} w_t}{\partial n}\right\rangle_{L_2(\Gamma_1)} d\tau = \mathcal{E}_w(s); \tag{6.6.6}$$

(iii) we have

$$w|_{\Gamma_1} = \frac{\partial \mathcal{A}_0^{-1} w_t}{\partial n} \in L_2(0,\infty; L_2(\Gamma_1)). \tag{6.6.7}$$

(iv) [L-T.12, Lemma 2.1]

$$\left\| \frac{\partial w}{\partial n} \right\|_{H^{-1}(\Sigma_1)} \le C_T \left\| w|_{\Sigma_1} \right\|_{L_2(\Sigma_1)}. \tag{6.6.8}$$

□

Next, we provide an observability-type estimate for the lower-level energy $\mathcal{E}_w(t)$ in (6.6.4), indeed, for the full Eqn. (6.1.1) with $f \equiv 0$.

THEOREM 6.6.2. *[L-T.12], [L-T-Y.3], [Ta.1–3] Assume (H.1′) = (3.1.2). Let* w *be a solution of Eqn. (6.1.1) [and with no boundary conditions imposed] within the class*

$$\begin{cases} \{w, w_t\} \in C([0,T]; L_2(\Omega)\times H^{-1}(\Omega)); & \text{(6.6.9a)} \\[2ex] w|_{\Sigma_1} \in L_2(\Sigma_1); \quad \left.\dfrac{\partial w}{\partial n}\right|_{\Sigma_1} \in H^{-1}(\Sigma_1). & \text{(6.6.9b)} \end{cases}$$

Assume, moreover, condition (6.6.3).

Then the following inequality holds true: If T *is large enough, then*

$$\int_0^T \mathcal{E}_w(t)dt \le C_T \left\{ \int_0^T \int_{\Gamma_1} w^2 d\Sigma_1 + \left\| \frac{\partial w}{\partial n} \right\|^2_{H^{-1}(\Sigma_{1,T})} + \|w\|^2_{H^{-1}(Q)} + \|f\|^2_{L_2(0,1;H^{-1}(\Sigma_T))} \right\}. \tag{6.6.10}$$

□

The notation on the norm of f means: L_2 in the normal direction and H^{-1} in the tangential direction and time [Hor.2].

We now return to the feedback problem (6.6.1) with $\{w_0, w_1\} \in Z$: By Theorem 6.6.1, we see that conditions (6.6.9) are satisfied. We can then apply estimate (6.6.10) supplemented by (6.6.8), where we use (6.6.1c) on Σ_1. We thus obtain the U.S.I. (0.5) polluted by the lower-order term: $w \in H^{-1}(Q)$. This *lot* is then absorbed by a standard compactness/uniqueness argument, where the uniqueness is provided by elliptic theory [Hor.2, p. 14] for the static problem corresponding to (6.6.1a). This way, the desired uniform stabilization problem is recovered as a special case when the dissipative B.C. (6.6.1c) on Σ_1 is invoked. We obtain

THEOREM 6.6.3. *In reference to problem (6.6.1), assume (H.1′) = (3.1.2) as well as (6.6.3) on* Γ_0. *Then, the s.c. contraction semigroup* $e^{A_D t}$ *guaranteed by Theorem 6.6.1 is uniformly stable on the space* Z *defined in (6.6.5): there exists constants* $M \geq 1$, $\omega > 0$ *such that*

$$\|e^{A_D t}\|_{\mathcal{L}(Z)} \leq Me^{-\omega t}; \text{ or } \mathcal{E}_w(t) \leq Me^{-\omega t}\mathcal{E}_w(0),\ t \geq 0. \tag{6.6.11}$$

□

Remark 6.6.2. Theorem 6.6.1 and Theorem 6.6.3 were first established in [L-T.1] in the Euclidean case $\Delta_g = \Delta =$ Euclidean Laplacian [this assumption is not critical for Theorem 6.6.1, which holds true with essentially the same proof, based on Lumer-Phillips]: in the case of Theorem 6.6.3 in the Euclidean case, $\Delta_g = \Delta$, the original proof in [L-T.1] required a geometrical condition: that $\Omega \subset \mathbb{R}^n$ be strictly convex (or the set difference of two strictly convex sets). The required shift of topology, described in the opening of this Subsection 6.6, was accomplished in [L-T.1] by using an operator-theoretic global change of variable $p = \mathcal{A}_0^{-1} w_t$ to shift the needed topology $L_2(\Omega) \times H^{-1}(\Omega)$ for $\{w, w_t\}$ into the more convenient topology $H^1(\Omega) \times L_2(\Omega)$ for $\{p, p_t\}$, where the energy method applies. Later, a pseudo-differential change of variable was introduced in [L-T.12] from $\{w, w_t\}$ to $\{p, p_t\}$ in the aforementioned topologies, which permitted one to eliminate the geometrical conditions of [L-T.1]. Lower-level energy inequalities such as (6.6.10) were obtained in [L-T.12], [L-T-Y.3] (in a form suitable for the feedback problem (6.6.9) of interest), [Ta.1–3], [E-L-T.1], [E-L-T.4], including the RHS forcing term f.

A more general (and more complicated) pseudo-differential method to obtain lower-order energy estimates (with no reference to any particular uniform stabilization problem was given in [Ta.7] for general evolution equations. See also [B-L-R-H-Z.1] for second-order hyperbolic equations.

6.7. Geometric optics and strictly convex functions. Since the work of [B-L-R.1], it has been understood in specialized PDE control circles that, even in the case of the Euclidean Laplacian, a 'minimal' controlled portion of the boundary—such as is provided by the sharp geometric optics condition [B-L-R.1]—may not be expected, in general, to be obtained by using classical multipliers with vector fields which are coercive, in particular the gradients of strictly convex functions in the Euclidean metric. Several examples may be given below.

Example #1. [G-L.2] An example is constructed in [G-L.2] of a domain in $\mathbb{R}^2$ with controls on the outer of the four boundary components and $g_{ij}(x) \equiv \delta_{ij}$, therefore with the Euclidean Laplacian $\Delta_g = \Delta$, so that *control is achieved in a finite time* T_0, but *there is no strictly convex function* v on Ω with *the additional property of nonpositive normal derivative* *on the uncontrolled boundary* (see Figure 2 taken from [G-L.2]). In fact, the analysis of [B-L-R.1], [Lit.1] shows that control may be achieved in a time equal to the maximum length of geodesics which reflect with equal angles at the uncontrolled boundary, before they cross the controlled boundary. In Figure 2 from [G-L.2], this will be achieved by a polygonal curve consisting of two line segments, each of which grazes the outer boundary curve, reflecting with equal angles at the upper boundary circle. On the other hand, a strictly convex function v on Ω must have positive outward normal derivative somewhere on one of the three uncontrolled boundary circles. Namely, let the black dot in Figure 2 from [G-L.2] indicate the origin. The segments l_1, l_2 and $l_3 \subset \Omega$ of the rays from the origin through the centers of the three circles meet the uncontrolled boundary at right angles at x_1, x_2 and x_3, resp. If the outward normal derivative of v at x_k is ≤ 0, then since the restriction of v to l_k is strictly convex, the derivative of v at the origin in the x_k direction is strictly negative. But x_1, x_2 and x_3 do not lie in any half-plane of $\mathbb{R}^2$, so this contradicts the differentiability of v at the origin.

On the other hand, the Euclidean Laplacian in $\mathbb{R}^n$ always admits infinitely many strictly convex functions, e.g., the quadratic functions $v(x) = \|x - x_0\|^2$, where x_0 is *any* fixed point in $\mathbb{R}^n$. Thus, by Theorem 6.5.4, the above example in $\mathbb{R}^2$ with the Euclidean Laplacian is always exactly controllable in optimal Sobolev spaces over a time $T > T_0$, provided that we apply control on an additional portion of the boundary. Namely, we have to retain control on the entire portion of the boundary where $\nabla v \cdot n \geq 0$, n being the unit outward normal. One symmetric way to do this is as follows. Call now x_0 the center of the domain (the dot in the picture). Then, apply Theorem 6.5.4 with the strictly convex function $v(x) = \|x - x_0\|^2$. This requires that we apply control not only on the exterior boundary, as done before, but also on the arcs of each circle illuminated by a light source at x_0.

Example #2. [B-L-R.1] An interesting 2-d example (with the Euclidean Laplacian) is given in [B-L-R.1, p. 1031, Fig. 4]. It displays a disconnected 'minimal' portion of a circumference, which is sufficient for control in light of the geometric optics criterion. As there is no discussion, however, on whether or not such example could also be obtained by a strictly convex function and Theorem 6.5.4, we warn the reader—who may be induced to the opposite conclusion by the last paragraph on p. 1031 of [B-L-R.1]—that an argument similar to the one provided for our own Example #1 shows likewise that Fig. 4 of [B-L-R.1] cannot be obtained by a strictly convex function.

Remark 6.7.1. A non-Euclidean example has been constructed in [Ga.1] in which *control on the whole boundary in finite time is possible, but there is no strictly convex function whatever.*

The example (the wiffle-ball) is as follows. Delete from the sphere a neighborhood of three closed arcs of the equator, each arc subtending an

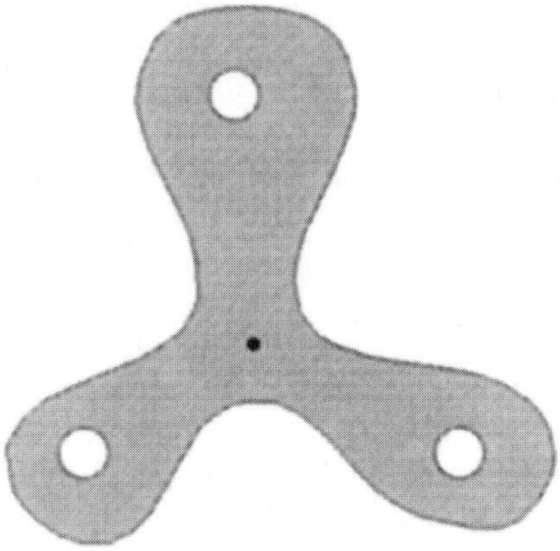

FIG. 2. *The outer boundary is subject to Dirichlet controls. The three inner boundary circles are uncontrolled; instead, homogeneous Dirichlet boundary conditions are imposed. No line segment inside the domain joins distinct inner boundary circles, which implies finite-time control. But the three inner boundary circles may be reached from the point at center, in directions not lying in a half space; this implies that there is no convex function which would ensure finite-time control.*

angle $\pi/3$ from the center of the sphere and spaced an angle $\frac{\pi}{3}$ from the other two arcs. The set that remains is the domain Ω. (Of course, this domain could be constructed as a planar domain, but with a non-Euclidean metric; it is easier to see the example on the sphere as a surface in $\mathbb{R}^3$). Inside Ω, there are no longer any full great circles; however, there are three meridians joining the north pole to the south pole (specifically: the meridians which cross the equators at the midpoints of the three remaining arcs of the equator), meeting at each pole with equal angles $\frac{2\pi}{3}$. The proof that there is no convex function whatever is analogous to, but more complicated than, the argument of Example #1.

How do the above exact controllability results, Theorem 6.5.4, Theorem 6.5.5, based on the strict convexity assumption (H.1) = (3.1.1) [or its relaxed version (H.1′) = (3.1.2)], compare—at least in the case of second-order hyperbolic equations—with the exact controllability results obtained by the geometric optics approach [B-L-R.1], [Lit.1]? Since [B-L-R.1], it has been known that even in the Euclidean canonical case $\Delta_g = \Delta$, $F(w) \equiv 0$, geometric optics is more refined and general, except for requiring C^∞-smoothness of both the Euclidean domain and the coefficients. (But see [B.1], only for the wave equation with Dirichlet control.) See Examples #1 and #2 and Remark 6.7.1 We point out here that paper [Ga.1] shows the following result:

THEOREM 6.7.1. *[Ga.1] (i) If the boundary $\partial\Omega$ of the Euclidean domain Ω is C^∞; (ii) if the coefficient of the second-order hyperbolic equation (6.3.3) are C^∞ and time independent; and (iii) if one knows* a-priori *that* *the boundary $\partial\Omega$ does not allow infinite order of contact with the geodesics, then: the existence of a strictly convex function $v(x)$ on Ω in the Riemannian metric g, with $\langle\nabla_g v(x)n(x)\rangle_g \leq 0$ on Γ_0 as in (3.2.10) or (6.5.1) implies the remaining assumptions of [B-L-R.1] and hence exact controllability on a time $T > T_0$, with T_0 defined by (3.2.2).* □

In this result, the assumption on $\partial\Omega$ having only finite order of contact with rays of geometric optics appears not to be easy to verify, in general; it holds true, however, if $\partial\Omega$ has positive definite second fundamental form. Moreover, via this route, the final exact controllability result rests with the highly technical apparatus of geometric optics and microlocal analysis in [B-L-R.1] or [Lit.1]. By contrast, the proof of Theorem 3.2.2, as specialized to the exact controllability of second-order hyperbolic equations (see Section 6.5), is entirely self-contained.

7. Second-order hyperbolic equations: Energy method in the Riemannian metric yielding Carleman and control estimates with no lower-order terms. Purely Neumann B.C. Global unique continuation. In this section we return to the setting of (6.1.1)–(6.1.3) of a general Riemannian wave equation defined on a bounded set Ω of a Riemannian manifold (M, g). We shall present a more flexible and more general method which, in contrast with the one of Section 6, yields *pointwise Carleman estimates with a controlled 'right' sign of the coefficient in front of the lower-order term*, so that, eventually, the lower-order term can be dropped, yielding the advantages pointed out in Remark 3.2.2.

We shall concentrate on the most challenging case: the purely Neumann B.C. case. This is Eqn. (6.1.1) with initial data $\{w_0, w_1\} \in H^1(\Omega) \times L_2(\Omega)$, and B.C.

$$\left.\frac{\partial w}{\partial n}\right|_{\Sigma} \equiv 0 \qquad \Sigma = (0, T] \times \Gamma, \tag{7.1}$$

subject to assumption (6.1.3) for $F(w)$. The point is that we seek the C.O.I. such as (0.3) (or (6.5.6)) or the U.S.I. such as (0.5) (or (6.5.16)), involving only the subportion Γ_1 of $\Gamma = \overline{\Gamma_0 \cup \Gamma_1}$, $\Gamma_0 \cap \Gamma_1 = \emptyset$. This corresponds to Neumann-control on Γ_1 and homogeneous *Neumann* B.C. on Γ_0. The easier case—typical of the literature—with homogeneous *Dirichlet* B.C. on Γ_0 is technically included as well in the present approach, indeed in a more relaxed setting: with the stringent assumption: $\langle Dd, n\rangle = 0$ on Γ_0 replaced by the more relaxed assumption: $\langle Dd, n\rangle \leq 0$ on Γ_0. See (A.3) = (7.21) below. Lack of space prevents us from giving a string of the main blocks of the approach yielding the sought-after final results. Accordingly, we must refer to the original papers: [L-T-Z.1] in the case of Eqn. (6.1.1) with $\Delta_g = \Delta$ = the Euclidean Laplacian on $\Omega \subset \mathbb{R}^n$, and its generalization [T-Y.2] to the Riemannian setting as in Eqn. (6.1.1), (6.1.3). The approach in [L-T-Z.1] was inspired by [L-R-S.1, Lemma 1, p. 124], which was also used in [K-K.1]. Both works refer to the Euclidean setting. Even in this case, the technical "pointwise estimate" in [L-R-S.1] works well for $H^{2,2}(Q)$-solutions to second-order hyperbolic equations with constant coefficient principal part and either Dirichlet control on Γ_1, or Neumann control on Γ_1, *but with Dirichlet homogeneous B.C. on* Γ_0. This is the case of the stability estimate paper [K-K.1]. The delicate passage from $H^{2,2}(Q)$- to $H^{1,1}(Q)$-solutions is not dealt with in [K-K.1] in the critical Neumann case. (Refer to Remark 6.5.1 on this delicate passage.) By contrast, in studying the purely Neumann B.C. case—Neumann control on Γ_1 *and Neumann homogeneous B.C. on* Γ_0—mere application of the same pointwise estimate in [L-R-S.1] yields yes results, but under too strong geometrical conditions [I-Y.1]. The simultaneous and independent paper [L-T-Z.1] manages to considerably weaken the geometrical conditions through various additional technical points, one of which is a new and more refined version of the pointwise estimate [L-T-Z.1, Lemma 3.1]. This is too lengthy to report here. Moreover, [L-T-Z.1] performs the delicate and technical passage from $H^{2,2}(Q)$-solutions to $H^{1,1}(Q)$-solutions in the (more challenging) Neumann case. Carleman estimates of the type reported below, for use in control-theoretic results—C.O.I. and U.S.I.—with no lower-order terms were obtained in [Ta.7] in a technical paper cast in the pseudo-differential language. By contrast, all prior control theory literature contains estimates with lower-order terms (see Remark 6.5.1). Below we report results from [T-Y.2], which generalized to the Riemannian setting the treatment in the Euclidean case of [L-T-Z.1]. Only some highlights may be given here.

Assumptions. We begin with the two main assumptions yielding Carleman estimates for solutions of Eqn. (6.1.1), with no B.C.

(A.1): This is assumption (H.1) = (3.1.1) which w.l.o.g. we normalize with $\rho = 1$

$$D^2 d(X, X) \equiv \langle D_X(Dd), X\rangle \geq 2|X|^2, \quad \forall\, x \in \Omega, \quad \forall\, X \in M_x, \tag{7.2}$$

for a strictly convex function $d : \overline{\Omega} \Rightarrow \mathbb{R}$, which in this section we take of class $C^3(\overline{\Omega})$.

(A.2): The assumption that d has no critical point on $\overline{\Omega}$ (in the worst case, on Γ_0)

$$\inf_{x\in\Omega} |Dd| = p > 0, \tag{7.3}$$

in which case we shall assume w.l.o.g. the following translation/rescaling conditions

$$\begin{aligned} &\min_{\overline{\Omega}} d(x) = m > 0; \qquad k = \inf_{\overline{\Omega}} \frac{|Dd|^2}{d} > 4, \\ &\min_{x\in\overline{\Omega}, t\in[t_0,t_1]} \phi(x,t) \geq \sigma, \qquad 0 < \sigma < m, \end{aligned} \tag{7.4}$$

where we think of m as being small, and we may choose $0 < \sigma < m$ such that the condition on $\phi(x,t)$ in (7.4) holds true, with $\phi(x,t)$ defined by (3.2.1) with v replaced by d. [This is a small technical variation over (3.2.5).]

Remark 7.1. Actually, assumption (A.2) = (7.3) may be dispensed with, say, in the worst case where the critical point of d falls on Γ_0. This, however, requires a complicated procedure [L-T-Z.1, Section 10], [T-Y.2, Section 10], of which we may only give a brief sketch. We begin by writing $\Omega = \Omega_1 \cup \Omega_2$, $\Omega_1 \cap \Omega_2 \neq \emptyset$, with Ω_i appropriately chosen. We then relax (7.3) by simply assuming the existence of two strictly convex functions d_i: $\overline{\Omega} \to \mathbb{R}$ as in (7.2), satisfying only the weaker condition:

$$\inf_{\Omega_i} |Dd_i| \geq p > 0. \tag{7.5}$$

The sets Ω_i need to be chosen, as to satisfy a second requirement, which is always possible to accomplish: the condition is technical, but it roughly means that each Ω_i is obtained by removing from Ω a small set containing the critical points of d_i on Γ_0 (worst case).

Complicated cut-off functions $\chi_i(t,x)$ are next introduced [L-T-Z.1, Section 10], which, however, are only *time-dependent* (but not space-dependent) on a small internal layer of the boundary Γ. This requirement is critical in the purely Neumann B.C. case, but is not needed in the case of purely Dirichlet B.C. The results preliminarily obtained under assumptions (A.1), (A.2) are then applied to the solution $(\chi_i w)$ of a perturbed equation, and the estimates for $i = 1, 2$, are then combined. □

Carleman estimate, first version.

THEOREM 7.1. *[L-T-Z.1], [T-Y.2] With $0 < c < 1 = \rho$ (chosen as in (7.2)), assume (A.1) = (7.2) and (A.2) = (7.3) [subject to the translation/scaling conditions (7.4)]. Let $\phi(x,t)$ be the pseudo-convex function defined by (3.2.1) with v there replaced by d now. Let $w \in C^2(R_t \times \Omega)$ be a solution of Eqn. (6.1.1) [and no B.C.] under the standing assumptions (6.1.3) for $F(w)$ and (6.1.1) for f. Then:*

(i) for all $\tau > 0$ sufficiently large and any $\epsilon > 0$ small, the following one-parameter family of estimates holds true, with $\rho = 1 - c > 0$ and $\beta > 0$, β depending on ϵ:

$$\begin{aligned} B_\Sigma(w) + 2\int_0^T\int_\Omega e^{2\tau\phi} f^2 dQ + C_{1,T} e^{2\tau\sigma}\int_0^T\int_\Omega w^2 dQ \\ \geq [\tau\epsilon\rho - 2C_T]\int_0^T\int_\Omega e^{2\tau\phi}[w_t^2 + |Dw|^2] dQ \\ +(2\tau^3\beta + \mathcal{O}(\tau^2) - 2C_T)\int_{Q(\sigma)} e^{2\tau\phi} w^2 dx\, dt - C_T\tau^3 e^{-2\tau\delta}[E(0) + E(T)]; \end{aligned} \tag{7.6}$$

$$Q(\sigma) = \{(x,t) :\ x \in \Omega,\ 0 \leq t \leq T;\ \phi(x,t) \geq \sigma > 0\}; \tag{7.7}$$

(ii) for all $\tau > 0$ sufficiently large and any $\epsilon > 0$ small,

$$B_\Sigma(w) + 2\int_0^T \int_\Omega e^{\tau\phi} f^2 dQ$$

$$\geq [\tau\epsilon\rho - 2C_T]e^{2\tau\sigma} \int_{t_0}^{t_1} \int_\Omega [w_t^2 + |Dw|^2] d\Omega\, dt \tag{7.8}$$

$$- C_{1,T} e^{2\tau\sigma} \int_0^T E(t)dt - C_T \tau^3 e^{-2\tau\delta}[E(0) + E(T)].$$

Here, $\delta > 0$, $\sigma > 0$, and $\sigma > -\delta$ are the constants in (3.2.4) (7.7), while C_T is a positive constant depending on T and d. Moreover, the boundary terms $B_\Sigma(w)$, $\Sigma = [0, T] \times \Gamma$, are defined by

$$B_\Sigma(w) = -2\tau \int_\Sigma (f_1 + \theta w f_2)\langle Dd, n\rangle] d\Sigma, \tag{7.9}$$

$$\begin{aligned} f_1 + \theta w f_2 = \theta^2 \Big\{ & [|Dw|^2 - w_t^2] + a - \tau^2(|Dd|^2 + \phi_t^2) \\ & + 2\tau \ell_t \phi_t] w^2 + 2[\ell_t - \tau\phi_t] w w_t \Big\}. \end{aligned} \tag{7.10}$$

Moreover, we have set

$$E(t) = \int_\Omega [w_t^2 + |Dw|^2 + w^2] d\Omega. \tag{7.11}$$

(iii) The above inequality may be extended to all $w \in H^{2,2}(Q)$. □

The proof of the critical part (i) is given in [L-T-Z.1] in the Euclidean case and in [T-Y.2] in the Riemannian case. Here we comment explicitly on how to go from (i) to part (ii), as Eqn. (7.8) in part (ii) is the first step leading, in (7.12) of Theorem 7.2 below, to the elimination of the lower-order term.

Proof of part (ii). We take τ sufficiently large so that, since $\beta > 0$, we then have that the term $[2\tau^3\beta + \mathcal{O}(\tau^2) - 2C_T]$ is positive, and we then drop the corresponding lower-order *interior* term involving w^2 in (7.6). Moreover, we invoke the critical property (3.2.5) on ϕ on the first integral term on the right side of (7.6). Finally, we majorize $\int_\Omega w^2(t) d\Omega$ by $E(t)$, see (7.11), so that the term $e^{2\tau\sigma} \int_0^T E(t)dt$ appears on the right side of (7.8). □

Remark 7.1. To obtain the final version in Theorem 7.2, it is very important that the *same* exponent $[2\tau\sigma]$ appears for the two exponentials outside the integrals in (7.8). This is reflected in the term $\exp(2\tau\sigma)$ in (7.12a). □

Carleman estimate, second version, without lower-order term.

THEOREM 7.2. *[L-T-Z.1], [T-Y.2] Assume the setting of Theorem 7.1, that is assumptions (A.1) = (7.2), (A.2) = (7.3). Let $w \in H^{2,2}(Q)$ be* *a solution of Eqn. (6.1.1) [and no B.C.] under the standing assumptions (6.1.3) for $F(w)$ and (6.1.1) for f. Then the following one-parameter family of estimates hold true: for all τ sufficiently large, $\tau > \tau_0$, and any $\epsilon > 0$ small, there exists a constant $k_{\phi,\tau_0} > 0$ such that*

$$\overline{B}_\Sigma(w) + \int_0^T \int_\Omega e^{2\tau\phi} f^2 dQ + const_\phi \int_0^T \int_\Omega f^2 dQ$$

$$\geq \left\{ \left[\frac{k_1}{2}(\epsilon\tau\rho - 2C_T)(t_1 - t_2)e^{-C_T T} - \frac{C_{1,T}k_2}{2k_1} T e^{C_T T} \right] e^{2\tau\sigma} - C_T \tau^3 e^{-2\tau\delta} \right\} [E(0) + E(T)] \tag{7.12a}$$

$$\geq k_{\phi,\tau_0}[E(0) + E(T)], \qquad k_{\phi,\tau_0} > 0, \tag{7.12b}$$

since $\sigma > -\delta$, $\delta > 0$, see (7.4), (3.2.3). Here, the boundary terms $\overline{B}_\Sigma(w)$ are given in terms of the boundary terms $B_\Sigma(w)$ in (7.9):

$$\overline{B}_\Sigma(w) = B_\Sigma(w) + c_T(\tau + 1)e^{2\tau\sigma} \left[\int_0^T \int_\Sigma \left| w_t \frac{\partial w}{\partial n} \right| d\Sigma + \int_0^T \int_{\Gamma_1} |ww_t| d\Sigma_1 + \int_{t_0}^{t_1} \int_{\Gamma_1} w^2 d\Gamma_1 dt \right]. \tag{7.13}$$

□

As the elimination of *lot* is the key goal of the present energy method approach (over the energy method approach of Section 6), we provide a sketch of the proof.

Proof. **Step 1.** First, we introduce

$$\mathcal{E}(t) \equiv \int_\Omega [w_t^2 + |Dw|^2] d\Omega + \int_{\Gamma_1} w^2 d\Gamma_1, \tag{7.14}$$

with $\Gamma_1 = \Gamma \setminus \Gamma_0$, where Γ_0 is the uncontrolled, or unobserved, part of the boundary. Moreover, we note the following equivalence:

$$k_1 E(t) \leq \mathcal{E}(t) \leq k_2 E(t), \tag{7.15}$$

for some positive constants $k_1 > 0$, $k_2 > 0$. Next, we return to estimate (7.8) of Theorem 7.1 add the term $(\tau\epsilon\rho - 2C_T)e^{2\tau\sigma} \int_{t_0}^{t_1} \int_{\Gamma_1} w^2 d\Gamma_1 dt$ to both sides, recall (7.15) for $\mathcal{E}(t)$ and obtain

$$B_\Sigma(w) + (\epsilon\tau\rho - 2C_T)e^{2\tau\sigma} \int_{t_0}^{t_1} \int_{\Gamma_1} w^2 d\Gamma_1 dt + 2 \int_0^T \int_\Omega e^{2\tau\phi} f^2 dQ$$

$$\geq (\tau\epsilon\rho - 2C_T)e^{2\tau\sigma} \int_{t_0}^{t_1} \mathcal{E}(t) dt - C_{1,T} e^{2\tau\sigma} \int_0^T E(t) dt - C_T \tau^3 e^{-2\tau\delta}[E(0) + E(T)]. \tag{7.16}$$

Step 2. By multiplying (6.1.1) by w_t and using the Gronwall inequality one obtains as in (6.4.8)–(6.4.9)

$$\mathcal{E}(t) \geq \frac{\mathcal{E}(T)+\mathcal{E}(0)}{2}\, e^{-C_T T} - N(T) \geq \frac{a}{2}[E(T)+E(0)]e^{-C_T T} - N(T); \tag{7.17}$$

$$N(T) = \int_0^T\int_\Omega f^2 dQ + 2\int_0^T\int_\Gamma \left|\frac{\partial w}{\partial n}\, w_t\right| d\Sigma + 2\int_0^T\int_{\Gamma_1} |ww_t| d\Sigma_1. \tag{7.18}$$

Similarly,

$$E(t) \leq \frac{1}{a}\left[\frac{b(E(0)+E(T))}{2} + N(T)\right] e^{C_T T}, \tag{7.19}$$

and hence

$$\begin{aligned} -C_{1,T} e^{2\tau\sigma} \int_0^T E(t)dt \ \geq & -\tfrac{C_{1,T} b}{2a}\, T e^{C_T T} e^{2\tau\sigma}[E(0)+E(T)] \\ & - \frac{C_{1,T} T}{a}\, e^{C_T T} e^{2\tau\sigma} N(T). \end{aligned} \tag{7.20}$$

Step 3. We insert (7.17) into the first integral on the right side of (7.16) and use (7.20) and readily obtain (7.13), (7.12a), by invoking (7.18) for $N(T)$.

Finally, we recall the critical relation $\sigma > 0$ from (7.4), so that $\sigma > -\delta$, with $\delta > 0$ defined by (3.2.3), and then $[\epsilon\tau e^{2\tau\sigma} - \tau^3 e^{-2\tau\delta}]$ is positive for all τ large enough. Thus (7.12a) yields (7.12b). □

So far, solutions of (6.1.1) were not required to satisfy B.C. We now introduce an additional assumption on the uncontrolled (unobserved) part Γ_0 of the boundary Γ. We assume

(A.3):

$$\begin{aligned} \left.\frac{\partial d}{\partial n}\right|_{\Gamma_0} &= \langle Dd, n\rangle = 0 \ \text{ on } \ \Gamma \text{ (Neumann case)}; \\ & \langle Dd, n\rangle \leq 0 \ \text{ on } \ \Gamma_0 \text{ (Dirichlet case)}. \end{aligned} \tag{7.21}$$

Assumption (A.3) = (7.21) in the Neumann case (Euclidean setting) was introduced in [Tr.1, Section 5].

Remark 7.2. The question arises: which triples $\{\Omega, \Gamma_0, \Gamma_1\}$ satisfy (A.1) = (7.2), (A.2) = (7.3) [or (7.5)] and (A.3) = (7.21) in the Neumann case? This issue is amply analyzed, in the Euclidean setting, in [L-T-Z.1, Appendices], with several classes of triples $\{\Omega, \Gamma_0, \Gamma_1\}$ given. A sufficient (but not necessary) set of conditions for a construction of a suitable $d(x)$, given in the Euclidean setting in [L-T-Z.1, Theorem A.1], admits a generalization to the Riemannian setting and is reported at the end.

Global uniqueness and continuous observability/(stabiliz -ation) inequalities in one shot. The purely Neumann case. Theorem 7.2 permits at once to get global uniqueness results for over-determined boundary conditions associated with Eqn. (6.1.1) with $f \equiv 0$. We consider the following over-determined problem with Γ_0 defined by (7.21) and $\Gamma_1 = \Gamma \setminus \Gamma_0$:

$$\begin{cases} w_{tt} - \Delta_g w = F(w) & \text{in } (0,T] \times \equiv Q; & (7.22a) \\ \left.\frac{\partial w}{\partial n}\right|_{\Sigma} \equiv 0 & \text{in } (0,T] \times \Gamma \equiv \Sigma; & (7.22b) \\ w|_{\Sigma_1} \equiv 0 & \text{in } (0,T] \times \Gamma_1 \equiv \Sigma_1, & (7.22c) \end{cases}$$

with F satisfying (6.1.3). Then, $\overline{B}_\Sigma(w) = 0$ by (7.22b–c) and (7.21). We obtain

THEOREM 7.3. *[L-T-Z.1], [T-Y.2] Assume hypotheses (A.1) = (7.2), (A.2) = (7.3), (A.3) = (7.21), with Γ_0 defined there in the Neumann case. Let $\Gamma_1 = \Gamma \setminus \Gamma_0$. Let $T > T_0$, with T_0 defined in (3.2.2). Let $w \in H^{1,1}(Q)$ be a solution of problem (7.21). Then, in fact, $w \equiv 0$ in Q; indeed, in $R_t \times \Omega$.* □

The passage from $H^{2,2}(Q)$- to $H^{1,1}(Q)$-solutions requires a non-trivial argument [L-T-Z.1, Section 8].

The purely Dirichlet case. Consider problem (7.21a) coupled with the B.C.

$$w|_\Sigma \equiv 0; \qquad \left.\frac{\partial w}{\partial n}\right|_{\Sigma_1} \equiv 0, \tag{7.23}$$

where Γ_0 and Γ_1 are defined below.

THEOREM 7.4. *Assume hypotheses (A.1) = (7.2), (A.2) = (7.3), (A.3) = (7.21) (Dirichlet), (6.1.3) for $F(w)$, where now Γ_0 is defined by: $\langle Dd, n\rangle \le 0$ on Γ_0, as in (7.21) or (3.2.10). Let $\Gamma_1 = \Gamma \setminus \Gamma_0$. Let $w \in H^{1,1}(Q)$ be a solution of problem (7.22a), (7.23). Then, in fact, $w \equiv 0$ in Q; indeed, in $\mathbb{R}_t \times \Omega$.*

Control-theoretic inequalities.

THEOREM 7.5. *[L-T-Z.1], [T-Y.2] Consider Eqn. (6.1.1) with F satisfying (6.1.3). Assume hypotheses (A.1) = (7.2), (A.2) = (7.3), and (A.3) = (7.21) (Neumann). Let Γ_0 be defined by (7.21) (Neumann), and let $\Gamma_1 = \Gamma \setminus \Gamma_0$. Let $T > T_0$, with T_0 defined by (3.2.2). Then the following continuous observability inequality holds true for $H^{1,1}(Q)$ solutions: there exists a constant $C_T > 0$ such that*

$$\int_0^T \int_{\Gamma_1} [w_t^2 + w^2] d\Sigma_1 + \int_0^T \int_\Omega f^2 dQ \ge c_T E(0), \tag{7.24}$$

with $E(\,\cdot\,)$ defined by (7.11).

The proof of Theorem 7.5 combines Theorem 7.2, Lemma 6.5.3, and Theorem 7.3, the latter providing the uniqueness result needed in a standard compactness/uniqueness argument to absorb the interior $lot(w)$ occurring in Lemma 6.5.3. The constant c_T in (7.24) is not explicit, a price to pay to avoid geometrical conditions of the literature on the observed/controlled portion of the boundary Γ_1.

Assumptions (A.1), (A.2), (A.3) (Neumann). In the case of Euclidean domains, several rather general classes of triples $\{\Omega, \Gamma_0, \Gamma_1\}$ are given in [L-T-Z.1, Appendices] which satisfy assumptions (A.1), (A.2), (A.3) (Neumann case).

Here we report one such class, in fact in its generalization to a Riemannian manifold, as given in [T-Y.2, Appendix B]. In the Euclidean case, the perturbation formula (7.31) was used by Tataru in the context of pseudo-convex functions by pseudo-differential methods. The Euclidean proof in [L-T-Z.1, Theorem A.4.1] is direct. It admits a Riemannian version [T-Y.2].

Setting. Let $\{M, g\}$ be an n-dimensional Riemannian manifold, with Levi-Civita connection D. Let $\Omega \subset M$ be an open, connected, compact subset of M, with boundary $\partial\Omega = \overline{\Gamma_0 \cup \Gamma_1}$, $\Gamma_0 \cap \Gamma_1 = \emptyset$. The portion Γ_0 of $\partial\Omega$ is defined as follows. Let $\ell : M \to \mathbb{R}$ be a function of class C^2. Then, we define

$$\left\{\begin{array}{l}\Gamma_0 = \{x \in \partial\Omega : \ \ell(x) = 0\}, \\ \text{with the further provision that } D\ell(= \nabla_g \ell) \neq 0 \text{ on } \Gamma_0.\end{array}\right. \tag{7.25}$$

THEOREM 7.6. *In the above setting, assume that*

(i)

$$D^2\ell(X,X)(x) \geq 0, \qquad \forall\, x \in \Gamma_0, \ \forall\, X \in M_x \tag{7.26}$$

(convexity of ℓ near Γ_0);

(ii) There exists a function $d_0 : \overline{\Omega} \to \mathbb{R}$ of class C^2, such that

(ii_1)

$$D^2 d_0(X,X)(x) \geq \rho_0 |X|_g^2, \quad \forall\, x \in \Gamma_0, \ \forall\, X \in M_x, \tag{7.27}$$

for some constant $\rho_0 > 0$ (strict convexity of d_0 near Γ_0);

(ii_2)

$$\left.\frac{\partial d_0}{\partial n}\right|_{\Gamma_0} = \langle Dd_0, n\rangle_g \leq 0 \qquad \text{on } \Gamma_0, \tag{7.28}$$

where n is the unit outward normal field to Γ_0 on M which points in the same direction as $\nabla_g \ell = D\ell$. Then: there exists a function $d : \overline{\Omega} \to \mathbb{R}$ of class C^2 [which is explicitly constructed in a layer (collar) of Γ_0, the critical set], such that it satisfies the following two conditions:

(a)

$$\left.\frac{\partial d}{\partial n}\right|_{\Gamma_0} = \langle Dd, n\rangle_g = 0 \qquad \text{on } \Gamma_0; \tag{7.29}$$

(b)

$$D^2 d(X,X)(x) \geq (\rho_0 - \epsilon)|X|_g^2, \quad \forall\, x \in \Gamma_0,\ \forall\, X \in M_x, \tag{7.30}$$

where $\epsilon > 0$ *is arbitrarily small.*

DEFINITION OF d. *The function* $d(x)$ *is explicitly constructed* near Γ_0, within Ω, *as a perturbation of the original function* d_0 *assumed in (ii) above, as follows:*

$$\begin{cases} d(x) \equiv d_0(x) + z(x) & x \text{ near } \Gamma_0 \text{ in } \Omega; \quad (7.31)\\ z(x) \equiv -\dfrac{\partial d_0}{\partial n}(\ell k) + \lambda \ell^2, & k \equiv \dfrac{1}{|D\ell|_g}, \quad (7.32)\end{cases}$$

where λ *is a sufficiently large parameter, to be selected below in the proof, while*

(7.33a) $\dfrac{\partial d_0}{\partial n}$ *denotes an extension of* $\left.\dfrac{\partial d_0}{\partial n}\right|_{\Gamma_0}$ *from the set* Γ_0 *(defined by* $\ell(x) = 0$*) to a layer (collar) of* Γ_0*, within* Ω*,*

which is defined by

$$\frac{\partial d_0}{\partial n} \equiv \langle Dd_0, n\rangle_g = \left\langle Dd_0, \frac{D\ell}{|D\ell|_g}\right\rangle_g = \langle Dd_0, kD\ell\rangle_g. \tag{7.33b}$$

$n \equiv \frac{D\ell}{|D\ell|_g}$ *(consistently with the statement below (7.28).*

8. Schrödinger equations on a Riemannian manifold: Energy methods in the Riemannian metric yielding Carleman/control estimates with lower-order terms.

Models. Assumptions. We use the same notation (M,g), $\langle\ \cdot\ ,\ \cdot\ \rangle$, $|\ \ |$, Ω, $\Gamma = \Gamma_0 \cup \Gamma_1$, Δ_g, D as in Sections 6 and 7. The philosophical strategy is the same as that of Section 6 for second-order hyperbolic equations. Thus, our pace of exposition will be brisker. However, the technicalities involved are different. In this section we study the following Schrödinger equation on a bounded set Ω of the Riemannian manifold M:

$$iw_t + \Delta_g w = F(w) + f \quad \text{in } (0,T] \times \Omega \equiv Q;\ f \in L_2(Q), \tag{8.1}$$

where Δ_g is the Laplace-Beltrami operator on M. For this equation, the '*energy level*' is $H^1(\Omega)$ for w. We define

$$E(t) = \int_\Omega |Dw(t)|^2 d\Omega. \tag{8.2}$$

Accordingly, we make the following assumption on the energy level differential term:

(A.1) $F(w) = \langle P(t, \cdot\,), Dw\rangle + p_0(t,x)w$ satisfies the following estimate

$$|F(w)|^2 \le c_T[|Dw|^2 + w^2], \quad \forall\, t, x \in Q \text{ a.e.} \tag{8.3}$$

A remark on the coefficients such as Remark 6.1.1 applies. Furthermore, we assume $f \in L_2(Q)$.

We note at the outset that the 'energy level' term $F(w)$ in (A.1) is an *unbounded* term for Eqn. (8.1), as seen in $H^1(\Omega)$; while the 'energy level' term $F(w)$ in (6.1.3) for the second-order hyperbolic equation (6.1.1), as viewed in $H^1(\Omega) \times L_2(\Omega)$ is, by contrast, a *bounded* term. Thus, $F(w)$ is expected to give rise to more troubles in well-posedness issues in the present Schrödinger equation case. This is reflected by the need of assumption (A.2) = (8.9) below.

Strictly convex function $v(x)$. As in Sections 6.2 and 7, we assume hypothesis (H.1) = (3.1.1) of Section 3.1: *there exists a strictly convex (non-negative) C^2-function $v : \overline{\Omega} \Rightarrow \mathbb{R}$*, so that (3.1.1) holds true.

We then define the same pseudo-convex function $\phi(x,t)$ as in (3.2.1), except that now $T > 0$ is arbitrary and $c = c_T$ is such that $4\sup_\Omega v(x) < cT^2$. Properties (i) and (ii) in (3.2.4) and (3.2.5) continue to hold true with these choices. As in Section 6.3, we see that model (8.1) includes the case of a Schrödinger equation with variable coefficients defined on an open bounded domain $\tilde{\Omega}$ of $\mathbb{R}^n$: $iw_t + \mathcal{A}w = F(w) + f$, on $(0,T] \times \tilde{\Omega}$, where $\mathcal{A}$ is the second-order elliptic differential operator defined by (6.3.1). We repeat the pattern of Section 6.

Carleman estimates with lower-order terms.

THEOREM 8.1. *[Tr.2], [Tr-Y.1] (Carleman estimate, first version) With reference to Eqn. (8.1), assume (A.1) = (8.3) on the energy level term $F(w)$; (H.1) = (3.2.1) on the strictly convex function $v(x)$; and $f \in L_2(Q)$ as in (8.1). Let w be a solution of (8.1) in the following class*

$$\begin{cases} w \in C([0,T]; H^1(\Omega)), & (8.4a) \\ w_t \in L_2(0,T; H^{-1}(\Gamma)); \ \nabla_{\tan} w, \dfrac{\partial w}{\partial n} \in L_2(0,T; L_2(\Gamma)). & (8.4b) \end{cases}$$

Let $\phi(x,t)$ be the function defined in (3.2.1), this time with $T > 0$ arbitrary and $c = c_T$ sufficiently large as to have $4\sup_\Omega v(x) < cT^2$. Let $E(t)$ be defined by (8.2). Then, for all values $\tau > 0$, the following one-parameter family of estimates holds true:

$$B_\Sigma(w) + \frac{2}{\tau}\int_Q e^{\tau\phi}|f|^2 dQ + C_{T,\tau}\|w\|^2_{C([0,T];L_2(\Omega))}$$

$$\ge \left(\rho - \frac{c_T}{\tau}\right)\int_Q e^{\tau\phi}|Dw|^2 dQ - \frac{e^{-\delta\tau}}{\tau}[E(T) + E(0)] \tag{8.5}$$

$$\ge \left(\rho - \frac{c_T}{\tau}\right) e^{-\frac{\tau\delta}{2}} \int_{t_0}^{t_1} E(t)dt - \frac{e^{-\delta\tau}}{\tau}[E(T) + E(0)], \tag{8.6}$$

where the boundary terms $B_\Sigma(w)$ over $\Sigma = (0,T] \times \Gamma$ are given by

$$
(8.7)\quad \begin{aligned} B_\Sigma(w) = Re\left(\int_\Sigma e^{\tau\phi}\frac{\partial w}{\partial n}Dv(\overline{w})d\Sigma\right) - \frac{1}{2}\int_\Sigma e^{\tau\phi}|Dw|^2\langle Dv, n\rangle d\Sigma \\ + \frac{1}{2}\left|\int_\Sigma \frac{\partial w}{\partial n}\overline{w}\ div(e^{\tau\phi}h)d\Sigma - i\int_\Sigma \overline{w}\, w_t e^{\tau\phi}\langle Dv, n\rangle d\Sigma\right|. \end{aligned}
$$

□

[Eqn. (8.5) is obtained also by use of property (3.2.4) of ϕ. To obtain (8.6) from (8.5), one further uses property (3.2.5) of ϕ.]

A related result is in [Ta.7], in the pseudo-differential language. Our Theorem 8.1 has a more friendly proof and the estimate is more explicit.

Remark 8.1. (Idea of proof) [T-Y.1] The proof of Theorem 8.1 is by energy method in the Riemannian metric: it uses the following main multipliers, where $Dv = D\phi$, recalling (3.2.1):

$$
(8.8)\qquad e^{\tau\phi(x,t)}\langle Dv, D\overline{w}\rangle \ \text{ and } \ \overline{w}\ \text{div}(e^{\tau\phi}Dv).
$$

These multipliers were inspired by the second-order hyperbolic equation case, for which we refer to the multipliers (6.4.4) and Remark 6.4.1. The multipliers (8.8) represent a far-reaching generalization of the energy methods (multipliers) used for the pure Euclidean Schrödinger equation in the late '80s. The counterpart of Remark 6.4.1 applies now. More precisely:

(1) Under the following two conditions, that (i) Eqn. (8.1) is defined on a Euclidean domain Ω (thus, $\Delta_g = \Delta =$ the usual Laplacian, $g_{ij} = \delta_{ij}$) and, moreover, (ii) $F(w) \equiv 0$ (no energy level terms), then we recover the 'classical' multipliers $h \cdot \nabla\overline{w}$ and $\overline{w}$ div h of [L-T.8] (Dirichlet case) with $h(x)$ a smooth coercive field on Ω, by taking in (8.8) the Euclidean topology and setting $\tau = 0$ (that is, no free parameter). See also [M.1] (Neumann case) with radial field.

(2) Generalizing case (1), if in (8.1) we still have $\Delta_g = \Delta$, but $F(w)$ is now a first-order operator satisfying the Euclidean version of (8.3), then we recover the multipliers used in [Tr.2], by taking the Euclidean metric in (8.8).

It was the case in [Tr.2] that prompted the multipliers (8.8) for the general case of (8.1), (8.3), as it was previously done for second-order hyperbolic equations (Remark 6.4.1). The proof in [T-Y.2] for the general case is a parallel development of that in [Tr.2]: computations in the Euclidean metric performed in [Tr.2] are replaced by counterpart computations in the Riemannian metric g, using the Levi-Civita connection D.

A perfect counterpart of the considerations made in Remark 6.4.1 apply now regarding *the control-theoretic inequalities*—the C.O.I. (0.3) and the U.S.I. (0.5)—versus the *reverse trace regularity inequality.* If we restrict to the Dirichlet-control case, the trace regularity inequality is the *reverse* (for all $T > 0$) of the C.O.I. (8.14) below, see Remark 8.2. The reverse inequality is proved in [L-T.8]; [L-T.24, Section 10.9]. The proof applies equally well for the case of variable coefficients (even in time and space) of appropriate regularity (as in the case of second-order hyperbolic equations). By stark contrast, variable coefficients in the principal part and/or the energy level terms introduce most serious additional difficulties over the constant coefficient, canonical case in showing the relevant control-theoretic inequalities such as (8.14) and (8.17). □

As in the case of second-order hyperbolic equations of Section 6, the bulk of the proof of the control-theoretic inequalities (8.14), (8.17) rests with the Carleman estimate, first version, of Theorem 8.1. It is particularly in achieving this result that the Riemannian geometric energy method, based on the multipliers in (8.8), plays a critical role. The improvement from the first to the second version of the Carleman estimate, as in Theorem 8.2 below, is definitely more computationally extensive and delicate than in the case of second-order hyperbolic equations (Section 6), since $F(w)$ is now an *unbounded* term for the Schrödinger Eqn. (8.1) in $H^1(\Omega)$. Moreover, more importantly, it requires an additional *structural* assumption on the energy level term $F(w)$, unlike the case of second-order hyperbolic equations. More specifically, in the present case of the Schrödinger equation (8.1), we need to impose the following structural assumption on the energy level term $F(w)$, as already pointed out in connection with estimate (3.2.7): namely that the first-order coefficient has real part that is the gradient of a scalar function (i.e., it is a conservative vector field). That is, we further assume that:

(A.2) the first-order term $F(w)$ in (8.3) is of the form

$$F(w) = R(w) + p_0 w, \quad R(w) = \langle P(t,x), Dw \rangle, \tag{8.9}$$

where Re $P(t,x) = \nabla\chi(t,x)$ on Ω for $t > 0$ for some $W^{1,\infty}$ function χ, while P has imaginary part, say, time independent with div(Im P) $\in L_\infty(\Omega)$ [while $p_0 : \Omega \to \mathbb{C}$ is a function which is in $L_\infty(Q)$ [or even in $L_p(Q)$, $p = \dim \Omega + 1$, by using a Sobolev embedding theorem], see (8.3)]. By a change of variable on the solution, we may achieve $\nabla\chi \equiv 0$ [L-T-Z.2].

Assumption (A.2) = (8.10) is invoked to obtain the well-posedness relationship (3.2.7). With this inequality at one's disposal, one then obtains the counterparts of inequalities (6.4.9a) and (6.4.10). Thus, one then obtains the following result from Theorem 8.1. The results below were obtained in [Tr.2] in the Euclidean case $\Delta_g = \Delta$ ($g_{ij} = \delta_{ij}$) while the generalization to the Riemannian setting is in [T-Y.2].

THEOREM 8.2. *[Tr.2], [Tr-Y.2] (Carleman estimate, second version) (a) Assume the hypotheses and setting of Theorem 8.1. In addition, assume (A.2) = (8.9) and $f \in L_2(0,T;H^1(\Omega))$. Then, for all $\tau > 0$ sufficiently large, $\tau \geq \tau_0 > 0$, there exists a constant $k_{\phi,\tau} > 0$ such that the following one-parameter family of estimates holds true:*

$$\overline{B}_\Sigma(w) + \frac{2}{\tau}\int_Q e^{\tau\phi}|f|^2 dQ + C\|f\|^2_{L_2(0,T;H^1(\Omega))} + C_{T,\phi,\tau}\|w\|^2_{C([0,T];L_2(\Omega))}$$

$$\geq e^{-\frac{\tau\delta}{2}}\left\{\left(\rho-\frac{c_T}{\tau}\right)\frac{e^{-kT}}{2}(t_1-t_0)-\frac{e^{-\frac{\delta\tau}{2}}}{\tau}\right\}[E(T)+E(0)] \tag{8.10}$$

$$\geq k_{\phi,\tau_0}[E(T)+E(0)], \tag{8.11}$$

where the boundary terms $\overline{B}_\Sigma(w)$ *are given by*

$$\overline{B}_\Sigma(w)=B_\Sigma(w)+const_{\phi,\tau,\rho}\int_\Sigma\left|\frac{\partial w}{\partial n}\right|[|W(w)|+|w_t|+|p_0w|]d\Sigma, \tag{8.12}$$

where $W(w)=\langle W,Dw\rangle$, *and where* $W(x)$ *is a vector field on the submanifold* Γ *such that* $W(x)\in\Gamma_x$ *(the tangent space to* Γ *at* x*) for* $x\in\Gamma$.

(b) Assume further that the solution w *of (8.1) satisfies:* $w|_{\Sigma_0}\equiv 0$, $\Sigma_0=(0,T]\times\Gamma_0$, *with* Γ_0 *defined by (6.5.1). Then, estimate (8.11) holds true for all* $\tau>0$ *sufficiently large and with the boundary terms evaluated only on* $\Sigma_1=(0,T]\times\Gamma_1$, $\Gamma_1=\Gamma\setminus\Gamma_0$, *while the boundary terms evaluated on* Σ_0 *are negative:* $\overline{B}_{\Sigma_0}(w)\leq 0$.

Control-theoretic results. We preliminarily let $\partial\Omega\equiv\Gamma=\overline{\Gamma_0\cup\Gamma_1}$, $\Gamma_0\cap\Gamma_1=\emptyset$, Γ_0, Γ_1 open in Γ, where Γ_0 is defined, as in (6.5.1) or (6.4.7), by $\Gamma_0=\{x\in\Gamma:\langle Dv,n\rangle\leq 0\}$, with $n(x)$ the unit outward normal field at $x\in\Gamma$, where v is the strictly convex function of assumption (H.1) = (3.1.1), in the Riemannian metric g.

Continuous observability inequalities.

THEOREM 8.3. *(Dirichlet case) Assume the above setting: hypothesis (H.1) - (3.1.1), (A.1) = (8.3), and (A.2) = (8.9). Let* $T>0$, *and let* Γ_0 *be defined as above (in (6.5.1)). Let* $f=0$. *Assume further that (8.1) possesses the following unique continuation property: if* ψ *is a solution of (8.1) in the class (8.4), which satisfies the over-determined B.C. (as in (6.5.2)):*

$$\psi|_\Sigma\equiv 0 \text{ and } \left.\frac{\partial\psi}{\partial n}\right|_{\Sigma_1}\equiv 0 \text{ on } \Gamma_1=\Gamma\setminus\Gamma_0, \tag{8.13}$$

then, in fact, $\psi\equiv 0$ *in* Q. *Finally, let* w *be a solution of (8.1) in the class (8.4) which satisfies the homogeneous Dirichelt B.C.:* $w|_\Sigma\equiv 0$. *Then, the following continuous observability inequality holds true: there exists* $C_T>0$ *such that*

$$\int_0^T\int_{\Gamma_1}\left(\frac{\partial w}{\partial n}\right)^2 d\Sigma_1\geq C_T\|w_0\|^2_{H_0^1(\Omega)}. \tag{8.14}$$

[This is the C.O.I. (0.3), as specialized to the present Dirichlet case.]

Remark 8.2. The converse (trace regularity) of inequality (8.14) always holds true, for any $T>0$ for (8.1), (8.3) with $w|_\Sigma\equiv 0$ [L-T.8], [L-T.24, Section 10.9].

Remark 8.3. In case of the pure Schrödinger equation, $F(w)\equiv 0$, in the Euclidean setting, the C.O.I. (8.14) is shown in [Le.2] under a geometric optics condition. □

THEOREM 8.4. *(Neumann case) We assume the same hypotheses of Theorem 8.3, [thus (H.1) = (3.1.1), (A.1) = (8.3), (A.2) = (8.9),* $T > 0$, Γ_0 *as in (6.5.1)], except that the present version of the unique continuation property is as follows: let* ψ *be a solution of (8.1) in the class (8.4) satisfying the over-determined B.C. (same as (6.5.4)):*

$$\left.\frac{\partial \psi}{\partial n}\right|_{\Sigma} \equiv 0 \text{ and } \psi|_{\Sigma_1} \equiv 0, \ \Gamma_1 = \Gamma \setminus \Gamma_0, \tag{8.15}$$

then, in fact, $\psi \equiv 0$ *in* Q. *Finally, let* w *be a solution of (8.1) in the class (8.4) which satisfies the homogeneous Neumann*

$$w|_{\Sigma_0} = 0; \qquad \left.\frac{\partial w}{\partial n}\right|_{\Sigma_1} \equiv 0, \qquad \Gamma_0 \neq 0. \tag{8.16}$$

Then, the following continuous observability inequality holds true: there exists $C_T > 0$ *such that*

$$\int_0^T \int_{\Gamma_1} |w_t|^2 d\Sigma_1 \geq C_T \|w_0\|^2_{H^1_{\Gamma_0}(\Omega)}. \tag{8.17}$$

[This is the C.O.I. (0.3), as specialized to the present Neumann case (8.16).]

For global uniqueness results such as the ones for ψ satisfying either (8.13) or (8.16), we refer to [L-T-Z.2–3].

In the statement of Theorem 8.4, key to the elimination of geometrical conditions on the controlled/observed (and also stabilized) portion Γ_1 of the boundary $\Gamma = \partial\Omega$ is a lemma such as Lemma 6.5.3 for a solution w of (8.1) in the class (8.4): see [Tr.2, Theorem 2.1.4], [T-Y.2, Lemma 7.2], which, in fact, can be reinforced [L-T-Z.2]. As a result of this lemma and of Theorem 8.2, one obtains the following main estimate (at the energy level).

THEOREM 8.5. *Assume the hypotheses and the setting of Theorem 8.2. Let* $f = 0$. *Let* w *be a solution of (8.1) in the class (8.4).*

(a) Then the following estimate holds true: there exists a constant $k_{\phi,\tau} > 0$ *for* τ_0 *sufficiently large,* $\tau \geq \tau_0 > 0$, *such that, for any* $\epsilon_0 > 0$:

$$\int_0^T \int_{\Gamma} \left[\left|\frac{\partial w}{\partial n}\right|^2 + |w_t|^2 \right] d\Sigma + c_{\phi,\epsilon_0} \|w\|^2_{L_2(0,T;H^{\frac{1}{2}+\epsilon_0}(\Omega))} \geq k_{\phi,\tau_0}[E(T) + E(0)]. \tag{8.18}$$

(b) Assume, further, that the solution w *satisfies also the conditions of Theorem 8.2(b), that is,* $w|_{\Sigma_0} \equiv 0$ *with* Γ_0 *defined by (6.5.1), then estimate (8.18) holds true with* $\int_\Gamma$ *replaced by* $\int_{\Gamma_1}$. □

Exact controllability. By duality, Theorems 8.3 and 8.4 give exact controllability results of the Schrödinger equation, Eqn. (8.1), (8.3) on the

Riemannian manifold M on any $T > 0$, with controls $L_2(0,T;L_2(\Gamma_1))$ on the space of optimal regularity $H^{-1}(\Omega)$ [Dirichlet case], or on the space of finite energy $H^1_{\Gamma_0}(\Omega)$ [Neumann case]. This is a parallel treatment of the case for second-order hyperbolic equations, Theorem 6.5.4 and Theorem 6.5.5. A counterpart of Remark 6.5.1 regarding the passage from $H^{2,2}(Q)$-solutions to $H^{1,1}(Q)$-solutions in the Dirichlet and Neumann cases still holds true for Schrödinger equation [L-T-Z.2].

Uniform stabilization. Neumann case. Consider the following Neumann feedback problem with $\Gamma_0 \neq \phi$:

$$\begin{cases} iw_t + \Delta_g w = 0 & \text{in } Q; & (8.19\text{a}) \\ w(0, \,\cdot\,) = w_0 & \text{in } \Omega; & (8.19\text{b}) \\ w|_{\Sigma_0} \equiv 0; \quad \dfrac{\partial w}{\partial n} = -w_t & \text{in } \Sigma_i. & (8.19\text{c}) \end{cases}$$

The problem is well-posed, in the semigroup sense in $H^1(\Omega)$. Then, inequality (8.18) permits to obtain a *uniform stabilization* result for the Neumann feedback problem (8.19), and with *no geometric* conditions imposed on the controlled boundary Γ_1.

THEOREM 8.6. *Assume (H.1′) = (3.1.2) and (A.1) = (8.3), (A.2) = (8.9), and the unique continuation property of Theorem 8.4 with* Γ_0 *defined by (6.5.1). Then:*

(a) For all $\tau > 0$ *sufficiently large,* $\tau \geq \tau_0 > 0$, *there exists a constant* $k_{\phi,\tau_0} > 0$ *such that*

$$\int_0^T \int_{\Gamma_1} \left| \frac{\partial w}{\partial n} \right|^2 d\Sigma_1 \geq k_{\phi,\tau_0} E(T). \tag{8.20}$$

[This is the U.S.I. (0.5), as specialized to the present Neumann case.]

(b) Equivalently, there exist constants $M \geq 1$, $\mu > 0$, *such that the energy (8.2) of problem (8.19) satisfies*

$$E(t) \leq Me^{-\mu t} E(0), \qquad \forall\, t \geq 0. \tag{8.21}$$

□

For uniform stabilization under Dirichlet feedback we refer to [L-T.8], [L-T-Z.2–3].

9. Schrödinger equations: Energy method yielding Carleman/ control estimates with no lower-order terms. Purely Neumann B.C. Global unique continuation. So far, the program encompassed by the title of this subsection has been accomplished only in the Euclidean setting; that is, for Eqn. (8.1) with $\Delta_g = \Delta$ (or $g_{ij} = \delta_{ij}$) [L-T-Z.2]. As for other evolution equations, analysis of the Euclidean case with $\Delta_g = \Delta$ is the first step of the investigative process. Once the appropriate approach has been found and the resulting technicalities have been resolved at the

Euclidean level, then the next step is to extend these approaches and techniques to the Riemannian setting, by replacing the Euclidean metric with the Riemannian metric g.

Paper [L-T-Z.2] is, in a sense, the 'perfect' counterpart in the Schrödinger case of the treatment for second-order equations in Section 7, as specialized to the Euclidean setting $\Delta_g = \Delta$ (that is, the 'perfect' counterpart of [L-T-Z.1]). The two evolutions—the second-order hyperbolic Eqn. (6.1.1) and the Schrödinger Eqn. (8.1) require vastly different technical treatments, while sharing a common philosophy. Since only the Euclidean case $\Delta_g = \Delta$ has been worked out so far (summer 2002), we shall confine ourselves to give here explicitly only the first version of the corresponding Carleman estimate. More results are in [L-T-Z.2], [L-T-Z.3]. Carleman estimates with no lower-order terms are given in the technical paper [Ta.7], in the pseudo-differential setting.

Let Ω be an open bounded domain in $\mathbb{R}^n$, with boundary $\partial\Omega = \Gamma = \overline{\Gamma_0 \cup \Gamma_1}$, $\Gamma_0 \cap \Gamma_1 = \emptyset$, Γ of class C^2. We consider Eqn. (8.1) in the Euclidean case, that is,

$$iw_t + \Delta w = F(w) + f \text{ in } Q = (0,T] \times \Omega,\; f \in L_2(Q), \tag{9.1}$$

$$F(w) = q_1(t,x) \cdot \nabla w + q_0(t,x) w, \tag{9.2}$$

with $|q_1|$, $q_0 \in L_\infty(Q)$ so that the following pointwise estimate holds true:

$$|F(w)|^2 \le c_T[|\nabla w|^2 + |w|^2], \qquad \forall\, (t,x) \in Q. \tag{9.3}$$

This is the setting of Section 8.1, as restricted to the Euclidean case. As in the case of Section 7, the emphasis of [L-T-Z.2] is on the most challenging case: the purely Neumann B.C. case: i.e., Eqn. (9.1) with initial condition $w_0 \in H^1(\Omega)$, and B.C.

$$\left.\frac{\partial w}{\partial n}\right|_{\Sigma} = 0 \quad \text{in } \Sigma = (0,T] \times \Gamma. \tag{9.4}$$

Again, the point is that we seek the C.O.I. such as (0.3) or the U.S.I. such as (0.5), involving only the subportion Γ_1 of Γ. This corresponds to Neumann-control on Γ_1 and homogeneous *Neumann* B.C. on Γ_0. The following assumptions yield Carleman estimates for solutions of (9.1), with no B.C.

(A.1) There exists a strictly convex (non-negative) function $d : \overline{\Omega} \Rightarrow \mathbb{R}$ of class $C^3(\overline{\Omega})$, so that the (symmetric) Hessian matrix $\mathcal{H}_d$ of $d(x)$ [i.e., the Jacobian matrix of $h(x) \equiv \nabla d(x)$] is strictly positive on $\overline{\Omega}$: $\mathcal{H}_d \ge \rho I$, $\rho > 0$, on $\overline{\Omega}$.

(A.2) [This can, in fact, be removed eventually as in [L-T-Z.1], as described in Section 7, Remark 7.1.]

$$\inf_{\Omega} |\nabla d(x)| = p > 0. \tag{9.5}$$

We next introduce the same function $\phi(x,t)$ defined in (3.2.1) with v replaced by d and with $T > 0$ arbitrary and $c = c_T$ sufficiently large as in (3.2.3), so that properties (3.2.4) and (3.2.5) hold true. We define

$$\mathbb{E}(t) = \int_\Omega [|\nabla w(t)|^2 + |w(t)|^2] d\Omega. \tag{9.6}$$

Carleman estimates.

THEOREM 9.1. *(first version) Assume (A.1) and (A.2) = (9.5) as well as the standing assumption (9.3) on $F(w)$ and (9.1) on f. Let $T > 0$, and let $\phi(t,x)$ be defined by (3.2.1) with v replaced by d. Let w be a solution of (9.1) [with no B.C.] in the following class*

$$\begin{cases} w \in C([0,T]; H^1(\Omega)); \quad \nabla_{\tan} w \quad \frac{\partial w}{\partial \nu} \in L_2(0,T; L_2(\Gamma)); & (9.7) \\ w_t \in L_2(0,T; H^{-1}(\Gamma)). & (9.8) \end{cases}$$

Then:

(i) the following one-parameter family of estimates holds true for all $\tau > 0$:

$$\begin{aligned} B_\Sigma(w) &+ 4 \int_0^T \int_\Omega e^{2\tau\phi} |f|^2 d\Omega\, dt \\ &\geq const_\tau \int_0^T \int_\Omega e^{2\tau\phi} |\nabla w|^2 d\Omega\, dt + [4\tau^3 \rho p + \mathcal{O}(\tau^2) - 4] \\ &\quad \times \int_0^T \int_\Omega e^{2\tau\phi} |w|^2 d\Omega\, dt - c_T \tau e^{-2\tau\delta} [\mathbb{E}(T) + \mathbb{E}(0)] \qquad (9.9) \\ &\geq \left[\tau \rho p - \frac{9}{2} c_T\right] e^{-\delta\tau} \int_{t_0}^{t_1} \mathbb{E}(t) dt - c_T \tau e^{-2\tau\delta} [\mathbb{E}(T) + \mathbb{E}(0)] \\ &\quad + [4\tau^3 \rho p + \mathcal{O}(\tau^2) - 4] \int_0^T \int_\Omega e^{2\tau\phi} |w|^2 d\Omega\, dt. \qquad (9.10) \end{aligned}$$

(ii) For all τ sufficiently large that $[4\tau^3 \rho p + \mathcal{O}(\tau^2) - 4] \geq \tau_0 > 0$, we then have

$$\begin{aligned} B_\Sigma(w) &+ 4 \int_0^T \int_\Omega e^{2\tau\phi} |f|^2 d\Omega\, dt \\ &\geq const_{\tau,p,\rho} e^{-\delta\tau} \int_{t_0}^{t_1} \mathbb{E}(t) dt - c_T \tau e^{-2\tau\delta} [\mathbb{E}(T) + \mathbb{E}(0)], \end{aligned} \tag{9.11}$$

with $c_{\tau,p,\rho} \nearrow \infty$ as $\tau \nearrow \infty$. In both (9.10) and (9.11), the boundary terms $B_\Sigma(w)$ are defined (with $h = \nabla d$) by:

$$
\begin{aligned}
B_\Sigma(w) = {} & 2\int_0^T \int_\Gamma e^{2\tau\phi}[2\tau^2|h|^2]|w|^2 h\cdot\nu \, d\Gamma\, dt \\
& - 2\int_0^T \int_\Gamma e^{2\tau\phi} 2c\tau \left(t - \frac{T}{2}\right)\left[\eta \frac{\partial \xi}{\partial \nu} - \xi \frac{\partial \eta}{\partial \nu}\right] d\Gamma\, dt \\
& - 2\int_0^T \int_\Gamma e^{2\tau\phi}[\xi_t \eta - \xi\eta_t]\tau h\cdot\nu \, d\Gamma\, dt \\
& + \int_0^T \int_\Gamma e^{2\tau\phi}[2\tau^2|h|^2 - \tau\Delta d]\left[\overline{w}\frac{\partial w}{\partial \nu} + w\frac{\partial \overline{w}}{\partial \nu}\right] d\Gamma\, dt \\
& + 2\int_0^T \int_\Gamma e^{2\tau\phi}\tau h\cdot\left[\nabla\overline{w}\frac{\partial w}{\partial \nu} + \nabla w\frac{\partial \overline{w}}{\partial \nu}\right] d\Gamma\, dt \\
& - 2\int_0^T \int_\Gamma e^{2\tau\phi}|\nabla w|^2 \tau h\cdot\nu\, d\Gamma\, dt,
\end{aligned}
\tag{9.12}
$$

where $\xi = Re\ w$ and $\eta = Im\ w$. □

As usual, establishing the first Carleman estimate is a main part of the entire proof leading to control-theoretic results. The critical novelty of Eqn. (9.10) over Eqn. (8.6) is that in (9.10) the lower-order term (last integral term) comes with a positive coefficient, for τ sufficiently large, and hence can *be absorbed within the* $\mathbb{E}(t)$*-term.* We then accrue the same benefits as in Section 7. The second (final) Carleman estimate (under further assumptions (8.9) as in Section 8) also comes with no lower-order terms, as do the C.O.I. and the U.S.I., whereby one then obtains *global unique continuation results* and *control-theoretic inequalities in one shot*, as part of the same flow of arguments. We refer to [L-T-Z.2] for more details. Such work with $\Delta_g = \Delta$ should form the basis of a generalization to the Riemannian setting, in the same way as the Euclidean paper [L-T-Z.1] led to the Riemannian paper [T-Y.2] for second-order hyperbolic equations.

10. Plate equations on the Riemannian manifold: Energy methods in the Riemannian metric yielding Carleman/control estimates with lower-order terms.

Model. Assumptions. We use the same notation of Sections 6, 7, 8. The philosophical strategy is the same as that of Sections 6 or 8. Accordingly, our exposition will be concise.

In this section we study the following plate equation, of Euler-Bernoulli type, on a bounded set Ω of the Riemannian manifold M, with Laplace-Beltrami operator Δ_g:

$$
w_{tt} + \Delta_g^2 w + F(w) + f = 0 \quad \text{in } (0,T]\times\Omega \equiv Q, \qquad f\in L_2(Q), \tag{10.1}
$$

where Δ_g is the Laplace-Beltrami operator on M. For this equation, the *'energy level'* is $H^3(\Omega) \times H^1(\Omega)$ for $\{w, w_t\}$. We define

$$(10.2) \qquad E(t) \equiv \int_\Omega \left[|D(\Delta w)|^2 + |Dw_t|^2\right] d\Omega.$$

Accordingly, we make the following assumptions on the energy level differential term:

(A.1): $F(w)$ satisfies the following estimate

$$(10.3) \quad |F(w)|^2 \le c_T\{|Dw|^3 + |D^2w|^2 + |Dw|^2 + w^2 + |Dw_t|^2 + w_t^2\}, \ \forall\, t, x \in Q.$$

A remark such as Remark 6.1.1 applies. Furthermore, we assume $f \in L_2(Q)$.

Strictly convex function $v(x)$. As in Sections 6.2, 7, and 8, we assume hypothesis (H.1) = (3.1.1) of Section 3.1: *there exists a strictly convex (non-negative) C^2-function $v : \overline{\Omega} \Rightarrow$*, so that (3.1.1) holds true.

We then define the same pseudo-convex function $\phi(x, t)$ as in (3.2.1), however, for $T > 0$ arbitrary and $c = c_T$ such that $4 \sup_\Omega d(x) \le cT^2$, precisely as in the case of the Schrödinger equations in Section 8. Thus, properties (i) and (ii) in (3.2.4) and (3.2.5) continue to hold true with these choices. As in Section 6.3, we see that model (10.1) includes the case of a plate-like equation of Euler-Bernoulli type with variable coefficients defined on an open bounded domain $\tilde{\Omega}$ of $\mathbb{R}^n$: $w_{tt} + \mathcal{A}^2(w) + F(w) + f = 0$ on $(0, T] \times \tilde{\Omega}$, where $\mathcal{A}$ is the second-order elliptic differential operator defined by (6.3.1).

The results below were obtained in [L-T.14] in the Euclidean case $\Delta_g = \Delta$ $(g_{ij} = \delta_{ij})$ and generalized in [L-T-Y.4] to the Riemannian setting. Related results are in [Ta.1–3] in the pseudo-differential setting.

Carleman estimates with lower-order terms. The present pattern repeats that of Sections 6 and 8.

THEOREM 10.1. *[L-T.14], [L-T-Y.4] (Carleman estimate, first version) With reference to (10.1), assume (A.1) = (10.3) on the energy level term $F(w)$; (H.1) = (6.2.1) on the strictly convex function $v(x)$; and (10.1) on f. Let w be a solution of (10.1) in the following class*

$$\begin{cases} (10.4a) \quad \{w, w_t\} \in L_2(0, T; H^3(\Omega) \times H^1(\Omega)), \\ (10.4b) \quad w_t, \Delta_g w, \dfrac{\partial \Delta_g w}{\partial n}, \ Dw_t, \ D(\Delta_g w) \in L_2(\Sigma). \end{cases}$$

Let $\phi(x, t)$ be the function defined in (3.2.1), this time with $T > 0$ arbitrary and $c = c_T$ sufficiently large as to have $4 \sup_\Omega v(x) \le cT^2$. Let $E(t)$ be defined by (10.2). Then, for all values $\tau > 0$, the following one-parameter family of estimates holds true:

$$B_\Sigma(w) + \frac{2}{\tau}\int_Q e^{\tau\phi} f^2 dQ + C_{\phi,\tau} \max_{[0,T]} \left\{ \int_\Omega [|D^2w|^2 + |Dw|^2 + w^2 + w_t^2] d\Omega \right\}$$

$$(10.5) \quad \geq \left(\rho - \frac{C}{\tau}\right) \int_Q e^{\tau\phi}[|Dw_t|^2 + |D(\Delta_g w)|^2] dQ - C\tau e^{-\delta\tau}[E(T) + E(0)]$$

$$(10.6) \quad \geq \left(\rho - \frac{C}{\tau}\right) e^{-\frac{\tau\delta}{2}} \int_{t_0}^{t_1} E(t)dt - C\tau e^{-\delta\tau}[E(T) + E(0)],$$

where the boundary terms $B_\Sigma(w)$ *on* $(0,T] \times \Gamma = \Sigma$ *are given by*

$$(10.7) \quad B_\Sigma(w) \equiv B^1_\Sigma(w) + B^2_\Sigma(w);$$

$$(10.8) \quad \begin{aligned} B^1_\Sigma(w) &= \int_\Sigma \left\{ [H(w_t) + w_t \; div \; H] \frac{\partial w_t}{\partial n} + H(\Delta_g w) \frac{\partial \Delta_g w}{\partial n} \right\} d\Sigma \\ &- \int_\Sigma \left\{ \left[w_t \Delta_g w_t + \frac{1}{2}(|Dw_t|^2 + |D(\Delta_g w)|^2) \right] \langle H, n \rangle \right. \\ &\left. + \frac{1}{2} w_t^2 \langle D(div \; H), n \rangle \right\} d\Sigma; \end{aligned}$$

$$(10.9) \quad \begin{aligned} B^2_\Sigma &\equiv \frac{1}{2} \int_\Sigma \left\{ \left[\Delta_g w \; \frac{\partial \Delta_g w}{\partial n} - w_t \; \frac{\partial w_t}{\partial n} \right] div \; H \right. \\ &\left. + \frac{1}{2} \left[w_t^2 - (\Delta_g w)|^2 \right] \langle D(div \; H), n \rangle \right\} d\Sigma, \end{aligned}$$

where H *is a vector field* $H = e^{\tau\phi} Dv$.

Remark 10.1. (Idea of proof) [L-T-Y.4] The proof of Theorem 10.1 is by energy methods in the Riemannian metric: it uses the following main multipliers, where $Dv = D\phi$:

$$(10.10) \quad H(\Delta_g w) = e^{\tau\phi(x,t)} Dv(\Delta_g w), \quad w \text{ div } H = \text{div}(e^{\tau\phi} Dv) w.$$

The multipliers (10.10) represent a far-reaching generalization of the energy methods (multipliers) used for the pure Euclidean Euler-Bernoulli equation in the late '80s [L-T.2], [L-T.4–5], [L.1]. Historical considerations similar to those already made for second-order hyperbolic equations, Eqn. (6.1.1), in Remark 6.4.1 and for the Schrödinger Eqn. (8.1), in Remark 8.1, apply now regarding the multipliers (10.10) in connection with the Euler-Bernoulli equation (10.1), (10.3). More precisely,

(1) If (i) Eqn. (10.1) is defined on a Euclidean domain Ω (thus $\Delta_g = \Delta$ = Euclidean Laplacian, $g_{ij} = \delta_{ij}$), and, moreover, (ii) $F(w) \equiv 0$ (no energy level terms), then we recover the 'classical' multipliers $h \cdot \nabla \Delta w$ and Δw div h of [L-T.2], [L-T.4–5] with $h(x)$ a coercive field in Ω (or the special case of [L.1] with a radial field $h(x) = (x - x_0)$), by taking in (10.10) the Euclidean topology and setting $\tau = 0$.

(2) Generalizing case (i), if in (10.1) we still have $\Delta_g = \Delta$, but $F(w)$ is now an energy level operator satisfying the Euclidean version of (10.3), then we recover the multipliers used in [L-T.14], by taking the Euclidean metric in (10.10).

It was the case in [L-T.14] that prompted the multipliers (10.10) for the general case (10.1), (10.3), under the inspiration of previously covered cases: second-order hyperbolic equations (Remark 6.4.1), and Schrödinger equations (Remark 8.1). The proof in [L-T-Y.4] for the general case is a parallel development of that in [L-T.14]: computations in the Euclidean metric performed in [L-T.14] are replaced by counterpart computations in the Riemannian metric g, using the Levi-Civita connection D. □

This time, as in the case of Schrödinger equations see (A.2) = (8.9) in Section 8, and unlike the case of second-order hyperbolic equations in Section 6, we need to impose a structural assumption on the energy level term $F(w)$. More precisely:

(A.2):

(10.11) Assume now that $F(w)$ is such that inequality (3.2.7) holds true,

with $E(t)$ defined by (10.2). See [L-T-Y.4], [Hor.3].

We then have, as a consequence of (A.2) = (10.11) used in Theorem 10.1:

THEOREM 10.2. *[L-T.14], [L-T-Y.4] (Carleman estimate, second version) Assume the hypotheses and setting of Theorem 10.1. In addition, assume (A.2) = (10.11). Then, for all $\tau > 0$ sufficiently large, $\tau \geq \tau_0 > 0$, there exists a constant $k_{\phi,\tau} > 0$, such that the following one-parameter family of estimates holds true:*

$$\begin{aligned}
&[1 + C_T(t_1 - t_0)]B_\Sigma(w) \\
&+ C_T(t_1 - t_0)\left\{BT_e(w) + \int_0^T \|f\|^2_{H^1(\Omega)}dt + lot(w)\right\} \\
&+ \frac{2}{\tau}\int_Q e^{\tau\phi}f^2 dQ + C_T e^{-\delta\tau} \max_{[0,T]} \int_\Omega [|D^2w|^2 + |Dw|^2 + w^2 + w_t^2] d\Omega \\
&\geq k_{\phi,\tau_0}[E(T) + E(0)],
\end{aligned} \tag{10.12}$$

where the boundary terms $B_\Sigma(w)$ and $BT_e(w)$ are defined by (10.8)–(10.10) and (3.2.7).

Control-theoretic results for (10.1), (10.3). The above Carleman estimates at the $H^3(\Omega) \times H^1(\Omega)$-level—in particular, estimate (10.12) of Theorem 10.2—form the key basic ground for obtaining continuous observability inequalities (hence, by duality, exact controllability results)/ stabilization inequalities for Eqn. (10.1), when this equation is accompanied by suitable boundary conditions (B.C.)

The $H^3(\Omega) \times H^1(\Omega)$-level of the preceding estimates are most directly of use when equation (10.1) is supplemented by *hinged* B.C. However, a full, sharp account even in the case of hinged B.C.—and surely more so for other B.C. such as *clamped* B.C.—requires additional tools [L-T.4-5], [L-T.14], [Ta.1-3] to deal with a few remaining issues. These include: (i) the issue of reducing the number of traces (see Remark 10.3) needed in the continuous observability estimates (that is, the issue of reducing the number of controls needed to obtain the corresponding (dual) exact controllability results); (ii) the issue of shifting (in particular, shifting down) the $H^3(\Omega)\times H^1(\Omega)$—level of topologies of the original estimate (10.12) [which is good, as we said, for *hinged* B.C.]. This shifting is required in order to obtain continuous observability estimates (exact controllability results) also in the case of *clamped* B.C. While lack of space induces us to leave this program and refer to the aforementioned references, we here derive the corresponding continuous observability/stabilization estimate for *hinged* B.C., in the case of two traces/two controls), since this follows readily from (10.12). More general consequences will be derived in a subsequent paper.

Uniqueness assumption. Below we shall need the following *uniqueness property for an over-determined problem* associated with (10.1): Let w be a sufficiently smooth solution of (10.1) as in (10.4) which satisfies the following B.C.

$$w|_\Sigma = \Delta_g w|_\Sigma \equiv 0 \qquad \text{on } (0,T]\times\Gamma\equiv\Sigma, \tag{10.13}$$

$$\left.\frac{\partial w}{\partial n}\right|_{\Sigma_1} = \left.\frac{\partial \Delta_g w}{\partial n}\right|_{\Sigma_1} \equiv 0 \qquad \text{on } (0,T]\times\Gamma_1\equiv\Sigma_1, \tag{10.14}$$

where the geometrical condition (6.5.1) or (6.4.7) for Γ_0:

$$\frac{\partial v}{\partial n} = \langle Dv, n\rangle \le 0, \qquad x\in\Gamma_0 \tag{10.15}$$

holds true on $\Gamma_0 = \Gamma/\Gamma_1$. Then, in fact: $w\equiv 0$ on $Q=(0,T]\times\Omega$.

We point out that our fourth-order operator is the iteration of a second-order operator.

Remark 10.2. The following are a few known cases, where the unique continuation property required above for problem (10.1), (10.3) holds true, say, in the Euclidean setting of Section 6.3, for $w_{tt}+\mathcal{A}^2 w + F(w) = 0$ with variable coefficients, on a bounded domain $\Omega\subset\mathbb{R}^n$:

(1) The case where the coefficients $a_{ij}(x)$ of the elliptic operator $\mathcal{A}$ in (6.3.1) satisfy $a_{ij}(x)\in C^1(\Omega)$, while the coefficients q_α and r_α of the energy level term F in (10.3) are *time-independent* and in $L_\infty(\Omega)$ in the space variable.

Indeed, in this case, the unique continuation property for the dynamical problem (10.1), (10.3) can be converted (via Laplace transform) into a unique continuation property for the corresponding fourth-order elliptic problem with space-variable coefficients. This latter problem has all four boundary conditions (Cauchy data) zero on the portion Γ_1 of the boundary. As a consequence, the solution of the corresponding elliptic problem has to vanish in a suitable neighborhood of Γ_1, interior to Ω. At this point, we can apply Hörmander's 1959-theorem [Hor.2] (of which a new proof has been recently given in [E.1, Theorem 1.1]) and conclude that, then, the solution of such elliptic problem must vanish on all of Ω: $w \equiv 0$ in Ω, as desired. [A stronger version of the unique continuation result in [Hor.2] for certain fourth-order elliptic equations, which are iterations of two second-order elliptic equations, is given in [S.1], under "weaker vanishing requirements," beyond our present needs. Papers [Hor.2] and [S.1] improve upon a prior result in [P.1], where the fourth-order equation has principal part restricted to Δ^2, followed by *all* third-order terms.]

(2) The case—given in [I.1, Theorem 5.1, p. 137]—of the equation in a Euclidean setting as in Section 6.3, where, however, the differential operator $\mathcal{A} = \Delta$ in (6.3.1) $[a_{ij} = \delta_{ij}]$, but with coefficients of F allowed to vary in both *time* and *space*:

$$F(w) = \sum_{|\beta|\leq 1} a_\beta \partial^\beta \Delta w + \sum_{|\alpha|\leq 2} a_\alpha \partial^\alpha w,$$

$a_\alpha \in L(Q)$, $Q = (0,T) \times \Omega$, plus some additional conditions. This result in [I.1] requires, however, all four zero B.C. on the *entire* boundary Γ (so, in (10.14), we must take $\Gamma_1 = \Gamma$). We remark that a readjustment of the proof for unique continuation result given in [E-L-T.1, Section 10], [E-L-T.2] for the corresponding Kirchhoff plate with all four B.C. zero, but this time only on an arbitrary common portion of the boundary, of positive measure, is likely to admit (*mutatis mutandis*) a counterpart version to the Euler-Bernoulli plate. [The constant $\gamma > 0$ for the Kirchhoff plate, which accounts for rotational inertia, becomes $\gamma = 0$ on the Euler-Bernoulli plate.] Should this be the case, the required $\Gamma_1 \subset \Gamma$ in (10.14) subject to (10.15) would be allowed. The unique continuation result in [E-L-T.1, Section 10], [E-L-T.2] for the Kirchhoff equation with all four zero B.C. on an arbitrary portion of the boundary extended the prior result [I.1, Theorem 1.2, p. 136], where vanishing of all four B.C. on the entire boundary was required. □

Continuous observability inequality.

THEOREM 10.3. *With reference to Eqn. (10.1) with $f = 0$, assume (A.1) = (10.3) on F and (H.1) = (3.1.1). In addition, assume: (i) the hinged boundary condition (10.13) on all of $\Sigma = (0,T] \times \Gamma$, as well as (ii) the geometrical condition (10.15) on Γ_0.*

(a) Then, given $T > 0$, there exists $C_T > 0$, such that the following estimate holds true:

$$\int_0^T\int_{\Gamma_1} \left[\left(\frac{\partial w_t}{\partial n}\right)^2 + \left(\frac{\partial \Delta w}{\partial n}\right)^2\right] d\Gamma_1\, dt + \ell ot(w) \geq C_T[E(0)+E(T)]. \tag{10.16}$$

(b) Assume further the uniqueness property stated before: (10.1), (10.13), (10.14), imply $w \equiv 0$ on Q. Then, estimate (10.16) simplifies to the following Continuous Observability Inequality*:*

$$\int_0^T\int_{\Gamma_1} \left[\left(\frac{\partial w_t}{\partial n}\right)^2 + \left(\frac{\partial \Delta w}{\partial n}\right)^2\right] d\Gamma_1\, dt \geq c_T[E(0)+E(T)]. \tag{10.17}$$ □

Remark 10.3. It is possible to reduce the number of traces arising in the C.O.I. or U.S.I. of plate equations. This topic—which is algebraic or analytic rather than geometric—will take us too far afield. In the case of the C.O.I. in the hinged case above, an ad hoc method was given in [Le.2] and is reproduced in [K.1]. In the more challenging stabilization problem of an Euler-Bernoulli plate with (true) "physical moment B.C. (rather than simply $\Delta w|_\Sigma$), [Las-Hor.1] also succeed in eliminating a trace from the U.S.I. A general algebraic method for expressing a trace in terms of others, rooted in PDE theory [Tay.1], was given in [Ta.1] for general evolution equation in the pseudo-differential language. Manifestations of it can be found in [L-T.14] for various single PDE classes, and in [Las.3] for the boundary stabilization of the 3-d structural acoustic model. At any rate, the Riemannian energy method of the present Part III merges well with the techniques of eliminating superfluous traces in the final C.O.I. and U.S.I. for plates. □

Exact controllability. As is well-known [L-T.2,4–5], by duality, the continuous observability inequality (10.17) is equivalent to the property of exact controllability of the corresponding non-homogeneous boundary control problem, with controls in the hinged B.C., in the space of 'optimal regularity' [L-T.2,4–5], [L.1].

Remark 10.4. The approach of Section 9 on Schrödinger equations accrues a benefit also in the case of the Euler-Bernoulli equation in the Euclidean case (i.e., Eqn. (10.1) with $\Delta_g = \Delta$, $g_{ij} = \delta_{ij}$, and $F(w) \equiv 0$) with hinged B.C. In fact, this case can be rewritten as the iteration of two pure Schrödinger equations, with Dirichlet B.C., to which the results of Section 9 can be applied. The result of this approach is that the constant c_T in (10.17) is thus *explicit*, contrary to the classical results [L-T.2, 4–5], [L.1], etc., where it is not explicit, due to the compactness uniqueness contradiction argument to absorb $l.o.t.(w)$. □

Part IV: A dynamic shell model.

11. Uniform stabilization of a shallow shell model with non-linear boundary feedbacks. Critical marriage between Riemannian geometry (continuous observability inequality) and microlocal analysis (boundary trace estimates). A dynamic shell: The need for differential geometry. A shell is a body in $\mathbb{R}^3$. We think of the middle surface of the shell as occupying a bounded region Ω of a smooth orientable surface M in $\mathbb{R}^3$. Then, the shell of thickness $h > 0$ ("small") is defined by

$$S = \left\{ p : \ p = x + zN(x), \quad x \in \Omega, \ -\frac{h}{2} < z < \frac{h}{2} \right\}.$$

Here $N(x)$ is the unit normal field on Ω. A dynamic model of a shell describes its displacement as a function of time. The displacement is a

3-dimensional vector $\{W, w\}$. Here, W is the 2-dimensional in-plane displacement, defined on the tangent plane of each point $x \in \Omega$; instead, w is the scalar normal displacement. The mathematical model of a dynamic shell consists of two *coupled* variable coefficient Partial Differential Equations (PDEs), both of hyperbolic type: a 2-d system of elasticity for the in-plane displacement W and a plate-like equation of Kirchhoff type for the scalar normal displacement w. The coefficients of the principal part of the operators are variable, due to the curved nature of the shell. Classically, the topic of *static* shells is covered by many books. They all assume the middle surface of a shell to be described by one coordinate patch: this is the image in $\mathbb{R}^3$ of a smooth function defined on a connected domain of $\mathbb{R}^2$. This view has geometrical limitations, as it forces the exclusion of interesting objects such as a half sphere, or a sphere. Moreover, the classical models use traditional geometry and end up with highly complicated analytical models. In these, the explicit presence of the Christoffel symbols Γ^k_{ij}, make them unsuitable for energy method computations, of the type needed for continuous observability/stabilization estimates, such as (0.3), (0.5), due to the presence of variable coefficients (in space) in the principal part of the differential operators, as well as in the energy level terms. Here, instead, in line with the main aim of the present paper, we shall present a recently proposed [Y.3], [Y.4], intrinsic model of a shallow shell viewed as a 2-d Riemannian manifold, within the intrinsic, coordinate-free setting of differential geometry. This approach allows for the use of a computational energy method in the Riemannian metric, as in Part III for single PDE equations.

Boundary feedback dissipative shell model: Stabilization. In this section, we first present the resulting analytic model of a shell, as it arises when suitable dissipative terms are imposed on the boundary conditions. The choice of the suitable dissipative terms is a non-trivial problem in its own merit, particularly in the case of the physically relevant, but mathematically challenging, free boundary conditions. Here we focus on free boundary conditions, and we then select the boundary feedback dissipative model proposed in [L-T.29]. This augments, with suitably chosen dissipative boundary terms, the original boundary homogeneous model in [Y.3]. We choose the boundary damping terms to be nonlinear. We then present a recent result [L-T.29] which claims that the aforementioned shallow shell model with nonlinear dissipative terms in the free boundary conditions is, in fact, uniformly stable.

Combination of differential geometry and microlocal analysis on traces as critical for the solution of the stabilization problem. The solution of the present stabilization problem relies critically on the combination or fusion of the differential geometric description of the shell—in particular, the continuous observability estimate in [Y.4] given in Section 11.3 below—with a delicate PDE-microlocal analysis, yielding sharp trace regularity of the solutions of the elastic waves (Section 11.5) and of the Kirchhoff plates (Section 11.6)—the two components of the displacement of a shell. This way, we first of all solve the problem, and, in the process, we achieve two main benefits: (i) we dispense altogether with restrictive geometrical conditions on the controlled part of the boundary of the shell, of the type used in wave and plate literature [Lag.3]; (ii) we avoid unnatural and mathematically undesirable terms in the boundary feedbacks of the elastic wave [P-T.1] even in the *flat* case, whose purpose was to cancel out boundary traces, which one would not control without sharp trace theory; the price paid this way was that of injecting boundary terms which are *not* in L_2. More explicitly, the sharp, microlocal trace theory of the plate component (w below in Section 11.6) is not strictly critical for achieving *some* solution of the present uniform stabilization problem: in fact, one could get a solution at the price of assuming, instead, restrictive and unnecessary geometrical conditions on the controlled part of the boundary Γ_1 as in prior literature [Lag.3]. By contrast, the contribution of a sharp, microlocal trace theory of the elastic wave component W in Section 11.5 is indispensible for the *very solution* of the present uniform stabilization problem.

11.1. A differential geometric model for a dynamic shallow shell's model in nonlinear, dissipative, feedback form. The middle surface of the shell is a bounded region Ω, which lies on a smooth orientable, two-dimensional surface M of $\mathbb{R}^3$. The regular boundary (on M) of Ω is denoted by $\Gamma = \Gamma_0 \cup \Gamma_1$, $\Gamma_0 \cap \Gamma_1 = \emptyset$ and consists of two disjoint portions: Γ_0 which will be the 'uncontrolled' part of the boundary; and Γ_1 which will be the 'controlled' part of the boundary; that is, the one where the dissipative feedback is active. The coupled system of two hyperbolic PDEs in the displacement vector $[W, w]$, which represent the dynamic model of a shallow shell in feedback form is [L-T.29],

$$
\begin{array}{ll}
(11.1.1a) & W_{tt} - [\Delta_\mu W + (1-\mu)kW + \mathcal{F}(w)] = 0 \quad \text{in } (0,\infty] \times \Omega \equiv Q_\infty; \\
(11.1.1b) & [I - \gamma\Delta]w_{tt} + \gamma[\Delta^2 w - (1-\mu)\delta(kdw)] \\
 & \quad + (H^2 - 2(1-\mu)k)w + \mathcal{G}(W) = 0 \quad \text{in } Q_\infty; \\
(11.1.1c) & W \equiv 0;\ w \equiv 0,\ \dfrac{\partial w}{\partial n} \equiv 0 \quad \text{in } (0,\infty) \times \Gamma_0 \equiv \Sigma_{0,\infty}; \\
(11.1.1d) & B_1(W,w) = g_1(\langle W_t, n\rangle);\ B_2(W,w) = g_2(\langle W_t, \tau\rangle) \quad \text{in } (0,\infty) \times \Gamma_1 \equiv \Sigma_{1,\infty}; \\
(11.1.1e) & \Delta w + (1-\mu)B_3(w) = -h_1\left(\dfrac{\partial w_t}{\partial n}\right) \quad \text{in } \Sigma_{1,\infty}; \\
(11.1.1f) & \dfrac{\partial \Delta w}{\partial n} + (1-\mu)B_4 w - \gamma\dfrac{\partial w_{tt}}{\partial n} = -\dfrac{\partial}{\partial \tau}h_2\left(\dfrac{\partial w_t}{\partial \tau}\right) \quad \text{in } \Sigma_{1,\infty}; \\
(11.1.1g) & \zeta(0,\,\cdot\,) \equiv [W(0,\,\cdot\,), w(0,\,\cdot\,)] \equiv \zeta_0 = [W_0, w_0]; \\
 & \zeta_t(0,\,\cdot\,) \equiv [W_t(0,\,\cdot\,), w_t(0,\,\cdot\,)] \equiv \zeta_1 = [W_1, w_1].
\end{array}
$$

We notice that model (11.1.1) has homogeneous boundary conditions on Γ_0, and suitable nonlinear dissipative feedback terms on Γ_1 involving tangential and normal components of W_t and w_t through the nonlinear scalar functions $g_i(\,\cdot\,)$ and $h_i(\,\cdot\,)$. In the boundary *homogeneous* case, where the boundary functions are all zero: $g_1 \equiv g_2 \equiv h_1 \equiv h_2 \equiv 0$, the mixed problem (11.1.1) specializes to the one considered in [Y.1–2]. In the *flat* case, the feedback problem (11.1.1) reduces to (a special case of) the fully nonlinear von Karman system considered in [Las.5], see also [Las.6]: here, the coupling between the W- and the w-equation is via nonlinear unbounded terms in the energy space; instead, problem (11.1.1) in the *flat* case yields *no* coupling terms: $\mathcal{F}(w) \equiv 0$, $\mathcal{G}(W) \equiv 0$.

Essential glossary of notation. Equations (11.1.1a–b). Essentially Eqn. (11.1.1a) is a system of elasticity in W, while (11.1.1b) is a Kirchhoff-like plate equation in w, both defined on the curved surface Ω. In (11.1.1a), we have that Δ_μ is a Hodge-Laplace type operator applied to 1-forms (equivalently, vector fields) on M and is defined by: $-\Delta_\mu = \frac{1-\mu}{2}\,\delta d + d\delta$. Here d is the exterior differential and δ is its formal adjoint. The constant μ, $0 < \mu < 1$ (physically $0 < \mu < \frac{1}{2}$), is the Poisson's coefficient of the material of the shell. Moreover, k and H are, respectively, the Gaussian curvature and the mean curvature of the shell's middle surface Ω. Furthermore, in (11.1.1a–b), Δ is the Laplace-Beltrami operator on the manifold M. In addition, the coupling terms $\mathcal{F}(w)$ and $\mathcal{G}(W)$ are first-order differential operators on w and W, respectively, whose structure is not essential (see [Y.3]). However, in the *flat* case, the second fundamental form is zero, and so $H \equiv k \equiv 0$, and these coupling terms vanish $\mathcal{F}(w) \equiv \mathcal{G}(W) \equiv 0$. We thus obtain the (usual) system of elasticity in 2-d and a Kirchhoff plate equation, this time uncoupled. Finally, $\gamma = \frac{h}{12}$, where h is the thickness of the shell.

Boundary equations. (11.1.1d-f) Set preliminarily $\frac{\partial\,\cdot}{\partial n} = \langle D\,\cdot\,, n\rangle$, $\frac{\partial}{\partial\tau} = \langle D\,\cdot\,,\tau\rangle$, where D denotes the Levi-Civita connection on M in the induced metric from $\mathbb{R}^3$. Finally, the boundary operators B_1, B_2, B_3, B_4 in (11.1.1d-f) are defined by

$$\left\{\begin{aligned} B_1(W,w) &= (1-\mu)\Upsilon(W,w)(n,n) + \mu(wH - \delta W) &\quad (11.1.2)\\ B_2(W,w) &= (1-\mu)\Upsilon(W,w)(n,\tau) &\quad (11.1.3)\\ B_3 &\equiv -D^2 w(\tau,\tau) &\quad (11.1.4)\\ B_4 &\equiv \frac{\partial}{\partial\tau}[D^2 w(\tau,n)] + k(x)\frac{\partial w}{\partial n} + \ell w,\ \ell \geq 0. &\quad (11.1.5)\end{aligned}\right.$$

In (11.1.2)–(11.1.5), $\Upsilon(\zeta) = \Upsilon(W,w)$ is the linearized 2-covariant strain tensor defined by [Y.3, (1.22)]

$$\Upsilon(W,w) = \frac{1}{2}(DW + D^*W) + w\Pi \tag{11.1.6}$$

in terms of the covariant differential DW of W and its transpose D^*W, as well as of the second fundamental form Π of the surface M. Moreover,

D^2w is the Hessian of w [$-D^2w$ denotes the change of curvature tensor of the middle surface Ω]. We note that the operator B_4 associated with the plate component of the shell is given in terms of normal and tangential coordinates, precisely as in [L-T.24, Chapter 3, Appendix D], which is a more convenient geometric and analytic representation of that arising in the variation model [Lag.2–3]. In (11.1.5), ℓ is a non-negative constant, whose role is seen in hypothesis (H.4) below (11.2.18). The present model appears to be the linearization of models by Bernadou and Ciarlet.

11.2. Main results: Well-posedness and uniform stabilization. Well-posedness of feedback problem (11.1.1a–g). The following well-posedness/regularity results are known for the feedback problem (11.1.1a–g).

THEOREM 11.2.1. *(a) [generalized (weak) solutions] Assume that: the non-linear functions h_i, g_i in (11.1.1d-f) are possibly multivalued, monotone, and that $0 \in h_i(0)$, $0 \in g_i(0)$. Then: there exists a* unique, *global solution of finite energy of problem (11.1.1a–g). This is to say that: for any initial data (see [He.1], [L-T.29], [Y.3], [Y.4], for these spaces)*

$$W_0, W_1 \in H^1_{\Gamma_0}(\Omega, \Lambda) \times L^2(\Omega, \Lambda); \;\; w_0, w_1 \in H^2_{\Gamma_0}(\Omega) \times H^1_{\Gamma_0}(\Omega), \tag{11.2.1}$$

that is, subject to the boundary conditions (B.C.) $W_0 = w_0 = \dfrac{\partial w_0}{\partial n} = w_1 = 0$ on Γ_0, there exists a unique solution

$$\begin{aligned} \{W, w\} &\in C([0,T]; H^1_{\Gamma_0}(\Omega, \Lambda) \times H^2_{\Gamma_0}(\Omega)); \\ \{W_t, w_t\} &\in C([0,T]; L_2(\Omega, \Lambda) \times H^1_{\Gamma_0}(\Omega)), \end{aligned} \tag{11.2.2}$$

where $T > 0$ is arbitrary. This solution is described by a non-linear semigroup acting on the finite energy space. The form of the generator is given in (11.3.3) below.

(b) [regular solutions] Assume that the boundary functions h_i, g_i satisfy, in addition to the above hypotheses of part (a), the following more specific hypotheses: h_i, g_i are single valued, and moreover $h_i, g_i \in C(\mathbb{R})$; $h_i', g_i' \in L_\infty(\mathbb{R})$. Then: for any initial data (see [L-T.29], [He.1], [Y.3–4] for these spaces)

$$W_0, W_1 \in H^2(\Omega, \Lambda) \times H^1(\Omega, \Lambda); \;\; w_0, w_1 \in H^3(\Omega) \times H^2(\Omega), \tag{11.2.3}$$

subject to the B.C. below (11.2.1), there exists a unique solution of (11.1.1a–g):

$$[W, w] \in C([0,T]; H^2(\Omega, \Lambda) \times H^3(\Omega)); \tag{11.2.4}$$

$$[W_t, w_t] \in C([0,T]; H^1(\Omega, \Lambda) \times H^2(\Omega)). \tag{11.2.5}$$

□

Henceforth, the boundary functions h_i, g_i are assumed to satisfy the condition of Theorem 11.2.1 unless otherwise stated.

Uniform stabilization. The main goal of the present section is to show that the solutions of problem (11.1.1a-g), asserted by Theorem 11.2.1, decay to zero at $t \to \infty$ at a uniform rate. Since the dissipative feedback terms are located on the portion Γ_1 of the boundary Γ of the mid-surface Ω on the surface M, and the dissipation needs to be propagated from the boundary onto the interior of the shell, then we surmise that the geometry of the shell is bound to play a critical role in the stabilization arguments. Indeed, we shall require geometric assumptions (H.1) and (H.2) below.

Preliminaries. Let $\zeta = [W, w]$ be the displacement field of the middle surface Ω of the shell and denote $\hat{\zeta} = [\hat{W}, \hat{w}]$. Introduce the bilinear form [Y.3, (1.25)]

$$(11.2.6) \quad B(\zeta, \hat{\zeta}) = a(\Upsilon(\zeta), \Upsilon(\hat{\zeta})) + \gamma a(D^2 w, D^2 \hat{w}), \quad \gamma = \frac{h^2}{12}.$$

See [B-B, p. 15] for (11.2.6). In (11.2.6), the 2-covariant tensor $\Upsilon(\,\cdot\,,\,\cdot\,)$ was defined by (11.1.6), while the 2-covariant tensor $D^2 w$ is the Hessian of w, which is defined in (3.1.1). Moreover, in (11.2.6), $a(\,\cdot\,,\,\cdot\,)$ is a bilinear form [Y.3, (1.26)]

$$(11.2.7) \quad a(T_1, T_1) = (1-\mu)\langle T_1, T_1\rangle_{T_x^2} + \mu(tr\, T_1)^2, \ x \in \Omega, \ T_1 \in T^2(\overline{\Omega}),$$

defined on 2-order tensors $T^2(\overline{\Omega})$ of $\overline{\Omega}$, see [Y.3–4], [L-T.29] for the inner product and the trace tr. Finally, with (11.2.6) we can associate the following symmetric bilinear form, defined directly on the middle surface Ω:

$$(11.2.8) \quad \begin{aligned} \mathcal{B}(\zeta, \hat{\zeta}) = \int_\Omega B(\zeta, \hat{\zeta}) dx; \qquad & \zeta(x) = W(x) + w(x) N(x); \\ & W(x) \in M_x, \quad x \in \Omega, \end{aligned}$$

M_x being the tangent space at $x \in M$. $N(x)$ is the unit normal field.

After these preliminaries, to state our stabilization result, we recall that the energy functional associated with model (11.1.1) is given by

$$(11.2.9) \qquad E(t) = E_k(t) + E_p(t),$$

where E_k is the kinetic energy

$$(11.2.10) \quad E_k(t) \equiv \int_\Omega \left\{ |W_t|^2_{T_x} + w_t^2 + \gamma |Dw_t|^2_{T_x} \right\} dx$$

$$(11.2.11) \qquad = \|W_t\|^2_{L^2(\Omega,\Lambda)} + \|w_t\|^2_{L^2(\Omega)} + \gamma \|Dw_t\|^2_{L^2(\Omega,\Lambda)}$$

(see [He.1 [Y.3–4], [L-T.29] for these spaces), and E_p is the potential energy, see (11.2.6)–(11.2.8):

$$(11.2.12) \quad E_p(t) \equiv \mathcal{B}(\zeta, \zeta) = \int_\Omega B(\zeta, \zeta) dx$$

$$(11.2.13) \qquad = \int_\Omega [a(\Upsilon(W,w), \Upsilon(W,w)) + \gamma\, a(D^2 w, D^2 w)] dx, \ \zeta = [W, w].$$

Next, in line with the statement above (11.2.6), we need to impose some geometric conditions on the shell.

Geometric assumptions. We shall assume the following hypotheses that were needed in [Y.4] to prove the observability estimate which we shall invoke in Section 11.3.

(H.1): ellipticity of the shell strain energy: there exists a constant $\lambda_0 \geq 1$ such that

$$\begin{cases} \lambda_0 \mathcal{B}(\zeta, \zeta) \geq \|DW\|^2_{L^2(\Omega, T^2)} + \gamma \|D^2 w\|^2_{L^2(\Omega, T^2)} \\ \text{for } \zeta = [W, w] \in H^1(\Omega, \Lambda) \times H^2(\Omega), \end{cases} \tag{11.2.14}$$

where the function spaces are defined in [He.1], [Y.3–4], [L-T.29] below.

In particular, a sufficient condition for (H.1) to hold true is that *both* Π and $D\Pi$ are small enough [B-O], where Π is the second fundamental form of M. A much weaker condition where (H.1) holds true is given in [Y.3, Theorem 3.2]: it basically says that the shell is "sufficiently shallow."

Main assumption (H.2). We assume that there exists a vector field $V \in \mathcal{X}(M)$ such that the covariant differential satisfies

$$DV(X, X) = b(x)|X|^2, \quad X \in M_x, \ x \in \overline{\Omega}, \tag{11.2.15}$$

where b is a function on Ω. Set

$$a(x) = \frac{1}{2}\langle DV, \mathcal{E}\rangle_{T^2_x}, \qquad x \in \overline{\Omega}, \tag{11.2.16}$$

where $\mathcal{E}$ is the volume element of M. Moreover, suppose that b and a satisfy the following inequality

$$2 \min_{x \in \overline{\Omega}} b(x) > \lambda_0 (1 + \mu) \max_{x \in \overline{\Omega}} |a(x)|. \tag{11.2.17}$$

Assumption (H.2) consists of (11.2.15) and (11.2.16). □

Illustrations where Assumption (H.2) holds true are given in [Y.3]. The include shells whose mid-surface lies on a surface of constant curvature or a surface of revolution.

THEOREM 11.2.2. *Assume (H.1) in (11.2.14) and (H.2) in (11.2.15), (11.2.16) above. In addition to the well-posedness assumptions on h_i, g_i in Theorem 11.1.1(a), assume further that:*

(H.3): there exist positive constants $0 < m < M$ and a sufficiently large constant $R > 0$, such that: for all $s \in \mathbb{R}$ with $|s| > R$, we have

$$m|s|^2 \leq g_i(s)s \leq M|s|^2; \ m|s|^2 \leq h_i(s)s \leq M|s|^2, \ i = 1, 2. \tag{11.2.18}$$

Next, assume that:

(H.4): either the coefficient ℓ in (11.1.1) is positive: $\ell > 0$ or else $\Gamma_0 \neq \emptyset$.

Let $[W, w]$ be a weak solution of the feedback problem (11.1.1a-g), as asserted by Theorem 11.2.1(a). Then: there exists a constant $T_0 > 0$ such that, with reference to the energy $E(t)$ in (11.2.9), the following estimate holds true:

$$E(t) \leq C(E(0))\, s\left(\frac{t}{T_0} - 1\right), \quad \forall\, t \geq T_0, \tag{11.2.19}$$

where $C(E(0))$ denotes a constant depending on the initial energy $E(0)$, and where $s(t)$ is a real-valued function converging to zero: $s(t) \to 0$ as $t \to \infty$, which is constructed as a solution of the following Cauchy problem

$$s_t(t) + q(s(t)) = 0, \quad s(0) = E(0) \tag{11.2.20}$$

involving a non-linear ordinary differential equation where the function $q(\,\cdot\,)$ is, in turn, constructed from the data of problem (11.1.1). More precisely, the non-linear monotone increasing function $q(\,\cdot\,)$ is determined entirely *from the behavior at the origin of the non-linear boundary functions g_i, h_i, according to the following algorithm [L-Ta.1].*

Step 1. *Due to the assumed monotonicity of the non-linear boundary functions h_i, g_i one can readily construct [L-Ta.1] functions $\tilde{g}_i$, $\tilde{h}_i$, concave and strictly increasing; vanishing at the origin: $\tilde{g}_i(0) = \tilde{h}_i(0) = 0$, such that the following inequalities are satisfied for $|s| \leq 1$:*

$$\tilde{g}_i(sg_i(s)) \geq |s|^2 + |g_i(s)|^2; \quad \tilde{h}_i(sh_i(s)) \geq |s|^2 + |h_i(s)|^2, \quad \forall\, |s| \leq 1. \tag{11.2.21}$$

We then define first the functions $r_0(\,\cdot\,)$ and its rescaled version $r(\,\cdot\,)$ by:

$$r_0(s) \equiv \sum_{i=1}^{2} \tilde{g}_i(s) + \tilde{h}_i(s); \qquad r(\,\cdot\,) = r_0\left(\frac{\cdot}{\text{meas } \Sigma_1}\right), \tag{11.2.22}$$

and next the function p,

$$p = (cI + r)^{-1}, \tag{11.2.23}$$

where c is a constant dependent on $\frac{1}{\text{meas } \Sigma_1}\left(\frac{1}{m} + M\right)$, where $\Sigma_1 = (0, T] \times \Gamma_1$.

Step 2. *Having constructed the function $p(\,\cdot\,)$ in (11.2.23) from the given boundary feedback functions h_i, g_i (data of the problem) via (11.2.21)-(11.2.23), we next introduce the function $q(\,\cdot\,)$ by [L-Ta.1]*

$$q = I - (I + p)^{-1}, \tag{11.2.24}$$

so that q is monotone increasing and $q(0) = 0$. It is such function q that defines the non-linear ordinary differential equation in the Cauchy problem (11.2.20), whose solution $s(\,\cdot\,)$ determines the decay rate of the energy $E(t)$ in (11.2.9) as $t \to \infty$. □

Remark 11.2.1. (i) Assume, in particular, that the nonlinear functions g_i, h_i are bounded from below by a linear function; that is, that (reinforcing (11.2.18) valid for $|s| > R$):

$$|h_i(s)| \geq c|s|, \qquad |g_i(s)| \geq c|s|, \ \forall\, s \in \mathbb{R}, \tag{11.2.25}$$

for some $c > 0$. Then, it can be shown that the decay rates predicted by Theorem 11.2.2 are exponential. That is, there exist positive constant C, ω—possibly depending on $E(0)$—such that

$$E(t) \leq Ce^{-\omega t}, \quad \forall\, t \geq T_0. \tag{11.2.26}$$

(ii) Assume, instead, that the functions h_i, g_i have polynomial growth at the origin; that is

$$\begin{gathered} h_i(s)s \geq a_i|s|^{p+1}; \qquad g_i(s)s \geq b_i|s|^{p+1} \\ \text{for } |s| \leq 1: \ a_i, b_i \text{ positive constants}, \ p > 1. \end{gathered} \tag{11.2.27}$$

Then, the decay rates predicted by Theorem 11.2.2 are algebraic:

$$E(t) \leq Ct^{\frac{2}{1-p}}, \qquad p > 1, \tag{11.2.28}$$

where $C = C(E(0))$ = a constant depending on $E(0)$.

11.3. Preliminary results.

Dissipativity equality. A starting point is, as usual, the dissipativity equality which states that the energy $E(t)$ in (11.2.9) of the entire system is nonincreasing. This is (0.6), as specialized to the present case. This fact alone does not prove, of course, that the energy is decaying, but it is a necessary preliminary step of the stability analysis.

LEMMA 11.3.1. *Let $[W, w]$ be a finite energy solution of system (11.1.1), as guaranteed by Theorem 11.2.1. Then, for any $s \leq t$, the following identity holds true for the energy $E(t)$ defined by (11.2.9),*

$$\begin{aligned} E(t) + 2 \int_s^t \int_{\Gamma_1} \Big[& g_1(\langle W_t, n\rangle)\langle W_t, n\rangle + g_2(\langle W_t, \tau\rangle)\langle W_t, \tau\rangle \\ & + h_1\left(\frac{\partial w_t}{\partial n}\right)\frac{\partial w_t}{\partial n} + h_2\left(\frac{\partial w_t}{\partial \tau}\right)\frac{\partial w_t}{\partial \tau}\Big] d\Gamma_1\, dt = E(s). \end{aligned} \tag{11.3.1}$$

Proof. The proof is standard and follows by a classical energy-type argument. We multiply the two equations (11.1.1a-b) by W_t, w_t, respectively, integrate over $\Omega \times (s, t)$ and apply the divergence theorem (as in [Y.3–4]) first to smooth solutions, and then we extend by density to all weak solutions. The main tool of our computations is the following Green's formula [Y.4, (3.1.35)] in the notation of [Y.4]:

$$
\begin{aligned}
(\mathcal{A}\eta, \hat{\eta})_{L^2(\Omega,\Lambda)\times L^2(\Omega)} &= \int_\Omega (\mathcal{A}\eta, \hat{\eta}) dx \\
&= \int_\Omega B(\eta, \hat{\eta}) dx - \int_\Gamma \Big\{ B_1(W, w)\langle \hat{W}, n\rangle \\
&\quad + B_2(W, w)\langle \hat{W}, \tau\rangle + \gamma\Big[\big(\Delta w + (1-\mu)B_0 w\big) \frac{\partial \hat{w}}{\partial n} \\
&\quad - \left(\frac{\partial \Delta w}{\partial n} + (1-\mu)B_4 w \right) \hat{w}\Big)\Big] \Big\} d\Gamma,
\end{aligned} \tag{11.3.2}
$$

where $\eta = [W, w]$, $\hat{\eta} = [\hat{W}, \hat{w}]$,

$$
\mathcal{A}\eta \equiv \begin{bmatrix} -\Delta_\mu W - (1-\mu)kW - \mathcal{F}(w) \\ \gamma\big[\Delta^2 w - (1-\mu)\delta(kdw)\big] + \big(H^2 - 2(1-\mu)k\big)w + \mathcal{G}(W) \end{bmatrix}, \tag{11.3.3}
$$

and where Δ_μ and $B(\,\cdot\,,\,\cdot\,)$ are defined below (11.1.1g) and by (11.2.6), while B_1, B_2, B_3, B_4 are defined by (11.1.2)–(11.1.5). Finally, the coupling $\mathcal{F}(w)$ and $\mathcal{G}(W)$ are first-order operators, defined in [Y.3, (1.33), p. 1733] whose precise structure is not essential in the present paper. However, $\mathcal{F}(w) = 0$, $\mathcal{G}(W) = 0$ in the *flat* case. □

In our next step we apply multipliers to (11.1.1). These are the same as those used in the flat non-linear case in [Las.5], except that now they are in differential geometric form. The corresponding calculations in the curved shell case are the Riemannian metric counterpart of those in the *flat* case, and thus follow the same philosophy as those in [Las.5] where they were used for the full non-linear von Karman system. More precisely, we apply:

(i) the multipliers $[D_V W, V(w)]$, in the usual notation, in order to handle the potential energy $E_p(t)$ in (1.2.12), where V is the vector field on M assumed in (H.2);

(ii) the multipliers $[W, w]$, in order to obtain an estimate for the difference between kinetic energy $E_k(t)$ and potential energy $E_p(t)$.

The actual computations are performed in [Y.3, particularly formulas (2.122), (2.123)] and lead to the following inequality, which is the counterpart, in the curved case, of a special case of the inequality in the *flat* case given in [Las.5, Lemma 3.2], when specialized to the linear model (11.1.1a-b).

PROPOSITION 11.3.2. *Assume (H.1), (H.2). With reference to strong solutions of the original problem (11.1.1a-g) as guaranteed by Theorem 11.2.1, the following inequality holds true for the energy $E(t)$ defined in (11.2.9): for $T > 0$ given, there exist constants $C > 0$, $C_T > 0$, such that*

$$
\begin{aligned}
\int_0^T E(t)dt &\le C[E(0) + E(T)] + C_T(BT^{\text{good}}) \\
&\quad + C_T(BT^{\text{bad}}) + LOT(W, w),
\end{aligned} \tag{11.3.4}
$$

where:

(i) $LOT(W, w)$ are lower terms with respect to the energy $E(t)$ in (11.2.9) where $E(t)$ is topologically equivalent to

$$H^1(\Omega, \Lambda) \times L_2(\Omega, \Lambda) \times H^2(\Omega) \times H^1(\Omega) \tag{11.3.5}$$

for $[W, W_t, w, w_t]$, see [He.1], [Y.3–4], [L-T.29] for these spaces;

(ii) if $BT = BT^{\text{good}} + BT^{\text{bad}}$ are the boundary terms, divided into 'good' and 'bad' terms, these are defined by

$$BT^{\text{good}} = \int_0^T \left[\|W_t\|^2_{L^2(\Gamma_1,\Lambda)} + \|Dw_t\|^2_{L^2(\Gamma_1,\Lambda)} \right] dt \tag{11.3.6a}$$

$$BT^{\text{bad}} = \int_0^T \|DW\|^2_{L^2(\Gamma_1,T^2)} dt + \int_{\Sigma_1} B(\zeta,\zeta) d\Sigma_1, \tag{11.3.6b}$$

where we recall $\zeta = [W, w]$ and $B(\zeta,\zeta) \equiv a(\Upsilon(\eta), \Upsilon(\eta)) + a(D^2 w, D^2 w)$ as in (11.2.6).

Remark 11.3.1. While all boundary terms B^{good} involving time derivatives in (11.3.6a) will be determined by the dissipation, the boundary integral in the term BT^{bad} in (11.3.6b) contains traces of the first order for W and of the second order for w, see (11.2.6), (11.1.6). These traces are *not* determined either by the energy or by the boundary conditions. In fact, the main challenge to, and contribution by, this paper is to provide an estimate for the traces in the BT-term. This will be done by extending microlocal estimates ([L-T.12] for scalar waves, hence [Hn.1] for elastic waves in Lame form, to the present W-component which is now in Lame form, and [L-T.13] for the plate w-component and applying arguments as in [Las.5] for the case of full von Karman model. □

11.4. First step of the proof: Local reduction to a Euclidean (flat) coordinate system. Throughout this section Assumptions (H.1) and Assumption (H.2) are in force. The following estimate is critical for the proof of Theorem 11.2.2.

THEOREM 11.4.1. *Let $0 < \alpha < T/2$. Then, the following trace estimate holds true for any regular solution of problem (11.1.1), as guaranteed by Theorem 11.2.1: there exists a constant $C_{\alpha T} > 0$ such that*

$$\begin{aligned} BT^{\text{bad}}[\alpha, T-\alpha] &\equiv \int_\alpha^{T-\alpha} \int_{\Gamma_1} \left[B(\zeta,\zeta) + |DW|^2_{T^2_x} \right] dx\, dt \\ &\le C_{\alpha T} \int_{\Sigma_1} \left[|Dw_t|^2_{T_x} + |W_t|^2_{T_x} + |g_1(\langle W_t, n\rangle)|^2 \right. \\ &\quad + |g_2(\langle W_t, \tau\rangle)|^2 + \left| h_1\left(\frac{\partial w_t}{\partial n}\right)\right|^2 \\ &\quad \left. + \left| h_2\left(\frac{\partial w_t}{\partial \tau}\right)\right|^2 \right] d\Sigma_1 + LOT(W, w). \end{aligned} \tag{11.4.1}$$

where the lower order terms LOT are below energy level and satisfy

$$LOT(W,w) \le C_\epsilon \sup_{t\in[0,T]} \Big[|W(t)|_{1-\epsilon,\Omega} + |W_t(t)|_{-\epsilon,\Omega} + |w(t)|_{2-\epsilon,\Omega} + |w_t(t)|_{1-\epsilon,\Omega} \Big]. \tag{11.4.2}$$

for any $\epsilon > 0$, see notation adopted in [ ] in 1-forms W, W_t or functions w, w_t.

Remark 11.4.1. The traces $B(\zeta,\zeta)$, DW in (11.4.1) are *not* bounded by the energy. □

Proof. By recalling the definition of $B(\zeta,\zeta)$ in (11.2.6) we see that it suffices to prove the following estimate for regular solutions of problem (11.1.1): there exists a constant $C_{\alpha T} > 0$, such that

$$\begin{aligned} \int_\alpha^{T-\alpha} \int_{\Gamma_1} \Big[|D^2 w|_{T^2_x} + |DW|^2_{T^2_x} \Big]\, dx\, dt & \\ \le C_{\alpha T} \int_{\Sigma_1} \Big[|Dw_t|^2_{T_x} + |W_t|^2_{T_x} + |g_1(\langle W_t, n\rangle)|^2 & \\ + |g_2(\langle W_t, \tau\rangle)|^2 + \left| h_1\left(\frac{\partial w_t}{\partial n}\right)\right|^2 & \\ + \left| h_2\left(\frac{\partial w_t}{\partial \tau}\right)\right|^2 \Big] d\Sigma_1 + LOT(W,w). & \end{aligned} \tag{11.4.3}$$

Orientation. The proof of estimate (11.4.3) requires the insertion of microlocal estimates. To carry this out, we shall apply a basic strategy similar to that already employed in [Las.5] in the *flat*, nonlinear case of the full von Karman system. Indeed, the proof of estimate (11.4.3) will comprise three main steps. In Step 1, one introduces a coordinate cover of a boundary layer of $\bar{\Omega}$ and a subordinate partition of unity, and reduce our task to prove estimate (11.4.3) for just one coordinate system. Next, in Step 2, see Subsection 11.5, one provides a sharp trace estimate for the linear model of dynamic elasticity, to be used for the in-plane W-components of the displacement vector ζ. Finally, in Step 3, see Subsection 11.6, one provides a sharp trace estimate for the linear Kirchoff plate model, to be used for the normal w-component of the displacement ζ.

Steps 2 and 3 are critical for the proof of the stabilizability estimates *without* assuming geometric conditions on the controlled portion Γ_1 of the boundary, as done in wave-plate literature [Lag.2–3] and *without* considering additional tangential components of the horizontal displacement W in the structure of the stabilizing feedback as done in [P-T.1] in the *flat* case. □

11.5. Trace regularity for elastic waves (W-component). Orientation. This subsection and the next provide sharp trace regularity results which are critical for the proof of stability estimates *without* assuming geometric conditions on Γ_1 and *without* considering artificial tangential components of the in-plane displacement W in the structure of the stabilizing feedback (as was done in [P-T.1] in the study of the von Karman problem) which are *not* in L_2. These estimates are based on the corresponding trace estimates valid for: (i) linear model of dynamic elasticity and (ii) linear Kirchhoff model. They are obtained by methods of microlocal analysis.

As to (i), we need to extend to the present non-Lame elastic W-component the analysis begun in [L-T.12] for second-order hyperbolic equations, which was then the basis for the analysis in [Hn.1] of Lame-type elastic systems. We cannot merely quote [Hn.1] as our W-system is not of Lame-type. As to (ii) instead, we shall invoke [L-T.13] for the sharp trace regularity of second-order derivatives for plates. The sharp trace regularity of first-order traces for W—to be given in Proposition 11.5.1 below—are critical for solving the stabilization problem in the first place. Instead, the sharp trace regularity of second-order traces for w—to be given in Proposition 11.6.1—merely avoid unnecessary and restrictive geometrical conditions on the controlled portion of the boundary Γ_1. The main idea is to obtain the estimates for the tangential derivatives on the boundary in terms of the velocity traces and lower order terms: see Proposition 11.5.1 for W and Proposition 11.6.1 for w. □

The constant C is a generic constant, different in various occurrences.

PROPOSITION 11.5.1. *Let W, w be a finite energy solution corresponding to the system (11.1.1) as guaranteed by Theorem 11.2.1(a). Then, for any $0 < \epsilon < 1/4$ and $0 < \alpha < T/2$, there exist constants $C > 0$ such that the following trace regularity takes place:*

$$\begin{aligned} \int_\alpha^{T-\alpha} \int_{\Gamma_1} |DW|^2_{T^2_x} d\Sigma_{1\alpha} \le C_{\alpha\epsilon T} \int_{\Sigma_1} & |W_t|^2 + |g_1(\langle W_t, n\rangle)|^2 \\ & + |g_2(\langle W_t, \tau\rangle)|^2 d\Sigma_1 \\ & + C_{\alpha T} \int_0^T [|w|^2_{2-\epsilon,\Omega} + |W|^2_{1-\epsilon,\Omega}] dt. \end{aligned} \tag{11.5.1}$$

Remark 11.5.1. Proposition 11.5.1 is a counterpart of [Las.5, Lemma 2.2], proved for the "flat" but nonlinear case. Notice that the regularity of the trace of DW, claimed by Proposition 11.5.1 (see also Proposition 11.6.1 for the Kirchhoff plate below) does not follow from the standard interior regularity of finite energy solutions via trace theory. These are independent regularity results which rely heavily on microlocal arguments applied to both: the dynamic system of elasticity and the dynamic Kirchhoff plate.

Remark 11.5.2: Orientation. Proposition 11.5.1 is established by introducing a coordinate cover of the boundary layer of $\overline{\Omega}$ and a subordinate partition of unity, and thus by proving a corresponding estimate in the

Euclidean setting just for one coordinate system. This latter estimate for the trace of the elastic space in the Euclidean case cannot be just quoted from available literature, since the resulting system is *not* Lame-type, as in [Hn.1]. Indeed, two additional technical lemmas need to be established for this more general elastic wave problem [L-T.29], which push further to the present level of generality arguments in the literature [L-T.12], [L-T.13]. □

11.6. Trace regularity for normal component w. Our next result deals with the improved trace regularity for the normal displacement w.

PROPOSITION 11.6.1. *Let W, w be a finite energy solution to problem (11.1.1) as guaranteed by Theorem 11.2.1. Then, there is a constant $C_{\alpha T\epsilon} > 0$ such that*

$$\int_{\alpha}^{T-\alpha}\int_{\Gamma_1} |D^2 w|^2_{T^2_x} d\Sigma_{1\alpha}$$

$$\leq C\left\{\int_{\Sigma_1}\left[|Dw_t|^2 + \left|h_1\left(\frac{\partial w_t}{\partial n}\right)\right|^2 + \left|h_2\left(\frac{\partial w_t}{\partial \tau}\right)\right|^2\right] d\Sigma_1 + \int_0^T \left[|w(t)|^2_{2-\epsilon,\Omega} + |w_t(t)|^2_{1-\epsilon,\Omega} + |W(t)|^2_{1-\epsilon,\Omega}\right] dt\right\}. \quad (11.6.1)$$

Remark 11.6.1. Again Proposition 11.6.1 is established (as Proposition 11.5.1) by introducing a coordinate cover of the boundary layer of $\overline{\Omega}$ and a subordinate partition of unity, and thus by proving a corresponding estimate in the Euclidean setting just for one coordinate system. This latter estimate for the corresponding Kirchhoff plate problem is then lifted from [L-T.13, Theorem 2.1, p. 279], except that we are giving it a sharper form. This is explained in [L-T.29, Remark 3.3.1]. □

11.7. Stabilizability estimate and completion of the proof of Theorem 11.2.2. The remaining part of the proof of Theorem 11.2.2 is contained (as a strict subset) in the arguments presented in [L-Ta.1]. For the reader's convenience, we shall outline the remaining steps.

Step 1. We proceed as in [L-T.12, Section 7.3], [Las-Ta.1]. We return to estimate (11.3.4) of Proposition 11.3.2, this time over the interval $[\alpha, T-\alpha]$ rather than over $[0, T]$. On the left-hand side, we obtain $(T-2\alpha)E(T) \leq \int_{\alpha}^{T-\alpha} E(t)dt$ by the dissipativity property in (11.3.1). On the right-hand side, we use: (i) estimate (11.4.1) of Theorem 11.4.1 for the boundary terms $BT^{\text{bad}}[\alpha, T-\alpha]$ over $[\alpha, T-\alpha]$; (ii) identity (11.3.1) again with $s = 0$, $t = T$, to express $E(0)$ in terms of $E(T)$ plus boundary terms, where we use: $2g_i(s)s \leq |g_i(s)|^2 + s^2$; $2h_i(s)s \leq |h_i(s)|^2 + s^2$ for these boundary terms, as well as the relevant definitions to estimate them by the norms of Dw_t and W_t. This way, using again (11.3.4), we readily obtain the counterpart of [L-Ta.1, Lemma 3.3].

PROPOSITION 11.7.1. *With reference to the energy in (11.2.9), the strong solutions of the original dynamics (11.1.1), guaranteed by Theorem*

11.2.1, satisfy the following estimate for all $T > 0$ sufficiently large, there is a constant $C_T > 0$ such that

$$\begin{aligned} E(T) + \int_0^T E(t)dt \le C_T \int_{\Sigma_1} \Big\{ &\|W_t\|^2_{L^2(\Gamma,\Lambda)} + \|Dw_t\|^2_{L^2(\Gamma,\Lambda)} \\ &+ |g_1(\langle W_t, n\rangle)|^2 |g_2(\langle W_t, \tau\rangle)|^2 + \left| h_1\left(\frac{\partial w_t}{\partial n}\right)\right|^2 \\ &+ \left| h_2\left(\frac{\partial w_t}{\partial \tau}\right)\right|^2 \Big\} d\Sigma_1 + LOT(W, w), \end{aligned} \tag{11.7.1}$$

where, as before, $LOT(W, w)$ are lower terms (with respect to the energy) as defined in (11.4.2).

Step 2. Lower-order terms are absorbed, as usual, by a compactness/uniqueness argument. This requires an appropriate unique continuation result (from the boundary). Since the coefficients are time-independent, it is far convenient to require a uniqueness result for the corresponding static problem. The latter is established in [Y.4, Proposition 2.3], by reducing the Cauchy problem to a system of three equations of the fourth-order with the same principal part Δ^2, where Δ is the Laplacian on the manifold M. For this latter problem, the result in [S.1] is then invoked to obtain uniqueness. (One could also use results of [I.]). Thus, a by-now standard compactness/uniqueness argument (see [L-Ta.1, Lemma 4.1] leads to the absorption of lower-order terms. We thus obtain

LEMMA 11.7.2. *With reference to the $LOT(W, w)$ in estimate (11.7.1) for the strong solutions of problem (11.1.1), there exists $T > 0$ large enough, so that:*

$$\begin{aligned} LOT(W, w) &\equiv \int_0^T \left[|W(t)|^2_{1-\epsilon,\Omega} + |w(t)|^2_{2-\epsilon,\Omega} \right] dt \\ &\le C_T \Bigg[\int_{\Sigma_1} \Big[|W_t|^2 + |Dw_t|^2 + |g_1(\langle W_t, n\rangle)|^2 \\ &\quad + |g_2(\langle W_t, \tau\rangle)|^2 + \left| h_1\left(\frac{\partial w_t}{\partial n}\right)\right|^2 + \left| h_2\left(\frac{\partial w_t}{\partial \tau}\right)\right|^2 \Big] d\Sigma_1. \end{aligned} \tag{11.7.2}$$

□

Step 3. By combining the results of Proposition 11.7.1 and Lemma 11.7.2, we obtain the final stabilization estimate.

PROPOSITION 11.7.3. *Let $[W, w]$ be a regular solution to the original system (11.1.1), as guaranteed by Theorem 11.2.1. Then, there exists a constant $T_0 > 0$ such that for any $T > T_0$, there is a constant $C_T > 0$ such that the following estimate holds true:*

$$\begin{aligned} E(0) + E(T) + \int_0^T E(t)dt \le C_T \int_{\Sigma_1} \Big[&|W_t|^2 + |Dw_t|^2 \\ &+ |g_1(\langle W_t, n\rangle)|^2 + |g_2(\langle W_t, \tau\rangle)|^2 \end{aligned} \tag{11.7.3}$$

$$+ \left| h_1 \left(\frac{\partial w_t}{\partial n} \right) \right|^2 + \left| h_2 \left(\frac{\partial w_t}{\partial \tau} \right) \right|^2 \Bigg] d\Sigma_1. \qquad \square$$

Step 4. Our next step is to express the boundary terms in terms of the feedbacks in (11.1.1d-f). To accomplish this, we shall use the growth conditions (11.2.18) imposed on the non-linear dissipation terms g_i, h_i at infinity, together with the consequent properties (11.2.21) of the "comparison" functions $\tilde{g}_i$, $\tilde{h}_i$ which contain information on the growth at the origin. It is *only at this point* that we use the *growth conditions* (11.2.18) imposed on the non-linear functions g_i, h_i given in assumption (H.3) = (11.2.18), together with the construction of the functions $\tilde{g}_i$, $\tilde{h}_i$ which capture the behavior of the non-linearity at the origin.

By splitting the integration on the boundary between "low" and "high" frequencies, and using Jensen's inequality in the same manner as it is used on [L-Ta.1, p. 1400-1401], we arrive at the following conclusion.

PROPOSITION 11.7.4. *Consider regular solutions of problem (11.1.1), as guaranteed by Theorem 11.2.1. Let $E(t)$ be defined in (11.2.9). Then: there exists $T > 0$ such that*

$$p(E(T)) + E(T) \leq E(0), \tag{11.7.4}$$

where the monotone function p is defined constructively in (11.2.23).

Step 5. The final conclusion of Theorem 11.2.2 now follows from (11.7.4) and [L-Ta.1, Lemma 3], which is based on a comparison theorem. $\square$

REFERENCES

[B-L-R.1] C. BARDOS, G. LEBEAU, AND J. RAUCH, Sharp sufficient conditions for the observation, control, and stabilization of waves from the boundary, *SIAM J. Control Optim.* **30** (1992), 1024–1065.

[B-L-R-H-Z.1] C. BARDOS, G. LEBEAU, J. RAUCH, L. HALPERIN, AND E. ZUAZUA, Stabilisation de l'equation des ondes au moyen d'un feedback portant sur la condition aux limites de Dirichlet, *Asymptotic Analysis* **4** (1991), 285–291.

[B-B.1] M. BERNADOU AND J.M. BOISSERIE, *The Finite Element Method in Thin Shell Theory*, Progress in Scientific Computing 1, Birkhäuser, Boston, 1982.

[B-O.1] M. BERNADOU AND J.T. ODEN, An existence theorem for a class of non-linear shallow shell problems, *J. Math. Pures et Appl.* **60** (1981), 258–298.

[B-G-L.1] S. BETELU, R. GULLIVER, AND W. LITTMAN, Curvature flow methods in boundary control of partial differential equations, *Appl. Math. & Optimiz.* **46** (Sept./Dec. 2002), 167–178. Special issue dedicated to J.L. Lions.

[B-C] R. BISHOP AND R. CRITTENDEN, *Geometry of Manifolds*, Academic Press, New York 1964.

[B-Z] YU. D. BURAGO AND V.A. ZALGALLER, *Geometric Inequalities*, Springer-Verlag, Berlin 1988.

[B.1] N. BURQ, Controlabilite exacte des ondes dans des ouverts peu reguliers, *Asymptotic Analysis* **14** (1997), 157–191.

[Ch.1] I. CHAVEL, *Riemannian Geometry, a Modern Introduction*, Cambridge University Press, Cambridge 1993.

[C-G] J. CHEEGER AND D. GROMOLL, On the structure of complete manifolds of nonnegative curvature. *Annals of Math.* **96** (1972), 413–443.

[C-Z.1] K.-S. CHOU AND X.-P. ZHU, *The Curve Shortening Problem*, Chapman and Hall, 2001.

[C-G.1] J. CHOE AND R. GULLIVER, Isoperimetric inequalities on minimal submanifolds of space forms, *Manuscripta Math.* **77** (1992), 169–189.

[C-H.1] R. COURANT AND D. HILBERT, *Methods of Mathematical Physics*, Vol. 2, Wiley Interscience, New York, 1962.

[DoC.1] M.P. DO CARMO, *Riemannian Geometry*, Birkhäuser, Boston 1992.

[Du.1] G.F.D. DUFF, *Partial Differential Equations*, University of Toronto Press, 1956.

[E.1] M. ELLER, Uniqueness of continuous theorems, in *Direct and Inverse Problems of Mathematical Physics*, Kluwer Academic, Dordrecht/Norwell, MA, 1999.

[E.2] M. ELLER, Carleman estimates with a second large parameter, *JMAA* **249** (2000), 491–514.

[E-I-N-T] M. ELLER, V. ISAKOV, G. NAKAMURA, AND D. TATARU, Uniqueness and stability in the Cauchy problem for Maxwell's and the elastic system, in *Nonlinear Partial Differential Equations*, Vol. 16, Chapman and Hall, London/New York, 2000.

[E-L-T.1] M. ELLER, I. LASIECKA, AND R. TRIGGIANI, Simultaneous exact/approximate boundary controllability of thermo-elastic plates with variable transmission coefficients, Marcel Dekker, *Lecture Notes Pure Appl. Math.*, J. Cagnol, M. Polis, and J.P. Zolesio (2001), 109–230.

[E-L-T.2] M. ELLER, I. LASIECKA, AND R. TRIGGIANI, Simultaneous exact/approximate boundary controllability of thermoelastic plates with variable thermal coefficient and moment controls, *J. Math. Anal. Appl.* **251** (2000), 452–478.

[E-L-T.3] M. ELLER, I. LASIECKA, AND R. TRIGGIANI, Unique continuation for over-determined Kirchoff plate equations and related thermoelastic systems, *J. Inv. Ill-Posed Prob.* **9**(2) (2001), 103–148.

[E-L-T.4] M. ELLER, I. LASIECKA, AND R. TRIGGIANI, Simultaneous exact/approximate controllability of thermoelastic plates with variable thermal coefficient and clamped controls, *Discrete Contin. Dynam. Systems* **7**(2) (2001), 283–301.

[F-I.1] A.V. FURSIKOV AND O.YU. IMANUVILOV, *Controllability of Evolution Equations*, Lecture Notes, Vol. 34, pp. 151–742, Research Institute of Mathematics, Seoul National University, Seoul, Korea.

[Ga.1] M. GALBRAITH, Geometric optics and convex functions in the boundary control of the wave equation, *Proceedings of 3rd ISAAC Conference*, 2001, World Scientific, to appear.

[Gr.1] M.A. GRAYSON, *Shortening embedded curves*, *Annals of Math.* **129** (1989), 71–111.

[G-W.1] R.E. GREENE AND H. WU, C^∞ convex functions and manifolds of positive curvature, *Acta Math.* **137** (1976), 209–245.

[Gri.1] P. GRISVARD, *Elliptic Problems in Nonsmooth Domains*, Monograph & Studies in Mathematics, Pitman, 1995.

[G.1] R. GULLIVER, On the variety of manifolds without conjugate points, *Trans. Amer. Math. Soc.* **210** (1975), 185–201.

[G-L.1] R. GULLIVER AND W. LITTMAN, Chord uniqueness and controllability: The view from the boundary, *Contemporary Mathematics*, AMS **268** (2000), 145–176.

[G-L.2] R. Gulliver and W. Littman, The use of geometric tools in the boundary control of partial differential equations, *Proceedings of 3rd International ISAAC Contress*, Berlin, August 2001.

[He.1] E. Hebey, *Sobolev Spaces on Riemannian Manifolds*, Lecture Notes in Mathematics, Springer, 1996.

[Ho.1] L.F. Ho, Observabilité frontière de l'equation des ondes, *C. R. Acad. Sci. Paris, Ser. I Math.* **202** (1986), 443–446.

[Hor.1] L. Hörmander, *Linear Partial Differential Operators*, Springer Verlag, 1963.

[Hor.2] L. Hörmander, *The Analysis of Linear Partial Differential Operators*, Vol. III, Springer Verlag, Berlin/New York, 1985.

[Hor.3] L. Hörmander, On the characteristic Cauchy problem, *Ann. Math.* **88** (1968),

[Hor.4] L. Hörmander, On the uniqueness of the Cauchy problem, II, *Math. Scand.* **7** (1959), 177–190.

[Hor.5] L. Hörmander, On the uniqueness of the Cauchy problem under partial analyticity assumptions, in *Geometrical Optics and Related Topics*, F. Colombini and N. Lerner (eds.), Birkhäuser, 1997.

[Hn.1] M.A. Horn, Implications of sharp trace regularity results on boundary stabilization of the system of linear elasticity, *J. Math. Anal. Appl.* **223** (1998), 126–150.

[Hn.2] M.A. Horn, Sharp trace regularity of the traces to solutions of dynamic elasticity, *J. Math. Systems, Estimation and Control* **8** (1998), 217–229.

[Koi] W.T. Koiter, On the foundations of the linear theory of thin elastic walls, *Proc. Kon. Ned. Akad. Wetensch.* B**73** (1970), 169–195.

[I.1] V. Isakov, A nonhyperbolic Cauchy problem for $\square_b\square_c$ and its applications to elasticity theory, *Comm. PDE's* XXXIX (1986), 747–767.

[I.2] V. Isakov, *Inverse Problems for Partial Differential Equations*, Springer Verlag, 1998.

[I.3] V. Isakov, Uniqueness of the continuation across a time-like hyperplane and related inverse problems for hyperbolic equations, *Comm. Partial Diff. Eqn.* **14** (1989), 465–478.

[I.4] V. Isakov, *Inverse Source Problems*, AMS, Providence, RI, 1990.

[I.5] V. Isakov, On the uniqueness of the continuation for a thermoelastic system, preprint, 1998.

[I.6] V. Isakov, On uniqueness on a lateral Cauchy problem with multiple characteristics, *J. Diff. Eqns.* **134** (1997), 134–147.

[I-Y.1] V. Isakov and M. Yamamoto, Carleman estimate with Neumann boundary condition and its application to the observability inequality and inverse hyperbolic problems, *Contemporary Mathematics*, AMS 268 (2000).

[K.1] V. Komornik, Exact controllability and stabilization: the multiplier method, *Research in Applied Mathematics*, 1994, Masson/John Wiley.

[K-K.1] M. Kazami and M.V. Klibanov, Stability estimates for ill-posed Cauchy problems involving hyperbolic equations and inequalities, *Applicable Analysis* **50** (1993), 93–102.

[Lag.1] J. Lagnese, Decay of solutions of the wave equation in bounded domains, *JDE* **50** (1983), 163–182.

[Lag.2] J. Lagnese, Uniform asymptotic energy estimates for solutions of the equations of dynamic plane elasticity with nonlinear dissipation at the boundary, *Nonlinear Anal.* **16** (1991), 35–54.

[Lag.3] J. Lagnese, *Boundary Stabilization of Thin Plates*, SIAM, Philadelphia, 1989.

[L-L.1] J. LAGNESE AND J.L. LIONS, *Modelling, Analysis and Control of Thin Plates*, Masson, 1988.

[Las.1] I. LASIECKA, Controllability of a viscoelastic Kirchoff plate, *Int. Series Num. Math.* **91** (1989), 237–247, Birkhäuser Verlag, Basel.

[Las.2] I. LASIECKA, Mathematical control theory in structural acoustic problems, *Mathematical Models and Methods in Applied Sciences* **8** (1998), 1119–1153.

[Las.3] I. LASIECKA, Boundary stabilization of a 3-dimensional structural acoustic model, *J. Math. Pure Appl.* **78** (1999), 203–232.

[Las.4] I. LASIECKA, Intermediate solutions to full von Kármán systems of dynamic nonlinear elasticity, *Appl. Anal.* **68** (1998).

[Las.5] I. LASIECKA, Uniform stabilizability of a full von Kármán system with nonlinear boundary feedback, *SIAM J. Control* **36**(4) (1998), 1376–1422.

[Las.6] I. LASIECKA, Uniform decay rates for full von Kármán system of dynamic elasticity with free boundary conditions and partial boundary dissipation, *Comm. PDE* **24** (1999), 1801–1849.

[Las.7] I. LASIECKA, *Mathematical Control Theory of Coupled PDEs*, NSF-CMBS Lecture Notes (2001), SIAM.

[L-M.1] I. LASIECKA AND R. MARCHAND, Uniform decay for solutions to nonlinear shells with nonlinear dissipation, *Nonlinear Analysis* **30** (1997), 5409–5418.

[Las-Hor.1] I. LASIECKA AND M.A. HORN, Asymptotic behavior with respect to thickness of boundary stabilizing feedback for the Kirchoff plate, *J. Diff. Eqns.* **114**(2) (1994), 396–433.

[L-O.1] I. LASIECKA AND J. ONG, Global solvability and uniform decays of solutions to quasilinear equation with nonlinear boundary dissipation, *Comm. PDE* **24** (11 & 12) (1999), 2069–2107.

[L-Ta.1] I. LASIECKA AND D. TATARU, Uniform boundary stabilization of semilinear wave equations with nonlinear boundary damping, *Diff. Int. Eqns.* **6** (1993), 507–533.

[L-T.1] I. LASIECKA AND R. TRIGGIANI, Uniform exponential energy decay of wave equations in a bounded region with $L_2(0,\infty; L_2(\Gamma))$-feedback control in the Dirichlet boundary conditions, *J. Diff. Eqns.* **66** (1987), 340–390.

[L-T.2] I. LASIECKA AND R. TRIGGIANI, Exact controllability of the Euler-Bernoulli equation with $L_2(\Sigma)$-control only in the Dirichlet B.C., *Atti Acad. Naz. Lincei Rend. Cl. Sc. Fis. Math. Nat.* Vol. LXXXII, No. 1 (1988), Rome.

[L-T.3] I. LASIECKA AND R. TRIGGIANI, Exact controllability of the wave equation with Neumann boundary control, *Appl. Math. Optimiz.* **19** (1989), 243–290.

[L-T.4] I. LASIECKA AND R. TRIGGIANI, Exact controllability of the Euler-Bernoulli equations with controls in the Dirichlet and Neumann B.C.: A nonconservative case, *SIAM J. Control Optim.* **27** (1989), 330–373.

[L-T.5] I. LASIECKA AND R. TRIGGIANI, Exact controllability of the Euler-Bernoulli equation with boundary controls for displacement and moment, *JMAA* **146** (1990), 1–33.

[L-T.6] I. LASIECKA AND R. TRIGGIANI, Sharp regularity for mixed second-order hyperbolic equations of Neumann type, Part I: The L_2-boundary case, *Annal. di Matemat. Pura e Appl. (IV)* **157** (1990), 285–367.

[L-T.7] I. LASIECKA AND R. TRIGGIANI, Exact controllability of the semilinear abstract system with application to waves and plates, *Appl. Math. Optimiz.* **23** (1991), 109–154.

[L-T.8] I. LASIECKA AND R. TRIGGIANI, Optimal regularity, exact controllability and uniform stabilization of the Schrödinger equation, *Diff. Int. Eqns.* **5** (1991), 521–535.

[L-T.9] I. LASIECKA AND R. TRIGGIANI, Exact controllability and uniform stabilization of Kirchoff plates with boundary controls only in $\Delta w|_{\Sigma}$, *J. Diff. Eqns.* **93** (1991), 62–101.

[L-T.10] I. LASIECKA AND R. TRIGGIANI, Regularity theory of hyperbolic equations with non-homogeneous Neumann boundary conditions, Part II: General boundary data, *J. Diff. Eqns.* **93** (1991), 112–164.

[L-T.11] I. LASIECKA AND R. TRIGGIANI, *Differential and Algebraic Riccati Equations with Applications to Boundary/Point Control Problems: Continuous Theory and Approximation Theory*, Springer Verlag LNICS, Vol. 164 (1991).

[L-T.12] I. LASIECKA AND R. TRIGGIANI, Uniform stabilization of the wave equation with Dirichlet or Neumann feedback control without geometric conditions, *Appl. Math. Optim.* **25** (1992), 189–224.

[L-T.13] I. LASIECKA AND R. TRIGGIANI, Sharp trace estimates of solutions to Kirchhoff and Euler-Bernoulli equations, *Appl. Math. Optimiz.* **28** (1993), 277–306.

[L-T.14] I. LASIECKA AND R. TRIGGIANI, *A-Priori Observability Inequalities*, Chapter 1, Lecture Notes, University of Virginia, May 1995.

[L-T.15] I. LASIECKA AND R. TRIGGIANI, Abstract models and semigroup well-posedness of spherical shells with boundary dissipation, invited paper, *Dynam. Sys. Appl.* **4** (1995), 453–471, World Scientific Series in Applicable Analysis.

[L-T.16] I. LASIECKA AND R. TRIGGIANI, Carleman estimates and uniqueness for the system of strongly coupled PDEs of spherical shells, special volume of *Zeitschrift für Angewandte Mathematik and Mechanik, ZAMM*, Akademie Verlag, Berlin, ICIAM 1995, 76, supp. **4** (1996), 277–280.

[L-T.17] I. LASIECKA AND R. TRIGGIANI, Carleman Estimates and Exact Boundary Controllability for a System of Coupled, Non-Conservative Second-Order Hyperbolic Equations, Lecture Notes in Pure and Applied Mathematics, Vol. 188, pp. 215–243, Marcel Dekker, New York, 1997, J.P. Zolesio and G. Da Prato (eds.).

[L-T.18] I. LASIECKA AND R. TRIGGIANI, Two direct proofs of analyticity of the s.c. semigroup arising in abstract thermo-elastic equations, *Advances in Diff. Eqns.* **3** (1998), 387–416.

[L-T.19] I. LASIECKA AND R. TRIGGIANI, Analyticity of thermo-elastic semigroups with coupled hinged/Neumann B.C., *Abstract Appl. Anal.* **3**(1–2) (1998), 153–169.

[L-T.20] I. LASIECKA AND R. TRIGGIANI, Analyticity of thermo-elastic semigroups with free B.C., *Annali Scuola Normale Superiore*, Pisa, cl. sci. (4), XXVII (1998).

[L-T.21] I. LASIECKA AND R. TRIGGIANI, Exact null controllability of structurally damped and thermo-elastic parabolic models, *Memoria, Academia dei Lincei* Rome (Italy), *Sezione Matematica* **9** (1998), 943–969.

[L-T.22] I. LASIECKA AND R. TRIGGIANI, Analyticity, and lack thereof, of thermo-elastic semigroups, *European Soc. Appl. Math. (ESIAM)* **4** (1999), 199–222.

[L-T.23] I. LASIECKA AND R. TRIGGIANI, sharp trace result on a thermo-elastic plate equation with coupled hinged/Neumann boundary conditions, *Discrete Cont. Dynam. Sys.* **5** (1999), 585–599.

[L-T.24] I. LASIECKA AND R. TRIGGIANI, *Control Theory for Partial Differential Equations*, Vol. 1 (680 pp.) and Vol. II (446 pp.), Cambridge University Press, 2000.

[L-T.25] I. Lasiecka and R. Triggiani, *Feedback Noise Control in an Acoustic Chamber: Mathematical Theory*, Gordon and Breach Scientific Publishers, 2000, 89–111.

[L-T.26] I. Lasiecka and R. Triggiani, Exact boundary controllability of a first order non-linear hyperbolic equation with non-local integral term arising in epidemic modeling, *Direct and Inverse Problems of Matheamtical Physics* R.P. Gilbert, J. Kajiwara, U. Xu (Editors), Kluwer (2000), 363–398.

[L-T.27] I. Lasiecka and R. Triggiani, Structural decomposition of thermoelastic semigroups with rotational forces, *Semigroup Forum* **60** (2000), 1–61.

[L-T.28] I. Lasiecka and R. Triggiani, A sharp trace result of Kirchof and thermo-elastic plate equations with free boundary conditions, *Rocky Mtn. J. Math.* **30**(3) (Fall 2000), 981–1024.

[L-T.29] I. Lasiecka and R. Triggiani, Uniform stabilization of a shallow shell model with nonlinear boundary feedbacks, *J. Math. Anal. Appl.* **269** (2002), 642–688.

[L-T.30] I. Lasiecka and R. Triggiani, $L_2(\Sigma)$-regularity of the boundary $\rightarrow$ boundary operator B^*L for hyperbolic and Petrowski PDEs, submitted.

[L-L-T.1] I. Lasiecka, J.L. Lions, and R. Triggiani, Nonhomogeneous boundary value problems for second-order hyperbolic operators, *J. Math. Pures Appl.* **65** (1981), 149–192.

[L-T-V.1] I. Lasiecka, R. Triggiani, and W. Valente, Uniform stabilization of spherical shells by boundary dissipation, *Advances Diff. Eqns.* **1**(4) (1996), 635–674.

[L-V.1] I. Lasiecka and W. Valente, Uniform stabilization of a nonlinear shallow and tin spherical cap, *JMAA* **22** (1996), 951–994.

[L-T-Y.1] I. Lasiecka, R. Triggiani, and P-F. Yao, Exact controllability for second-order hyperbolic equations with variable coefficient-principal part and first-order terms, *Nonlinear Anal.* **30** (1997), 111–122.

[L-T-Y.2] I. Lasiecka, R. Triggiani, and P-F. Yao, Inverse/observability estimates for second-order hyperbolic equations with variable coefficients, *J. Math. Anal. Appl.* **235** (1999), 13–57.

[L-T-Y.3] I. Lasiecka, R. Triggiani, and P-F. Yao, An observability estimate in $L_2 \times H^{-1}$ for second-order hyperbolic equations with variable coefficients, *Control of Distributed Parameter and Stochastic Systems*, S. Chen, X. Li, J. Yong, X. Zhou (eds.), Kluwer Academic Publishers, Dordrect, 1999.

[L-T-Y.4] I. Lasiecka, R. Triggiani, and P-F. Yao, Carleman estimates for a plate equation on a Riemannian manifold with energy level terms, *Analysis and Applications*, Chapter 15, 199–236 ISAAC, 2001; Kluwer, to appear in 2003.

[L-T-Z.1] I. Lasiecka, R. Triggiani, and X. Zhang, Nonconservative wave equations with unobserved Neumann B.C.: Global uniqueness and observability in one shot, *Contemporary Mathematics*, AMS **268** (2000), 227–326.

[L-T-Z.2] I. Lasiecka, R. Triggiani, and X. Zhang, Nonconservative Schrödinger equations with unobserved Neumann B.C.: Global uniqueness and observability in one shot, Spring 2002. Announcement, under the same title, to appear in the Proceedings of the Conference, "Analysis and Optimization of Differential Systems," held at Ovidius University, Constanta, Romania, Sept. 10–15, 2002.

[L-R-S.1] M.M. Lavrent'ev, V.G. Romanov, and S.P. Shishat·skii, *Ill-Posed Problems of Mathematical Physics and Analysis*, English translation. (The original: Nauka Moscow, 1980), American Mathematical Society, Providence, RI, 1986.

[Le.1] G. LEBEAU, Contrôl analytique, I, Estimations á priori, *Duke Math. J.* **68** (1992), 1–30.

[Le.2] G. LEBEAU, Controle de l'equation de Schrödinger, *J. Math. Pures Appl.* **71** (1992), 267–291.

[Le.3] G. LEBEAU, Controle des ondes amorties, Prepublications de Université de Paris-Sud **19** (1994).

[L.1] J.L. LIONS, *Controllabilite Exacte, Stabilization et Perturbation des Systemes Distribues*, Vol. 1, Masson, Paris, 1988.

[Lit.1] W. LITTMAN, Near-optimal-time boundary controllability for a class of hyperbolic equations, *Control Problems for Systems Described by Partial Differential Equations and Applications* (Gainesville, FL, 1986), 307–312, *Lecture Notes in Control and Inform. Sci.* 97, Springer, Berlin-New York, 1987.

[Lit.2] W. LITTMAN, Remarks on global uniqueness theorems for partial differential equations, *Contemporary Mathematics*, AMS **268** (2000).

[Lit.3] W. LITTMAN, Boundary controllability for polyhedral domains, *LNCIS*, Springer Verlag, **178** (1992), 272–284.

[Lit-Li.1] W. LITTMAN AND B. LIU, On the spectral properties and stabilization of acoustic flow, *SIAM J. Appl. Math.* **1** (1998), 17–34.

[L-H.1] W. LITTMAN AND M. A. HORN, Boundary control of a Schrödinger equation with nonconstant principal part, *Lecture Notes in Pure and Applied Mathematics*, Marcel Dekker, **174** (1996), 101–106.

[Lit-Ta.1] W. LITTMAN AND S.W. TAYLOR, Smoothing evolution equations and boundary control theory, *J. d'Analyse Math.* **59** (1992), 201–215.

[M.1] E. MACHTYNGIER, Controlabilite exacte et stabilizativ frontiere de l'equation de Schrödinger, *C. R. Acad. Sci. Paris*, Ser. I, Math **310** (1990), 801–806.

[Miz.1] MIZOHATA, personal communication.

[P-T.1] J. PUEL AND M. TUCSNAK, Boundary stabilization for the von Karman equations, *SIAM J. Control* **33** (1996), 255–273.

[Ra.1] J. RALSTON, Gaussian beams and the propagation of singularities, *Studies in Partial Differential Equations*, W. Littman (ed.), *MAA Studies in Mathematics* **23** (1982), 206–248.

[R.1] D.L. RUSSELL, Controllability and stabilizability theory for linear partial differential equations, recent progress and open questions, *SIAM REVIEW* **20** (1978), 639–739.

[S.1] T. SHIROTA, A remark on the unique continuation theorem for certain fourth order elliptic equations, *Proc. Japan Acad.* **36** (1960), 571–573.

[S-Y] R. SCHOEN AND S.-T. YAU, *Lectures on Differential Geometry*, International Press, Boston 1994.

[Ta.1] D. TATARU, A-priori pseudoconvexity energy estimates in domains with boundary applications to exact boundary controllability for conservative PDEs, Ph.D. Thesis, University of Virginia, May 1992.

[Ta.2] D. TATARU, Boundary controllability for conservative PDEs, *Appl. Math. & Optimiz.* **31** (1995), 257–295.

[Ta.3] D. TATARU, A-priori estimates of Carleman's type in domains with boundaries, *J. Math. Pures et Appl.* **73** (1994), 355–387.

[Ta.4] D. TATARU, Unique continuation for solutions to PDEs, between Hörmander's theorem and Holmgren's theorem, *Comm. Partial Diff. Eqns.* **20**(5 & 6) (1995), 855–884.

[Ta.5] D. TATARU, On the regularity of boundary traces for the wave equation, *Ann. Scuola Norm. Sup. Pisa Cl. Sci.* **26**(4) (1998), 185–206.

[Ta.6] D. TATARU, Unique continuation for partially differential operators with partial analytic coefficient, *J. Math. Pure Appl.* **78** (1999), 505–521.

[Ta.7] D. TATARU, Carleman estimates and unique continuation for solutions to boundary value problems, *J. Math. Pure Appl.* **75** (1996), 367–408.

[Ta.8] D. TATARU, Unique continuation problems for PDEs, this volume.

[T-L.1] A.E. TAYLOR AND D.C. LAY, *Introduction to Functional Analysis*, 2nd ed., John Wiley, New York, 1980.

[Tay.1] M. TAYLOR, *Partial Differential Equations*, Vols. 1, 2, Springer-Verlag, 1991.

[Tay.2] M. TAYLOR, *Tools for P.D.E.: Pseudodifferential Operators, Paradifferential Operators, and Layer Potentials.*

[Tr.1] R. TRIGGIANI, Exact controllability on $L_2(\Omega) \times H^{-1}(\Omega)$ of the wave equation with Dirichlet B.C., *AMO* (1988), 241–247.

[Tr.2] R. TRIGGIANI, Carleman estimates and exact boundary controllability for a system of coupled non-conservative Schrödinger equations, *Rend. Istit. Math. Univ. Trieste* (Italy), suppl. Vol. XXVIII (1997), 453–504; volume in memory of P. Grisvard.

[Tr.3] R. TRIGGIANI, Regularity theory, exact controllability and optimal quadratic cost problem for spherical shells with physical boundary controls. Invited paper in special issue for *Control and Cybernetics*, Polish Academy of Sciences, *Control Problems for Partial Differential Equations* **25**(3) (1996), 553–569.

[T-Y.1] R. TRIGGIANI AND P.F. YAO, Inverse/observability estimates for Schrödinger equations with variable coefficients, *Control & Cybernetics* **28** (1999), 627–664. Special volume on control of PDEs.

[T-Y.2] R. TRIGGIANI AND P.F. YAO, Carleman estimates with no lower-order terms for general Riemannian wave equations. Global uniqueness and observability in one shot, *Appl. Math. & Optim.* **46** (Sept./Dec. 2002), 331–375. Special issue dedicated to J.L. Lions.

[Wu.1] H. WU, The Bochner technique in differential geometry, *Math. Reports* 3 (Part 2), Hardwood Academic Publishers, London, Paris, 1988.

[Y.1] P.F. YAO, On the observability inequalities for exact controllability of wave equations with variable coefficients, *SIAM J. Control Optim.* **37** (1999), 1568–1599.

[Y.2] P.F. YAO, Observability inequalities for the Euler-Bernoulli plate with variable coefficients, *Contemporary Mathematics*, Amer. Math. Soc. **268** (2000), 383–406.

[Y.3] P.F. YAO, On the shallow shell equations, manuscript, 1999.

[Y.4] P.F. YAO, Observability inequalities for shallow shells, *SIAM J. Control* **38** (2000), 1729–1756.

[Z.1] E. ZUAZUA, Exact boundary controllability for the semilinear wave equation, in Non-Linear PDEs and their Applications, *Research Notes in Mathematics*, Pitman, Boston (19), 1265–12.

ENERGY MEASUREMENTS AND EQUIVALENCE OF BOUNDARY DATA FOR INVERSE PROBLEMS ON NON-COMPACT MANIFOLDS

A. KATCHALOV*, Y. KURYLEV†, AND M. LASSAS‡

1. Introduction. Formulation of results.

1.1. The goal of this paper is to consider inverse problems with different types of boundary data given as boundary forms. In particular, the boundary measurements considered in this paper are related to the measurements of energy needed to force the boundary value of a physical field to a given one.

In various applications it is often not possible to measure the Cauchy data on the whole boundary, as one can not attach sources and measurement devices to the same locations. A perfect example is given by the seismic measurements when the boundary sources are often made of explosives. Similarly, in many cases it is difficult to measure both the amplitude and phase of a field. On the contrary, the total energy of a wave is often easily accessible and, in particular, it is possible to find the energy which is required to force the boundary value of the field to a given one, that is, to measure the energy needed to do a given measurement. In theoretical inverse problems this idea goes back to A. Calderón who in his 1980 seminal paper [Cl] considered the inverse problem for the conductivity equation from the point view of energy measurements. Similarly, the measurements based on energy or interference of waves have been used in many applications, e.g. in impedance tomography (see e.g. [CIN]) and near field optical tomography (see e.g. [SM]). The work of Calderón was extended by J. Sylvester and G. Uhlmann who developed a method of complex geometric optics to solve fully non-linear inverse problems [SU].

In the paper we consider also the question of the equivalence of different boundary data used in inverse problems. Our interest in the equivalence of inverse problems with various boundary data comes from the fact that there are numerous examples when the solution of a particular inverse problem is used to solve an inverse problem of another type. For example, A. Nachman, J. Sylvester, and G. Uhlmann [NSU] solved the inverse boundary spectral problem for a Schrödinger operator by reducing it to the inverse boundary value problem in fixed frequency and then using the method of the complex geometric optics. Similarly, inverse problems in the time-domain are often reduced to problems in the frequency domain (see e.g. [Is]). The equivalence is also useful for applications as it makes

*Steklov Mathematical Institute, RAN, Fontanka 27, 191011, St. Petersburg, Russia.
†Department of Mathematical Sciences, Loughborough University, Loughborough, LE11 3TU, UK.
‡Rolf Nevanlinna Institute, University of Helsinki, P.O.Box 4, FIN-00014, Finland.

possible the use of reconstruction algorithms developed for some type of inverse problems to other inverse problems.

In this paper we concentrate mainly on the case of non-compact domains and manifolds and Robin boundary conditions. Our intention in doing so stems from the existence of many rather detailed expositions of these type of results for the compact case, e.g. [KaL], [KKLM] and, in particular, [KKL]. In addition, the non-compact case differs significantly from the compact one and we believe that it is worth a separate discussion. As [KKL] deals with the Dirichlet boundary condition we have decided to concentrate here on the Robin one.

1.2. We start with necessary definitions. Let M be a smooth, complete (possibly non-compact) m-dimensional Riemannian manifold with boundary ∂M and let $a(x,D)$ be an elliptic 2nd-order partial differential operator,

$$a(x,D)v(x) = -\Delta_g v(x) + (V(x), \nabla v(x))_g + c(x)v(x), \tag{1}$$

where $\Delta_g = g^{-1/2}\partial_j g^{1/2} g^{jk}(x)\partial_k$ with $g = \det(g_{ij})$, $[g^{ij}] = [g_{ij}]^{-1}$ is the Laplace-Beltrami operator corresponding to the metric g_{ij}, V is a smooth real vector field (1st order operator) and c is a smooth real valued function. As usual we use Einstein's summation over repeated upper and lower indices.

Next we consider an operator A related to $a(x,D)$,

$$Av = a(x,D)v, \quad \mathcal{D}(A) = \{v \in H^2(M) : \ Bv|_{\partial M} = \partial_\nu v + \eta v|_{\partial M} = 0\}, \tag{2}$$

where ν is the exterior unit normal vector to ∂M and η is a smooth real valued function on ∂M. As usual $H^2(M)$ stands for the Sobolev space of functions having square integrable derivatives up to the second order. We assume that there is a smooth measure $d\mu$ on M,

$$d\mu = \rho\, dV_g, \quad dV_g = g^{1/2}\, dx^1 \cdots dx^m, \tag{3}$$

where dV_g is Riemannian volume on (M,g), so that A is self-adjoint with respect to $d\mu$. In particular,

$$\int_M vAu\, d\mu = \int_M uAv\, d\mu \tag{4}$$

for $u, v \in \mathcal{D}(A)$. Then

$$a(x,D)v(x) = -\rho^{-1}g^{-1/2}(\partial_i g^{1/2} g^{ij}\rho\partial_j v(x)) + q(x)v(x). \tag{5}$$

The fact that A of form (2), (5) is self-adjoint as well as just an invariant definition of the Sobolev spaces $H^s(M)$ puts some restrictions on $(M, g; \partial M)$, ρ, q, and η. For our purposes it is sufficient to assume that $(M, g; \partial M)$ is (possibly) a non-compact manifold of (finitely) bounded geometry and $\rho, \rho^{-1} \in C^\infty(M) \cap C^2_b(M)$, $q \in C^\infty(M) \cap C^0_b(M)$ and

$\eta \in C^\infty(\partial M) \cap C_b^1(\partial M)$. Here C_b^l is the class of functions having uniformly bounded derivatives upto order l. For the explanation of these facts as well the definition of a manifold of bounded geometry see Appendix.

As A is self-adjoint, it has a spectral resolution,

$$A = \int_{\mathbb{R}} \lambda dE(\lambda), \tag{6}$$

where $E(\lambda)$ is the spectral projector on the interval $(-\infty, \lambda]$.

Since $\mathcal{D}(A^n) \subset H^{2n}_{loc}(M)$ and $A \geq \lambda_0 I$ due to $q(x) \in C_b^0$, we see that for any $\lambda \in \mathbb{R}$ the projection $E(\lambda)$ is an infinitely smoothing operator, $E(\lambda) : H^s_{\mathrm{comp}}(M) \to C^\infty(M)$ for any $s \in \mathbb{R}$. Thus $E(\lambda)$ has a smooth Schwartz kernel and there is a measure $dp(\cdot, \cdot; \lambda)$ on $M \times M$ such that

$$(u, E(\lambda)v)_{L^2(M,d\mu)} = \int_{M\times M} u(x)\overline{v(y)} d_{x,y} p(x, y; \lambda). \tag{7}$$

Here, we use the notation $d_{x,y}p(x, y; \lambda) = dp(x, y; \lambda)$ to indicate that integration variables are x and y and λ is a parameter. Clearly, the Radon-Nikodym derivative $dp_{x,y}(x, y; \lambda)/(d\mu(x)d\mu(y))$ is a real valued $C^\infty(M \times M)$-function for any λ. In particular, if M is compact and λ_j and φ_j are the eigenvalues and the normalized eigenfunctions of A, then

$$dp(x, y; \lambda) = \sum_{\lambda_j \leq \lambda} \varphi_j(x)\varphi_j(y) d\mu(x) d\mu(y).$$

Now we consider the measure $dP(x, y; \lambda)$ on $\partial M \times \partial M$,

$$dP_{x,y}(x, y; \lambda) = \frac{dp_{x,y}(x, y; \lambda)}{dV_g(x) dV_g(y)} dS_g(x) dS_g(y),$$

where dS_g is the Riemannian volume of the boundary.

Let $\delta_{\partial M}$ is the surface delta-measure with respect to dV_g. Then

$$\int_M h(x, y)\delta_{\partial M}(x)\delta_{\partial M}(y) dp(x, y; \lambda) = \int_{\partial M} h(x, y) dP(x, y; \lambda). \tag{8}$$

Remark 1. In the compact case, if the measure ρdS_g on the boundary is known, then λ_j and $\varphi_j|_{\partial M}$ determine $dP(x, y; \lambda)$. Operating as in [KKL], we also see that $dP(x, y; \lambda)$ determines λ_j and $\varphi_j|_{\partial M}$ upto a unitary transformation of the eigenfunctions corresponding to the same eigenvalue.

Therefore, it is natural to give the following generalization of the Gel'fand data to the case of a possibly non-compact manifold:

DEFINITION 1. *We define the Gel'fand boundary spectral data of A to be the measure $dP(x, y; \lambda)$ given on $\partial M \times \partial M$ for all $\lambda \in \mathbb{R}$.*

Other objects on ∂M related to the spectral properties of A are the Robin-to-Dirichlet maps

$$\Lambda^z \phi = u_z^\phi|_{\partial M}, \tag{9}$$

where u_z^ϕ is the solution of the Robin problem

$$a(x,D)u_z^\phi = zu_z^\phi, \quad Bu_z^\phi|_{\partial M} = \phi \tag{10}$$

where $z \in \mathbb{C}$ is not an eigenvalue.

DEFINITION 2. *We define the Calderon-Gel'fand boundary form $\Lambda^z[\phi,\psi]$ related to problem (10) by the formula*

$$\Lambda^z[\phi,\psi] = \int_{\partial M} \Lambda^z\phi(x)\,\overline{\psi(x)}\,\rho(x)dS_g \tag{11}$$

which is equivalent to a Dirichlet-type form for u_z^ϕ and u_z^ψ,

$$\Lambda^z[\phi,\psi] = \int_{\partial M} \eta u_z^\phi \overline{u_z^\psi}\rho dS_g + \int_M \left((\nabla u_z^\phi, \nabla u_z^\psi)_g + (q-\overline{z})u_z^\phi\overline{u_z^\psi}\right)\rho dV_g, \tag{12}$$

where $(\nabla u, \nabla v)_g = g^{ij}\partial_i u \overline{\partial_j v}$ (compare with e.g. with [Cl], [SU]).

1.3. We consider also the hyperbolic initial boundary value problem corresponding to the elliptic operator A,

$$\begin{gathered} \partial_t^2 + a(x,D))u^f(x,t) = 0 \quad \text{in} \quad M \times \mathbb{R}_+, \\ Bu^f|_{\partial M\times\mathbb{R}_+} = f \in C_0^\infty(\partial M \times \mathbb{R}_+), \quad u^f|_{t=0} = 0, \quad \partial_t u^f|_{t=0} = 0. \end{gathered} \tag{13}$$

For initial boundary value problem (13) we define the non-stationary Robin-to-Dirichlet map (response operator) Λ,

$$\Lambda f = u^f|_{\partial M\times\mathbb{R}_+}. \tag{14}$$

The operator Λ gives rise to the hyperbolic form $\mathcal{B}[f,h]$,

$$\begin{aligned} \mathcal{B}[f,h] &= \int_0^\infty \int_{\partial M} \left(\partial_\nu u^f\,\overline{u^h} - u^f\,\overline{\partial_\nu u^h}\right)\rho dS_g\,dt \\ &= \int_0^\infty \int_{\partial M} (f\overline{\Lambda h} - \Lambda f\overline{h})\,\rho dS_g\,dt. \end{aligned} \tag{15}$$

There is a natural concept of energy for the wave equation (13) given by

$$\begin{aligned} E(u,t) &= \frac{1}{2}\int_{\partial M} \eta(x)|u(x,t)|^2\rho(x)dS_g \\ &+ \frac{1}{2}\int_M \left(|\nabla u(x,t)|_g^2 + q(x)|u(x,t)|^2 + |\partial_t u(x,t)|^2\right)d\mu(x). \end{aligned} \tag{16}$$

When $f \in C_0^\infty(\partial M \times [0,T])$ we see that $E(u^f,t)$ is constant for $t > T$, that is, the energy is conserved. Therefore, the energy $E(u^f,T)$ is brought into M through the boundary $\partial M \times \mathbb{R}_+$. We define the total energy flux $\Pi(f)$ through the boundary as

$$\Pi(f) = \lim_{t\to\infty} E(u^f,t). \tag{17}$$

Actually, $\Pi(f)$ is given by a quadratic form of f,

$$\Pi(f) = \text{Re} \int_{\partial M} \int_0^\infty f(x,t)\partial_t \overline{\Lambda f(x,t)}\, \rho(x) dS_x dt. \tag{18}$$

Because differential equation (13) is translation invariant in time, we extend the map Λ, the form $\mathcal{B}$, and the energy flux Π to $f \in C_0^\infty(\partial M \times \mathbb{R})$ by setting e.g. $\Lambda f(x,t) = (\Lambda(f(\cdot, \cdot - T)))(x, t+T)$.

1.4. Before formulating inverse problems, we need to introduce the notion of a gauge-transformation. We will consider all operators and boundary data in such a way that our considerations do not depend on a particular choice of the scale of measurements. For instance, if the change of the scale of measurements is described by a function $\kappa(x)$, $\kappa|_{\partial M} = 1$, that is at a point $x \in M$ the physical quantity $u(x)$ is replaced with $\kappa(x)u(x)$, this change of the scale of measurements does not affect the physical model or the measurements of this quantity on ∂M. However, it does change its mathematical description, i.e., the differential equation which describes the process. For this reason we formulate all our statements so that they are invariant in gauge transformations $u(x) \to \kappa(x)u(x)$.

DEFINITION 3. *Let $\kappa \in C^\infty(M)$, $\kappa(x) \geq \kappa_0 > 0$ for $x \in M$. The gauge transformation generated by the function κ is the transformation*

$$S_\kappa : L^2(M, d\mu) \to L^2(M, d\mu_\kappa), \quad S_\kappa u(x) = \kappa(x)u(x),$$

with $d\mu_\kappa = \kappa^{-2}(x)d\mu$. If $\kappa|_{\partial M} = 1$ the gauge transformation S_κ is normalized on ∂M.

Each gauge transformation determines the corresponding gauge transformation A_κ of the operator A,

$$A_\kappa u = \kappa A(\kappa^{-1} u).$$

If A is an elliptic differential operator in $L^2(M, d\mu)$ of form (2), (5) with $\rho, \rho^{-1} \in C_b^2(M)$, $q \in C_b^0(M)$ and $\eta \in C_b^1(\partial M)$, then A_κ is also an elliptic differential operator in $L^2(M, d\mu_\kappa)$ of form (2), (5) with ρ_κ, η_κ and q_κ from the same classes as soon as $\kappa, \kappa^{-1} \in C_b^2(M)$. Furthermore,

$$\mathcal{D}(A_\kappa) = \{v \in H^2(M) : \ B_\kappa v|_{\partial M} = \partial_\nu v + \eta_\kappa v|_{\partial M} = 0\}, \quad \eta_\kappa = \eta + \kappa^{-1}\partial_\nu \kappa,$$

and $a_\kappa(x, D)u(x)$ is given by the formula

$$a_\kappa(x, D)u(x) = \kappa^{-1}(x)a(x, D)(\kappa(x)^{-1}u(x)).$$

The gauge transformations $S_\kappa : L^2(M) \to L^2(M)$ parametrized by κ : $\kappa \in C^\infty(M)$, $\kappa, \kappa^{-1} \in C_b^2(M)$ form an Abelian group $\mathcal{G}$ with respect to composition

$$S_{\kappa_1} \circ S_{\kappa_2} = S_{\kappa_1 \kappa_2}.$$

The action of this group on the set of the second order elliptic differential operators is given by $S_\kappa(A) = A_\kappa = \kappa A \kappa^{-1}$. For any A

$$\sigma A = \{S_\kappa(A) : \quad S_\kappa \in \mathcal{G}\}$$

is the orbit of the group $\mathcal{G}$ through A. The gauge transformations normalized on ∂M form a subgroup $\mathcal{G}_{\partial M}$ and the corresponding orbit is denoted by $\sigma_{\partial M} A$.

Although gauge transformations change $a(x, D)$, the metric tensor $g^{ij} = a^{ij}$ associated with the operator A remains invariant, $g^{ij}_\kappa = g^{ij}$.

An important fact related to the gauge transformations is that any orbit σA of a self-adjoint operator A of form (2), (5) contains a unique Schrödinger operator which is called the Schrödinger operator corresponding to A.

LEMMA 1. *i. Let A be an elliptic differential operator of form (2), (5). There is a unique Schrödinger operator $-\Delta_g + \widetilde{q}$ in the orbit σA, that is, for a given A there is a unique κ such that $A = \kappa(-\Delta_g + \widetilde{q})\kappa^{-1}$ and $d\mu = \kappa^{-2} dV_g$.*
ii. A is a Schrödinger operator if and only if $d\mu = dV_g$.

Proof. The assertion is proven in [KKL], see also [K1], [KK]. The basic idea is to consider the corresponding Dirichlet quadratic form in M and observe that a gauge transformation is equivalent to changing the measure in this form. □

In gauge transformations, the hyperbolic form $\mathcal{B}$, energy flux Π and Calderon-Gel'fand form Λ^z are also changed. Indeed, if for example $u^f(x, t)$ is a solution of problem (13) for the operator $a(x, D)$, then $v(x, t) = \kappa(x) u^f(x, t)$ is the solution of the problem

$$\text{(19)} \qquad \begin{aligned} &(\partial_t^2 + a_\kappa(x, D)) v(x, t) = 0 \quad \text{in} \quad M \times \mathbb{R}_+, \\ &B_\kappa v|_{\partial M \times \mathbb{R}_+} = \kappa f, \quad v|_{t=0} = 0, \ \partial_t v|_{t=0} = 0. \end{aligned}$$

Thus, if $\mathcal{B}, \Pi$ correspond to the operator A and $\mathcal{B}_\kappa, \Pi_\kappa$ – to its gauge transformation A_κ then

$$\text{(20)} \qquad \Lambda_\kappa[f, h] = \Lambda[\kappa f, \kappa h], \quad \Pi_\kappa[f, h] = \Pi[\kappa f, \kappa h].$$

Similarly, if Λ^z and $dP(x, y, \lambda)$ are the Calderon-Gel'fand forms and boundary spectral data of A and Λ^z_κ, $dP_\kappa(x, y, \lambda)$ – of A_κ, then

$$\text{(21)} \quad dP_\kappa(x, y; \lambda) = \kappa(x)^{-1} \kappa(y)^{-1} dP(x, y; \lambda), \quad \Lambda^z_\kappa[\phi, \psi] = \Lambda^z[\kappa\phi, \kappa\psi].$$

The forms Λ^z and Λ^z_κ, $\mathcal{B}$ and $\mathcal{B}_\kappa$, etc. which satisfy (20), (21) are called gauge equivalent.

1.5. Now we are in the position to formulate various inverse problems on M related to the above concepts.

Inverse problems. Determine (M, g) and A upto a normalized gauge transformation, i.e., determine $\sigma_{\partial M} A$ when we are given one of the following data:

i. The Gel'fand boundary spectral data $dP(x, y; \lambda)$ on $\partial M \times \partial M$ for all $\lambda \in \mathbb{R}$.
ii. The Calderon-Gel'fand forms Λ^z for all $z \in \mathbb{C} \setminus \sigma(A)$, where $\sigma(A)$ is the spectrum of A.
iii. The hyperbolic form $\mathcal{B}$.
iv. The energy flux Π.

Our aim is to show that, in particular, the energy measurements can be used to obtain the other data. Moreover, we will show:

THEOREM 1. *Inverse problems i.-iv. are equivalent, i.e., any of the data i.-iv. determine all other data.*

Thus all inverse problems *i.-iv.* can be reduced to solving one of them. Instead of solving problem *iii.* we will solve a more general problem which also answers the above problems. To formulate this more general problem denote by $\mathcal{B}^{2T}$ the hyperbolic form $\mathcal{B}$ restricted to the set of sources $f, h \in C_0^\infty(\partial M \times (0, 2T))$. We will show that

THEOREM 2. *Assume that we are given the hyperbolic form $\mathcal{B}^{2T}$ of an operator A of form (2), (5). This data determines uniquely the manifold $M^T = \{x \in M : \ d(x, \partial M) < T\}$ and the metric tensor g on M^T. Moreover, we can find $a(x, D)$ on M^T upto a normalized gauge transformation, i.e., we can find the orbit*

$$\sigma_{\partial M} A|_{M^T} = \{a_\kappa(x, D)|_{M^T} : \ \kappa > 0, \ \kappa|_{\partial M} = 1\}.$$

In particular, if any of data $i. - iv.$ is given, it is possible to determine the whole manifold (M, g) and the orbit $\sigma_{\partial M} A$. If, in addition, we have a priori knowledge about the structure of the operator, we can in many cases solve the inverse problem uniquely. For instance, we have:

COROLLARY 1. *Let $M \subset \mathbb{R}^m$ is given. Assume the metric g to be conformally Euclidean, that is $g_{jk}(x) = \sigma(x)\delta_{jk}$ where $\sigma(x) > 0$. Moreover, assume that we know any of the data i.-iv. for a Schrödinger operator $A = -\Delta_g + q$. Then we can determine g, q and η uniquely.*

At last we consider the case when the data is given only on an open subset $S \subset \partial M$. In this case we can define the Gel'fand boundary spectral data dP_S

$$dP_S(x, y; \lambda) = dP(x, y; \lambda), \quad x, y \in S, \tag{22}$$

the Calderon-Gel'fand form Λ_S^z

$$\Lambda_S^z[f, h] = \Lambda^z[f, h], \quad f, h \in C_0^\infty(S), \tag{23}$$

the hyperbolic boundary forms $\mathcal{B}_S$ and even $\mathcal{B}_S^{2T}$

$$\mathcal{B}_S^{2T}[f, h] = \mathcal{B}[f, h], \quad f, h \in C_0^\infty(S \times (0, 2T)), \tag{24}$$

and the energy flux Π_S,

$$\Pi_S(f) = \Pi(f), \quad f \in C_0^\infty(S \times \mathbb{R}_+). \tag{25}$$

Then the analogs of Theorems 1 and 2 remain valid for dP_S, Λ_S^z, etc:

THEOREM 3. *i. Assume that we are given $S \subset \partial M$. Then any data (22)–(25) determine all others.*

ii. Assume that we are given $S \subset \partial M$ and $\mathcal{B}_S^{2T}$. Then this data determine uniquely the manifold $M(S,T) = \{x \in M : \ d(x,S) < T\}$, the metric tensor g and the operator $a(x,D)$ on $M(S,T)$ upto a gauge transformation normalized on S, i.e., we can find $\sigma_S A|_{M(S,T)}$.

1.6. This paper gives a concise review of some of the results obtained in the multidimensional inverse boundary value problems, especially those regarding the equivalence of various types of data, gauge equivalence and also reconstruction procedures and uniqueness for hyperbolic inverse problems. The used techniques are based on various variants of the BC-method (for the original paper see [B1]). There are currently several extended expositions of this method, e.g. [B2], [KK], [KaL]. The monograph [KKL] is particularly close to our exposition in this paper and we refer the interested reader to this monograph for further details. Having said so we should stress that the majority of works on the BC-method deal with the case of inverse boundary problems on compact manifolds. More precisely, due to the local in time nature of the method, the treatment of hyperbolic inverse problems is essentially the same for compact and non-compact cases. This makes possible to closely follow in our proof of Theorem 2 the method described in [KKL], Ch. 4.2. An alternative approach to hyperbolic inverse problems also based on the BC-method and technique of Gaussian beams is given in [BKa] which deals with the wave equation for the Laplace operator on a Riemannian manifold. However, when coming to inverse spectral problems, i.e., problems *i.* and *ii.*, non-compact manifolds differ rather significantly from the compact ones due to a more complicated nature of the spectral properties of elliptic operators on non-compact manifolds. To our knowledge the only paper where the BC-method is applied to an inverse boundary spectral problem on a non-compact domain is [BKu1] where $M = \mathbb{R}^m_+$. In particular, the definition of the boundary spectral data (Definition 1) differs from that for the compact case. Moreover, the proof of the equivalence, although ideologically close to that in [KKL], Ch. 4.1 and [KKLM], implies some technical ideas absent in [KKL]. There are some other differences in our exposition as compared to the previous ones. For example, in the proof of Lemma 4 dealing with the inner products of waves we use variational technique which, we believe, is more appropriate for the numerical realization of the method. Furthermore, the step by step reconstruction of the manifold from a part of the boundary is based on the direct continuation of Green's function for the wave equation which, to our knowledge, has been unknown. And, of course, we deal with boundary forms rather then the corresponding operators. We believe that the invariance properties possessed by the forms better reflect the nature of the problem. This approach is essentially similar to that in, e.g. [SU], [LU], [Sy] which use differential forms rather then functions.

The plan of the paper is as follows: In section 2 we prove Theorem 1 about the equivalence of data *i.-iv.* in the case of the whole boundary, $S = \partial M$. In section 3 we describe the procedure of the reconstruction of a Riemannian manifold (M^T, g) and the Schrödinger operator on it from the hyperbolic form $\mathcal{B}^{2T}$ and give the proof of Theorem (2). An alternative approach based on Gaussian beams is given in Section 4. Section 5 is devoted to the generalization of the above results to the case $S \neq \partial M$. At last, in Appendix we collect some necessary results about properties of manifolds of bounded geometry and elliptic operators on such manifolds. Our exposition is rather concise, especially in Sections 4 and 5 and when the construction used is similar to those for the compact case. Nevertheless, we provide (at least brief) proofs of the main ingredients of the method, namely the Blagovestchenskii identity and controllability results.

2. Proof of equivalence of the boundary data.

2.1. We start with the observation that since $a(x, D)$ of form (5) is real,

$$\overline{u_z^{\phi}} = u_{\overline{z}}^{\overline{\phi}}, \quad \overline{u^f(x,t)} = u^{\overline{f}}(x,t), \quad (A - \overline{z})^{-1}\overline{\Phi} = \overline{(A - z)^{-1}\Phi}, \tag{26}$$

where $z \notin \sigma(A)$ and u_z^{ϕ} and $u^f(x,t)$ are solutions of problems (10) and (13), correspondingly. This allows us to work not with the inner product in $L^2(M)$ but with the complex bilinear pairing $\langle \cdot, \cdot \rangle$, namely, $\langle \Phi, \Psi \rangle = \int_M \Phi(x)\Psi(x) d\mu$, or the corresponding distribution duality. Using (26) we see that the forms $\Lambda^z[\phi, \psi]$, $\mathcal{B}[f, h]$ and $\Pi[f]$ determine complex bilinear forms

$$\begin{aligned} \Lambda^z_{\mathbb{C}}[\phi, \psi] &= \int_{\partial M} (\Lambda^z \phi)(x)\psi(x)\, \rho(x) dS_g = \Lambda^z[\phi, \overline{\psi}] \\ &= \int_{\partial M} \eta u_z^{\phi} u_z^{\psi} \rho dS_g + \int_M \left(\langle \nabla u_z^{\phi}, \nabla u_z^{\psi} \rangle_g + (q - z) u_z^{\phi} u_z^{\psi}\right) \rho dV_g, \end{aligned} \tag{27}$$

where $\langle \nabla u, \nabla v \rangle_g = g^{jk} \partial_j u\, \partial_k v$,

$$\mathcal{B}_{\mathbb{C}}[f, h] = \int_0^{\infty} \int_{\partial M} (f \Lambda h - \Lambda f h)\, \rho dS_g\, dt = \mathcal{B}[f, \overline{h}], \tag{28}$$

$$\Pi_{\mathbb{C}}[f, h] = \frac{1}{2} \int_{\partial M} \int_0^{\infty} (f(t) \partial_t \Lambda h(t) + \partial_t \Lambda f(t) h(t))\, \rho dS_x dt \tag{29}$$

with

$$\Pi_{\mathbb{C}}[f, h] = \lim_{t \to \infty} E_{\mathbb{C}}[u^f(t), v^h(t)],$$

where

$$\begin{aligned} E_{\mathbb{C}}[u(t), v(t)] &= \frac{1}{2} \int_{\partial M} \eta u(t) v(t)\, \rho dS_g \\ &+ \frac{1}{2} \int_M \left(\langle \nabla u(t), \nabla v(t) \rangle_g + q u(t) v(t) + \partial_t u(t) \partial_t v(t)\right) d\mu. \end{aligned} \tag{30}$$

In the following, we will prove the equivalence for the complex forms. Clearly, due to (27)–(30) this will prove Theorem 1.

2.2. Spectral data. We start with the equivalence of the Gel'fand boundary data $dP(x,y;\lambda)$ and Calderon-Gel'fand forms $\Lambda^z_{\mathbb{C}}$.

<u>*i.* → *ii.*</u> Consider the complex bilinear form

$$\langle (A-z)^{-1}\Phi, \Psi\rangle_{L^2(M)}, \quad z \notin \sigma(A). \tag{31}$$

As $H^{-1}(M) = \mathcal{D}((A+c)^{1/2})$, this form can be continued onto $H^{-1}(M)$, in particular, onto Ψ, Φ of the form

$$\Psi = \psi(x)\delta_{\partial M}(x), \quad \Phi = \phi(x)\delta_{\partial M}(x), \tag{32}$$

where ψ, $\phi \in C_0^\infty(\partial M)$ or, more generally, in $H^{-1/2}(\partial M)$. Next we use the fact that

$$u_z^\phi = (A-z)^{-1}\Phi = \int_{\mathbb{R}} \frac{d_\lambda E(\lambda)\Phi}{\lambda - z}. \tag{33}$$

Then

$$\Lambda^z_{\mathbb{C}}[\phi,\psi] = \langle (A-z)^{-1}\Phi, \Psi\rangle_{L^2(M)} = \int_{\sigma(A)} \frac{dm_{\phi,\psi}(\lambda)}{\lambda - z}, \tag{34}$$

where $m_{\phi,\psi}(\lambda)$ is the complexified spectral measure

$$m_{\phi,\psi}(\lambda) = \langle E(\lambda)\Phi,\, \Psi\rangle = \big(E(\lambda)\Phi,\, \overline{\Psi}\big) = \int_{\partial M} \phi(y)\psi(x)\, dP(x,y;\lambda). \tag{35}$$

Thus the boundary spectral data uniquely determine the Calderon-Gel'fand forms, i.e., *i.* determines *ii.*

<u>*ii.* → *i.*</u> To prove the opposite we use the Pleijel-Stone formula (e.g. [Ko], [RS]) which together with (33)–(35) shows that

$$\begin{aligned} &\lim_{\varepsilon\to+0} \frac{1}{2\pi i} \int_{-\infty}^{\lambda} \langle [(A-k-i\varepsilon)^{-1} - (A-k+i\varepsilon)^{-1}]\Phi, \Psi\rangle_{L^2(M)} dk \\ &\quad = \frac{1}{2}\int_{\partial M} \phi(x)\psi(x)\, (dP(x,y;\lambda-0) + dP(x,y;\lambda)), \end{aligned} \tag{36}$$

where $dP(x,y;\lambda-0) = \lim_{\varepsilon\to 0-} dP(x,y;\lambda-\varepsilon)$ is considered as a distribution limit. Because $\phi,\psi \in C_0^\infty(\partial M)$ are arbitrary, equations (34), (36) imply that the forms $\Lambda^z_{\mathbb{C}}$ determine $dP(x,y;\lambda-0) + dP(x,y;\lambda)$. Since the spectral projectors $E(\lambda)$ are continuous from the right we see that, in the sense of distributions,

$$\begin{aligned} dP(x,y;\lambda) &= \lim_{\varepsilon\to+0} dP(x,y;\lambda+\varepsilon) \\ &= \frac{1}{2}\lim_{\varepsilon\to+0} (dP(x,y;\lambda+\varepsilon) + dP(x,y;\lambda+\varepsilon-0)), \end{aligned}$$

Thus we can also determine $dP(x, y; \lambda)$, i.e., *ii.* determines *i.*

2.3. Hyperbolic data. Next we return to the time-domain problem. For $f \in C_0^\infty(\partial M \times \mathbb{R})$, let $\widehat{f}$ be its Fourier transform with respect to time,

$$\widehat{f}(x,k) = \int_{\mathbb{R}} e^{-ikt} f(x,t)\, dt.$$

If $f = 0$ for $|t| \geq c_0$, the function $\widehat{f}(x,k)$ is analytic in $k \in \mathbb{C}$ and C^∞–smooth in x. Moreover, by the Paley-Wiener theorem this functions satisfies

$$(37) \quad ||\widehat{f}(\cdot\,,k)||_{C^m(\partial M)} \leq C_{m,N}(1+|k|)^{-N} \exp(c_0|\operatorname{Im} k|), \quad m, N > 0.$$

The solution $u^f(x,t)$ of problem (13) lies in $C^\infty(M \times \mathbb{R})$ and, using spectral resolution (6),

$$(38) \qquad u^f(t) = \int_{\mathbb{R}} \int_{-\infty}^{t} \frac{\sin\sqrt{\lambda}(t-t')}{\sqrt{\lambda}}\, d_\lambda(E(\lambda)F(t'))\, dt',$$

where $F(x,t) = f(x,t)\delta_{\partial M}(x)$. Using this representation,

$$(39) \qquad \|u^f(t)\|_{L^2(M)} \leq C_f e^{\tau_0 t}, \quad t > 0,$$

where

$$(40) \qquad \tau_0 = \sqrt{\max(0, -\lambda_0)},$$

and λ_0 is the bottom of the spectrum $\sigma(A)$. Then the Fourier transform $\widehat{u^f}(x,k)$ of $u^f(x,t)$ is well-defined for $\operatorname{Im} k < -\tau_0$ and is there the solution of elliptic problem (10) with $z = k^2$ and $\phi(x) = \widehat{f}(x,k)$. Thus, when $\operatorname{Im} k < -\tau_0$

$$(41) \qquad \widehat{\Lambda f}(k) = \widehat{u^f}(k)|_{\partial M} = \Lambda^{k^2} \widehat{f}(k).$$

However, the right-hand side of (41) is analytic when $k^2 \notin \sigma(A)$ which determines an analytic continuation of $\widehat{\Lambda f}(k)$ onto $k \in \mathbb{C}$, $k^2 \notin \sigma(A) \subset [\lambda_0, \infty) \subset \mathbb{R}$.

After these preparations we can show that the hyperbolic data *iii.* and *iv.* are equivalent to each other and to *i.*

<u>*i.* → *iii.*</u> We will show that $\int_0^\infty \int_{\partial M} \Lambda f\, h\, \rho dS_g\, dt$, where $f, h \in C_0^\infty(\partial M \times \mathbb{R})$, may be represented in terms of the Gel'fand boundary spectral data $dP(x, y; \lambda)$. Due to (15) and (28) this will prove that *i.* determines *iii.* Indeed, by the Parseval identity and formula (41)

$$(42) \qquad \begin{aligned} \int_0^\infty \int_{\partial M} \Lambda f\, h\, \rho dS_g\, dt &= \langle e^{-\tau t}\Lambda f, e^{\tau t} h\rangle_{L^2(\partial M \times \mathbb{R}_+)} \\ &= \int_{\mathbb{R}} \Lambda_{\mathbb{C}}^{(k-i\tau)^2} [\widehat{f}(k-i\tau),\, \widehat{h}(-k+i\tau)]\, dk. \end{aligned}$$

Using formulae (34), (35) we obtain from (42) that

$$
(43)\quad \begin{aligned}
&\int_0^\infty \int_{\partial M} \Lambda f h\, \rho dS_g\, dt \\
&= \int_{\mathbb{R}} \left[\int_{\mathbb{R}} \int_{\partial M} \int_{\partial M} \frac{\widehat{f}(x,k-i\tau)\widehat{h}(y,-k+i\tau)}{\lambda-(k-i\tau)^2}\, d_{x,y,\lambda} P(x,y;\lambda) \right] dk.
\end{aligned}
$$

Hence $dP(x,y;\lambda)$ determines the forms Λ and $\mathcal{B}$.

<u>iii. → iv.</u> Let $f \in C_0^\infty(\partial M \times \mathbb{R}_+)$. It follows from (30) that

$$
\partial_t E_{\mathbb{C}}(u^f,t) = \int_{\partial M} f(t) \partial_t \Lambda f(t)\, \rho dS.
$$

As $\partial_t \Lambda f = \Lambda \partial_t f$, integrating by parts and using definitions (15), (17) we obtain

$$
(44)\qquad \Pi_{\mathbb{C}}(f) = \int_0^\infty \int_{\partial M} f(t)\partial_t \Lambda f(t)\, \rho dS dt = \frac{1}{2}\mathcal{B}_{\mathbb{C}}[f,\partial_t f],
$$

which implies that the hyperbolic form determines the energy flux.

<u>iv. → i.</u> Let us consider the form $\Pi_{\mathbb{C}}[f,h]$ when f,h are of the form

$$
(45)\qquad f(x,t) = f_0(t)\phi(x), \quad h(x,t) = h_0(t)\psi(x),
$$

with $f_0, h_0 \in C_0^\infty(\mathbb{R})$ and $\phi, \psi \in C_0^\infty(\partial M)$.

We intend to use the Parseval formula as in (42) to represent the integral in the rhs of (29) in terms of the Fourier transforms $\widehat{f}$, $\widehat{h}$. When f,h are of form (45), then $\widehat{f}(x,k) = \widehat{f_0}(k)\phi(x)$, $\widehat{h}(x,k) = \widehat{h_0}(k)\psi(x)$ and satisfy (37). Thus using partial integration we see that

$$
\Pi_{\mathbb{C}}[f,h] = -\frac{1}{2}\int_0^\infty \int_{\partial M} \left[e^{\tau t}\partial_t f(t)(e^{-\tau t}\Lambda h(t)) + (e^{-\tau t}\Lambda f(t))e^{\tau t}\partial_t h(t) \right] \rho dS_x dt.
$$

As Λ^{k^2} is analytic for $k^2 \notin \sigma(A)$, applying the Parseval formula we see that

$$
(46)\quad \begin{aligned}
\Pi_{\mathbb{C}}[f,h] &= \frac{1}{4\pi i}\int_{\mathbb{R}}\int_{\partial M} (k+i\tau)\langle \widehat{f}(k+i\tau), \widehat{\Lambda h}(-k-i\tau)\rangle_{L^2(\partial M)} dk \\
&\quad - \frac{1}{4\pi i}\int_{\mathbb{R}}\int_{\partial M} (k-i\tau)\langle \widehat{\Lambda f}(k-i\tau), \widehat{h}(-k+i\tau)\rangle_{L^2(\partial M)} dk
\end{aligned}
$$

Using the above formulae together with (34) and (41) we obtain that

$$
(47)\quad \begin{aligned}
\Pi_{\mathbb{C}}[f,h] &= \frac{1}{4\pi i}\int_{\Gamma_\tau} \Lambda_{\mathbb{C}}^{k^2}[\phi,\psi]\widehat{f_0}(k)\,\widehat{h_0}(-k) k dk \\
&= \frac{1}{4\pi i}\int_{\Gamma_\tau} \langle (A-k^2)^{-1}\Phi,\, \Psi\rangle_{L^2(M)} \widehat{f_0}(k)\widehat{h_0}(-k)\, k dk,
\end{aligned}
$$

where Γ_τ is the boundary of the strip $\{k \in \mathbb{C} : |\operatorname{Im} k| \le \tau\}$ and Φ, Ψ are given by formula (32). Because $|\langle (A-k^2)^{-1}\Phi, \Psi\rangle| \le c_{\phi,\psi}$ for $k \in \Gamma_\tau$ when $\tau > \tau_0$ (40) and $\Phi, \Psi \in H^{-1}(M)$ while $\widehat{f_0}(k), \widehat{h_0}(-k) \in \mathcal{S}(\mathbb{R})$, integral (47) converges absolutely. Using again (34) we see that

$$\Pi_{\mathbb{C}}[f,h] = \frac{1}{4\pi i}\int_{\sigma(A)}\left[\int_{\Gamma_\tau} \widehat{f_0}(k)\widehat{h_0}(-k)\,\frac{kdk}{\lambda - k^2}\right] dm_{\phi,\psi}(\lambda), \tag{48}$$

Further considerations are based on the following lemma.

LEMMA 2. *For any z, $Im\, z > \tau_0$ there is a sequence $f_0^n \in C_0^\infty(\mathbb{R})$ such that*

$$\lim_{n\to\infty} \Pi_{\mathbb{C}}[f^n] = \frac{1}{2}\int_{\sigma(A)} \frac{dm_\phi(\lambda)}{\lambda - z^2}, \tag{49}$$

where $f^n = f_0^n(t)\phi(x)$ and $dm_\phi(\lambda) = dm_{\phi,\phi}(\lambda)$.

Proof. i. We note that if $f_0(t) = h_0(t) = \exp(izt)H(t)$, where $H(t)$ is the Heaviside function, then $\widehat{f_0}(k)\widehat{h_0}(-k) = (k^2 - z^2)^{-1}$. Therefore, using formally (48) we obtain by means of the residue theorem that

$$\Pi_{\mathbb{C}}[f] = \frac{1}{2}\int_{\sigma(A)} \frac{dm_\phi(\lambda)}{\lambda - z^2}.$$

Next, let $u^f(t)$, $t > 0$ solve (13) with $f = H(t)\exp(izt)\phi(x)$. Then,

$$u^f(x,t) = \exp(izt)u^\phi_{z^2}(x) + w(x,t), \tag{50}$$

where $u^\phi_{z^2} = (A - z^2)^{-1}\Phi$ exists since $\operatorname{Im} z > \tau_0$ and w satisfies $(\partial_t^2 + A)w = 0$, $w|_{t=0} = -u^\phi_{z^2}$, $w_t|_{t=0} = -izu^\phi_{z^2}$. Using spectral resolution (6) and formula (33), we obtain the representation

$$\begin{aligned} w(t) &= -\int_{\mathbb{R}} W(\lambda, z, t)d_\lambda(E(\lambda)u^\phi_{z^2}), \\ W(\lambda, z, t) &= \cos(\sqrt{\lambda}t) + iz\frac{\sin(\sqrt{\lambda}t)}{\sqrt{\lambda}} \end{aligned} \tag{51}$$

Since by (34),

$$d_\lambda\langle E(\lambda)u^\phi_{z^2}, u^\phi_{z^2}\rangle = \frac{d_\lambda\langle E(\lambda)\Phi, \Phi\rangle}{(\lambda - z^2)^2} = \frac{dm_\phi(\lambda)}{(\lambda - z^2)^2},$$

formula (51) implies that

$$\begin{aligned} E_{\mathbb{C}}(w,t) &= \frac{1}{2}\int_{\sigma(A)} \left[\lambda W^2 + (\partial_t W)^2\right] d_\lambda\langle E(\lambda)u^\phi_{z^2}, u^\phi_{z^2}\rangle \\ &= \frac{1}{2}\int_{\sigma(A)} \frac{dm_\phi(\lambda)}{\lambda - z^2}. \end{aligned} \tag{52}$$

In particular, $E_{\mathbb{C}}(w,t) = E_{\mathbb{C}}(w,0)$. Moreover, it follows from (51) that

$$\|w(t)\|^2_{H^1(M)} + \|\partial_t w(t)\|^2_{L^2(M)} < c_\phi \exp(2\tau_0 t), \quad \text{when } t > 0, \tag{53}$$

while

$$\|\exp(izt)u^\phi_{z^2}\|^2_{H^1(M)} + \|\partial_t(\exp(izt)u^\phi_{z^2})\|^2_{L^2(M)} \le c_\phi \exp[-2(\operatorname{Im} z)t]. \tag{54}$$

Because $\operatorname{Im} z > \tau_0$, representation (50) together with (52)–(54) imply that $E_{\mathbb{C}}(u^f, t)$ is defined for $t \ge 0$ and

$$\Pi_{\mathbb{C}}[f] = \lim_{t\to\infty} E_{\mathbb{C}}(u^f, t) = E_{\mathbb{C}}(w, 0). \tag{55}$$

Comparing formulae (52), (55) and (34) we see that

$$\Pi_{\mathbb{C}}[f] = \lim_{t\to\infty} E_{\mathbb{C}}(u^f, t) = \frac{1}{2}\Lambda^{z^2}_{\mathbb{C}}[\phi, \phi].$$

Next we consider $u_{\varepsilon,N}(x,t)$ that is the solution of (13) with $f = f_{\varepsilon,N}(t)\phi(x)$, where $f_{\varepsilon,N}(t)$ is given by

$$f_{\varepsilon,N}(t) = f_\varepsilon(t)\widetilde{\chi}_N(t), \;\; f_\varepsilon(t) = (H(\cdot)\exp(iz\cdot) * \chi_\varepsilon)(t).$$

Here $*$ stands for the convolution in time, $\chi_\varepsilon(t)$ is a usual mollifier with $\operatorname{supp}(\chi_\varepsilon) \subset (-\varepsilon, \varepsilon)$ and $\widetilde{\chi}_N$ is a smooth cut-off function, $\widetilde{\chi}_N = 1$ for $t \le N$ and 0 for $t \ge N+1$. Then

$$u_{\varepsilon,N}(x,t) = f_{\varepsilon,N}(t)u^\phi_{z^2}(x) + w^1_\varepsilon(x,t) + w^2_{\varepsilon,N}(x,t), \tag{56}$$

where, for $t > 1$,

$$w^1_\varepsilon(t) = -\int_{\sigma(A)} (W(\lambda, z, \cdot) * \chi_\varepsilon)(t)\, d_\lambda(E(\lambda)u^\phi_{z^2}), \tag{57}$$

and, for $t > N+1$,

$$\begin{aligned} w^2_{\varepsilon,N}(t) &= -\int_{\sigma(A)} W^2_{\varepsilon,N}(\lambda, z, t) d_\lambda(E(\lambda)u^\phi_{z^2}), \\ W^2_{\varepsilon,N}(t) &= \int_0^t \frac{\sin(\sqrt{\lambda}(t-t'))}{\sqrt{\lambda}}\,[2\partial_t\widetilde{\chi}_N(t')\partial_t f_\varepsilon(t') + \partial_t^2\widetilde{\chi}_N(t')f_\varepsilon(t')]\,dt'. \end{aligned} \tag{58}$$

Clearly, $(W(\lambda, z, \cdot) * \chi_\varepsilon(\cdot))(t) \to W(\lambda, z, t)$ uniformly on any compact set of λ, t and are uniformly bounded when $\varepsilon \in (0,1)$, $\lambda \in \sigma(A)$ on any compact set of t. Because $\int_{\sigma(A)}(1+|\lambda|)d(E(\lambda)u^\phi_{z^2}, u^\phi_{z^2}) < \infty$ this implies that

$$\lim_{\varepsilon\to 0} E_{\mathbb{C}}(w^1_\varepsilon, t) = E_{\mathbb{C}}(w, t)$$

for bounded t. However, $E_{\mathbb{C}}(w^1_\varepsilon, t)$ does not depend on t for $t > 1$.

Furthermore, $E_{\mathbb{C}}(w_\varepsilon^1 + w_{\varepsilon,N}^2, t)$ does not depend on t for $t > N+1$ and, as is seen from (58),

$$\lim_{N\to\infty} E_{\mathbb{C}}(w_\varepsilon^1 + w_{\varepsilon,N}^2, N+1) = \lim_{N\to\infty} E_{\mathbb{C}}(w_\varepsilon^1, N+1) = E_{\mathbb{C}}(w_\varepsilon^1, 1). \tag{59}$$

In addition, $w_\varepsilon^1(t)$, $w_{\varepsilon,N}^2(t)$ satisfy estimate (53) uniformly for ε and N while $f_{\varepsilon,N}(t)u_{z^2}^\phi$ satisfies estimate (54). Thus,

$$\Pi_{\mathbb{C}}[f_{\varepsilon,N}] = \lim_{t\to\infty} E_{\mathbb{C}}(u_{\varepsilon,N}, t) = E_{\mathbb{C}}(w_\varepsilon^1 + w_{\varepsilon,N}^2, N+1). \tag{60}$$

Combining (59) and (60) with (52), we can choose a sequence ε_n, N_n so that $f_n = f_{\varepsilon_n,N_n}$ satisfy

$$\lim_{n\to\infty} \Pi_{\mathbb{C}}(f_n) = \lim_{\varepsilon\to 0} E_{\mathbb{C}}(w_\varepsilon^1, 1) = \frac{1}{2}\int_{\sigma(A)} \frac{dm_\phi(\lambda)}{\lambda - z^2}.$$

□

Lemma 2 shows that $\Pi_{\mathbb{C}}[f]$ determines $F(\omega) = \int_{\sigma(A)} \frac{dm_\phi(\lambda)}{\lambda-\omega}$ for any $\omega = z^2$, $\operatorname{Im} z > \tau_0$. Because $F(\omega)$ is analytic outside $\sigma(A)$ we can continue it to $\mathbb{C}\setminus\sigma(A)$. Then the Pleijel formula (compare with (36)) may be used to find $m_\phi(\lambda)$ and, henceforth, using polarization, $m_{\phi,\psi}(\lambda)$. Clearly, this also determines $dP(x,y,\lambda)$.

Remark 2. When $\sigma(A) \subset \mathbb{R}_+$, $F(\omega)$, $\omega \notin \sigma(A)$ can be directly found from $\Pi_{\mathbb{C}}$ without analytic continuation. Thus, the step *iv.* $\to$ *i.* does not require analytic continuation.

3. Reconstructions.

3.1. Blagovestchenskii identity. In this section we will describe a procedure to reconstruct the manifold and the Schrödinger operator on it. Here the given data is the hyperbolic form $\mathcal{B}^{2T}$ that is gauge equivalent to the form $\widetilde{\mathcal{B}}^{2T}$ of the Schrödinger operator. In Section 5 we will generalize our results to the case when data is given both on a finite part $S \subset \partial M$ and finite time-interval. To this end we present our constructions so that they can be easily extended to this general case. Remarks in the text often give generalizations of results which are used later in Section 5. As in [KKL], we will actually construct an isometric copy of (M, g) and an operator on it.

For $x \in M$ and $\xi \in T_x(M)$, $|\xi|_g = 1$, we denote by $\gamma_{x,\xi}(s)$ the geodesic parametrized by its path length which starts at x in the direction ξ.

By Lemma 1 there is a gauge transformation S_κ which makes A into a Schrödinger operator $A_\kappa = -\Delta_g + q$. We denote by $\widetilde{\Lambda}^{2T}$, $\widetilde{\mathcal{B}}^{2T}$ the Robin-to-Dirichlet and hyperbolic forms for $-\Delta_g + q$, and by Λ^{2T}, $\mathcal{B}^{2T}$ – those forms for A. By (20), $\mathcal{B}^{2T}[f,h] = \widetilde{\mathcal{B}}^{2T}[\kappa|_{\partial M} f, \kappa|_{\partial M} h]$, i.e., we are given the form

$$f, h \to \widetilde{\mathcal{B}}^{2T}[\kappa|_{\partial M} f, \kappa|_{\partial M} h],$$

where $\kappa|_{\partial M}$ is unknown. (In the future, when it does not cause confusion we will write κ instead of $\kappa|_{\partial M}$).

Consider the initial boundary value problem for the Schrödinger operator,

$$\begin{aligned} &\partial_t^2 u^f - \Delta_g u^f + q u^f = 0 \quad \text{in} \quad M\times\mathbb{R}, \\ &Bu^f\big|_{\partial M\times\mathbb{R}_+} = f, \quad u^f\big|_{t=0} = 0, \quad \partial_t u^f\big|_{t=0} = 0. \end{aligned} \tag{61}$$

The reconstruction is based on two main ingredients, namely the computation of the inner products of solutions to problem (61) and controllability results.

We start with the inner products, that is the Blagovestchenskii identity. Denote by $\mathring{C}^\infty(\Gamma\times\mathbb{R}_+), \Gamma\subset\partial M$ the class of functions

$$f\in C_0^\infty(\Gamma\times\mathbb{R}), \quad f=0 \text{ when } t<0.$$

LEMMA 3. *Let* $f,h\in\mathring{C}^\infty(\partial M\times\mathbb{R}_+)$. *Then*

$$\int_M u^{\kappa f}(T)\overline{u^{\kappa h}(T)}\,dV_g = \frac{1}{2}\int_{-T}^{T} sign(\tau)\mathcal{B}^{2T}[Y_{T+\tau}(f), Y_{T-\tau}(h)]\,d\tau, \tag{62}$$

where Y_τ *is the delay operator,* $(Y_\tau f)(x,t) = f(x,t-\tau)$.

Proof. Let $w(t,s) = \int_M u^f(t)\overline{u^h(s)}\,dV_g$. Integrating by parts and using (61) we see that

$$\begin{aligned} (\partial_t^2-\partial_s^2)w(t,s) &= -\int_M [(-\Delta_g+q)u^f(t)\overline{u^h(s)} - u^f(t)\overline{(-\Delta_g+q)u^h(s)}]dV_g \\ &= -\int_{\partial M}[\partial_\nu u^f(t)\overline{u^h(s)} - u^f(t)\overline{\partial_\nu u^h(s)}]\,dS_g \\ &= \int_{\partial M}[f(t)\overline{\widetilde\Lambda^{2T}h(s)} - \widetilde\Lambda^{2T}f(t)\overline{h(s)}]\,dS_g. \end{aligned} \tag{63}$$

Moreover,

$$w|_{t=0} = w|_{s=0} = 0, \quad \partial_t w|_{t=0} = \partial_s w|_{s=0} = 0.$$

Thus, for $s\ge t$,

$$w(t,s) = \frac{1}{2}\int_{L(t,s)}\left[\int_{\partial M}[f(t')\overline{\widetilde\Lambda^{2T}h(s')} - \widetilde\Lambda^{2T}f(t')\overline{h(s')}]\,dS_g\right]dt'ds',$$

where $L(t,s)$ is the triangle bounded by $s'+t'=s+t$, $s'-t'=s-t$ and $t'=0$. Introducing $\tau=\frac{1}{2}(s'-t')$, $\theta=\frac{1}{2}(s'+t')$, we have

$$\begin{aligned} w(t,s) &= \int_{(s-t)/2}^{(s+t)/2}\int_\tau^{(s+t)/2}\left[Y_\tau f(\theta)\,\overline{\widetilde\Lambda^{2T}Y_{-\tau}h(\theta)} - \widetilde\Lambda^{2T}Y_\tau f(\theta)\,\overline{Y_{-\tau}h(\theta)}\right]d\theta d\tau \\ &= \int_{(s-t)/2}^{(s+t)/2}\widetilde{\mathcal{B}}^{2T}[Y_{\delta+\tau}f,\, Y_{\delta-\tau}h]\,d\tau, \quad \text{where } \delta = 2T-\frac{1}{2}(s+t). \end{aligned} \tag{64}$$

Similar formula can be written for $s \le t$. When $s = t$ we can use both to obtain a symmetrized one. Taking κf, κh instead of f, h and using relation (20), this symmetrized formula with $s = t = T$ takes the form (62). □

Remark 3. It is clear from the proof, e.g. formula (64) that $\mathcal{B}^{2T}$ determines $\big(u^{\kappa f}(t), u^{\kappa h}(s)\big)$ for $s + t \le 2T$.

3.2. Approximate controllability. Let $t > 0$ and $\Gamma \subset \partial M$ be open and

$$M(\Gamma, t) = \{x \in M \ : \ \mathrm{dist}(x, \Gamma) < t\}, \tag{65}$$

be the domain of influence of Γ at time t. When $f \in C_0^\infty(\Gamma \times \mathbb{R}_+)$,

$$u^{\kappa f}(t) \in L^2(M(\Gamma, t)) = \{u \in L^2(M) : \ supp(u) \subset M(\Gamma, t)\}.$$

The controllability result we need is based on the celebrated Tataru's Holmgren-John unique continuation theorem [Ta1] (see also [Ta2], [Ho]).

THEOREM 4. *Let u be a solution of wave equation (61). Assume that*

$$u|_{\Gamma \times (0, 2\tau)} = 0, \ \ \partial_\nu u|_{\Gamma \times (0, 2\tau)} = 0 \tag{66}$$

where $\Gamma \subset \partial M$, $\Gamma \neq \emptyset$ is open. Then,

$$u(x, \tau) = 0, \ \ \partial_t u(x, \tau) = 0 \ \mathit{for}\ x \in M(\Gamma, \tau).$$

This result yields the following controllability result.

THEOREM 5. *Let $\Gamma \subset \partial M$ be open and $\tau > 0$. Then the linear subspace,*

$$\{u^f(\tau) \in L^2(M(\Gamma, \tau)) : \ f \in C_0^\infty(\Gamma \times [0, \tau])\},$$

is dense in $L^2(M(\Gamma, \tau))$.

Sketch of the proof. Let $\psi \in L^2(M(\Gamma, \tau))$ be such that

$$\langle u^f(\cdot\,, \tau), \psi\rangle = 0 \tag{67}$$

for all $f \in C_0^\infty(\Gamma \times [0, \tau])$. We need to show that $\psi = 0$. To this end, consider the following initial boundary value problem,

$$(\partial_t^2 - \Delta_g + q)e = 0, \quad Be|_{\partial M \times \mathbb{R}} = 0, \quad e|_{t=\tau} = 0, \quad \partial_t e|_{t=\tau} = \psi. \tag{68}$$

Integrating by parts and using equations (67) and (68), we obtain that

$$\begin{aligned} 0 &= \int_{M \times [0,\tau]} [u^f\, \overline{(\partial_t^2 - \Delta_g + q)e} - (\partial_t^2 - \Delta_g + q)u^f\, \overline{e}]\, dV_g\, dt \\ &= \int_{\partial M \times [0,\tau]} f\, \overline{e}\, dS_g\, dt. \end{aligned}$$

Since $f \in C_0^\infty(\Gamma \times [0,\tau])$ is arbitrary, $e|_{\Gamma\times[0,\tau]} = 0$. Together with boundary conditions in (68), this yields that the Cauchy data of e vanish on $\Gamma \times [0,\tau]$. Moreover, since $e(x,t) = -e(x, 2\tau - t)$ due to $e|_{t=\tau} = 0$, the Cauchy data of e vanish on $\Gamma \times [0, 2\tau]$. Therefore, $\psi = 0$ due to Theorem 4. □

3.3. Inner products of waves. Now our main ingredients are obtained and there have been various ways to proceed. In this section we will consider a method based on a minimization algorithm. In the next section we will briefly present a method based on Gaussian beams (for more detail see e.g. [BKa], [KK], [KKL]). Alternatively, one can use methods based on propagation of singularities, e.g. [B2] or methods based on a wave approximation of delta-distributions [KL2].

We start with projections onto domains of influences. Actually, there are several ways to obtain these projections. Here we describe an approach based on minimization because we want to use in the reconstruction as few unstable procedures as possible. A more explicit construction based on the Gram-Schmidt orthgonalization procedure may be found in e.g. [B1] or [KKL].

Let $P_{\Gamma,\tau} : L^2(M) \to L^2(M(\Gamma,\tau))$ be the orthoprojection,

$$P_{\Gamma,\tau}u(x) = \chi_{M(\Gamma,\tau)}(x)u(x),$$

where $\chi_{M(\Gamma,\tau)}$ is the characteristic function of the set $M(\Gamma,\tau)$.

LEMMA 4. *Let $f, h \in C_0^\infty(\partial M \times \mathbb{R}_+)$, $T > 0$, $t, s, \tau_1, \tau_2 \in [0, T]$. Let also $\Gamma_1, \Gamma_2 \subset \partial M$ be open sets. Assume that we are given the form $\mathcal{B}^{2T}$. Then it is possible to find the inner products*

$$(69)\quad \begin{aligned} &\left(P_{\Gamma_1,\tau_1}u^{\kappa f}(t), P_{\Gamma_2,\tau_2}u^{\kappa h}(s)\right)_{L^2(M)} \\ &\qquad = \int_{M(\Gamma_1,\tau_1)\cap M(\Gamma_2,\tau_2)} u^{\kappa f}(x,t)\, \overline{u^{\kappa h}(x,s)}\, dV_g. \end{aligned}$$

Proof. Using Theorem 5 we see that

$$(70)\quad \begin{aligned} &\|u^{\kappa f}(t)\|^2 - \|P_{\Gamma_1,\tau_1}u^{\kappa f}(t)\|^2 \\ &\qquad = \|(1 - P_{\Gamma_1,\tau_1})u^{\kappa f}(t)\|^2 \\ &\qquad = \inf\{\|u^{\kappa f}(t) - u^{\kappa\eta}(\tau_1)\|^2 :\ \eta \in C_0^\infty(\Gamma_1 \times [0,\tau_1])\}. \end{aligned}$$

Since

$$(71)\quad \begin{aligned} \|u^{\kappa f}(t) - u^{\kappa\eta}(\tau_1)\|^2 = \|u^{\kappa f}(t)\|^2 - 2\,\mathrm{Re}\left(u^{\kappa f}(t), u^{\kappa\eta}(\tau_1)\right) \\ + \|u^{\kappa\eta}(\tau_1)\|^2, \end{aligned}$$

the rhs of (70) can be computed by Lemma 3 and Remark 3. Therefore, we can choose a sequence $\eta_j \in C_0^\infty(\Gamma_1 \times [0,\tau_1])$ such that $\lim u^{\kappa\eta_j}(\tau_1) = P_{\Gamma_1,\tau_1}u^{\kappa f}(t)$. Similarly, we find $\widetilde{\eta}_k \in C_0^\infty(\Gamma_2 \times [0,\tau_2])$ with $\lim u^{\kappa\widetilde{\eta}_k}(\tau_2) = P_{\Gamma_2,\tau_2}u^{\kappa h}(s)$. As $\left(u^{\kappa\eta_j}(\tau_1), u^{\kappa\widetilde{\eta}_k}(\tau_2)\right)$ can be found by Lemma 3 and Remark 3 this proves the result. □

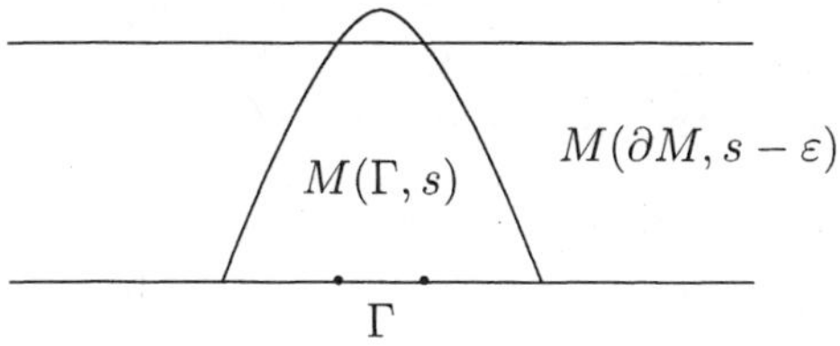

FIG. 1. *The set* $M(\Gamma, s) \setminus M(\partial M, s-\varepsilon)$.

Remark 4. Lemma 4 remains valid if

$$t+\tau_1 \leq 2T, \quad s+\tau_2 \leq 2T$$

This is true due to the remark after Lemma 3.

Let $z_1, z_2 \in \partial M$ and denote $M(z_1, \tau) = \{x \in M : \ d(x, z_1) < \tau\}$. If sequences of open sets $\Gamma_i^j \to z_i$, $i = 1, 2$, then $M(\Gamma_i^j, \tau_i) \to M(z_i, \tau_i)$. Hence we obtain

COROLLARY 2. *Let $\mathcal{B}^{2T}$ be given. Then for $f, h \in C_0^\infty(\partial M \times \mathbb{R}_+)$, $t, s, \tau_1, \tau_2 \in [0, T]$ we can find*

$$\left(P_{z_1,\tau_1} u^{\kappa f}(t), u^{\kappa h}(s)\right), \quad \left(P_{z_1,\tau_1} u^{\kappa f}(t), P_{z_2,\tau_2} u^{\kappa h}(s)\right),$$

where $P_{z,\tau}$ is the projection to $M(z, \tau)$.

3.4. Reconstruction of the manifold.

LEMMA 5. *Let $y \in \partial M$, $s \in (0, T)$, and $\gamma_{y,\nu}([0, s])$ be a normal geodesic. Then given $\mathcal{B}^{2T}$ we can determine whether $\gamma_{y,\nu}$ is the shortest geodesic between $\gamma_{y,\nu}(s)$ and the boundary ∂M or not. Moreover, when this geodesic is minimal, $\mathcal{B}^{2T}$ determines* $\min(d(\gamma_{y,\nu}(s), z), T)$ *for any $z \in \partial M$.*

Proof. i. The geodesic $\gamma_{y,\nu}$ is the shortest geodesic between $\gamma_{y,\nu}(s)$ and ∂M if and only if for any $\varepsilon > 0$ and any neighborhood Γ of y (see Fig. 1.),

$$M(\Gamma, s) \setminus M(\partial M, s-\varepsilon) \neq \emptyset. \tag{72}$$

By Theorem 5 property (72) is true if and only if for some $h \in C^\infty(\Gamma \times [0, s])$

$$||u^{\kappa h}(s)|| > ||P_{M(\partial M, s-\varepsilon)} u^{\kappa h}(s)||$$

However, this inequality can be checked when $\mathcal{B}^{2T}$ is given.

ii. Let now $\gamma_{y,\nu}$ be the shortest geodesic between $\gamma_{y,\nu}(s)$ and ∂M, $z \in \partial M$ and $t \in (0, T)$. Then $t \geq d(\gamma_{y,\nu}(s), z)$ if and only if there is a neighborhood $\Gamma \subset \partial M$ of y such that for sufficiently small $\varepsilon > 0$,

$$M(\Gamma, s) \subset M(\partial M, s-\varepsilon) \cup M(z, t). \tag{73}$$

Now, property (73) is true if and only if for any $h \in C_0^\infty(\Gamma \times [0, s])$,

$$||u^{\kappa h}(s)||^2 = ||P_{M(\partial M, s-\varepsilon)} u^{\kappa h}(s)||^2 + ||P_{M(z,t)} u^{\kappa h}(s)||^2$$
$$- (P_{M(\partial M, s-\varepsilon)} u^{\kappa h}(s), P_{M(z,t)} u^{\kappa h}(s))_{L^2(M)}.$$

This again can be checked using $\mathcal{B}^{2T}$. Thus taking infimum of all t for which property (73) is satisfied we find $\min(d(\gamma_{y,\nu}(s),z),T)$. □

Next we introduce the truncated boundary distance functions

$$r_x^T(z) = \min(d(x,z),T), \quad z \in \partial M, \; x \in M,$$

and the truncated boundary distance map

$$R^T : M \mapsto C(\partial M), \; x \mapsto r_x^T.$$

Since each x has a closest boundary point z_x and the shortest geodesic from z_x to x is normal, Lemma 5 implies

THEOREM 6. *The hyperbolic form $\mathcal{B}^{2T}$ determines the set $R^T(M(\partial M,T))$.*

One can show (e.g. [K2], [KKL]) that the map $R : M(\partial M,T) \to R^T(M(\partial M,T))$ is a homeomorphism.

Our further constructions use the evaluation functions, E_z, $z \in \partial M$,

$$E_z : R^T(M(\partial M,T)) \to \mathbb{R}, \qquad E_z(r_x^T) = r_x^T(z) = \min(d(x,z),T). \tag{74}$$

It can be shown that for any $r_{\widetilde{x}}^T \in R^T(M(\partial M,T))$ we can choose $z_1, \cdots, z_m \in \partial M$ so that E_{z_j}, $j = 1, \cdots, m$, form a system of coordinates near $r_{\widetilde{x}}^T$ which makes $R^T(M(\partial M,T))$ diffeomorphic to $M(\partial M,T)$. Let $\widetilde{g}$ be the metric on $R^T(M(\partial M,T))$ which makes it isometric to $(M(\partial M,T),g)$, i.e., $\widetilde{g} = ((R^T)^{-1})^{\cdot} g$. Then the differentials of E_z are covectors of length 1 on $(R^T(M(\partial M,T)),\widetilde{g})$. Using this observation it is possible to find infinitely many covectors of length 1 at any point and reconstruct the metric $\widetilde{g}$ (for details see e.g. [KKL]).

Identifying $(M(\partial M,T),g)$ with its isometric copy $(R^T(M(\partial M,T)),\widetilde{g})$ and using definition , we come to the following result:

LEMMA 6. *Let the form $\mathcal{B}^{2T}$ be given. Then it possible to construct the Riemannian manifold $(M(\partial M,T),g) = (M^T,g)$.*

3.5. Construction of the gauge class of the operator. So far we have constructed an isometric copy of $(M(\partial M,T),g)$, namely, the manifold $(R^T(M(\partial M,T)),\widetilde{g})$. Our next goal is to find the potential q and impedance η on this manifold.

LEMMA 7. *The form $\mathcal{B}^{2T}$ determines $\eta|_{\partial M}$ and $q|_{M(\partial M,T)}$, where η, q is the impedance and potential of the corresponding Schrödinger operator. Moreover, it determines also $\kappa|_{\partial M}$.*

Proof. We recall that any point $x \in M(\partial M,T)$ is represented as an endpoint of a shortest normal geodesic to ∂M.

Let $y \in \partial M$ and s be such that normal geodesic $\gamma_{y,\nu}([0,s])$ is shortest geodesic between its endpoints and let Γ be a neighborhood of y. Let $u^{\kappa f}(x,t)$ be the solution of (61) with $f \in C_0^\infty(\Gamma \times [0,T])$. By using Lemma 4, we can compute the inner products and find

$$\lim_{\varepsilon \to 0} \frac{||(P_{y,s} - P_{\Gamma,s-\varepsilon})u^{\kappa f}(t)||^2}{V_g(M(y,s) \setminus M(\partial M, s-\varepsilon))} = |u^{\kappa f}(x_0,t)|^2, \quad t \le T, \tag{75}$$

where $x_0 = \gamma_{y,\nu}(s)$. Indeed, the sets $\overline{M(y,s) \setminus M(\partial M, s-\varepsilon)}$ converge to x_0 (see Fig. 1), the waves $u^{\kappa f}(x_0,t)$ are smooth while the denominator can be found since the metric g is already known. Thus we can find $|u^{\kappa f}(x_0,t)|$ at any t.

Next, let $x_0 \in \operatorname{int}(M(\partial M, T))$ be a point for which there is real-valued f with $u^{\kappa f}(x_0,T) \neq 0$. These points form an open dense subset of $\operatorname{int}(M(\partial M,T))$ as can be seen from Theorem 4 and openness of the sets $\{x : u^f(x,T) \neq 0\}$. Then,

$$q(x_0) = \frac{-\partial_t^2 u^{\kappa f}(x_0,T) + \Delta_g u^{\kappa f}(x_0,T)}{u^{\kappa f}(x_0,T)} = \frac{-\partial_t^2 |u^{\kappa f}(x_0,T)| + \Delta_g |u^{\kappa f}(x_0,T)|}{|u^{\kappa f}(x_0,T)|}$$

and the rhs can be found from (75). Thus q can be found in a dense subset, and by smoothness on the whole $M(\partial M, T)$.

Finally, varying f and t and using the fact that $\kappa > 0$ we can find $\kappa|_{\partial M}$ and $\eta|_{\partial M}$ from

$$Bu^f = [\partial_\nu u^{\kappa f}(t) + \eta u^{\kappa f}(t)]|_{\partial M} = \kappa|_{\partial M} f(t).$$

□

Remark 5. Formula (75) and Remark 4 show that given $\mathcal{B}^{2T}$ and f we can find $|u^{\kappa f}(x,t)|$ when $d(x,\partial M) \leq T$ and $t + d(x,\partial M) \leq 2T$. Moreover, as $u^{\kappa f}(x,t) = 0$ when $d(x,\partial M) > T$ and $t < T$ we thus can find $|u^{\kappa f}(x,t)|$ when $t + d(x,\partial M) \leq 2T$.

4. Alternative reconstruction via Gaussian beams.

4.1 Gaussian beams Exposition in this section is very concise. It is based on properties of Gaussian beams (see formula (76) below) which are essentially identical for compact and non-compact manifolds. We refer an interested reader to [KKL], Ch. 2.4 or [KK] where the necessary properties of Gaussian beams are discussed in detail.

Gaussian beams, called also "quasiphotons", are a special class of solutions of the wave equation depending on a parameter ε. They can be described as an asymptotic sum

$$(76)\quad U_\epsilon(x,t) = M_\epsilon \exp\{-(i\epsilon)^{-1}\theta(x,t)\} \sum_{n=0}^{N} u_n(x,t)(i\epsilon)^n, \quad x \in M,\ t \in [t_-, t_+],$$

where $M_\epsilon = (\pi\epsilon)^{-m/4}$ is the normalization constant. The function $\theta(x,t)$ is called the phase function and $u_n(x,t)$, $n = 0,1,\dots,N$ – the amplitude functions. A phase function $\theta(x,t)$ is associated with a geodesic $t \mapsto \gamma(t) \in M$ so that

$$(77)\quad \operatorname{Im}\theta(\gamma(t),t) = 0,$$

$$(78)\quad \operatorname{Im}\theta(x,t) \geq C_0 d(x,\gamma(t))^2,$$

for $t \in [t_-, t_+]$. These conditions guarantee that the absolute value of $U_\epsilon(x,t)$ looks like a Gaussian distribution in x which moves in time along the geodesic $\gamma(t)$. The phase function satisfies the eikonal equation

$$(\partial_t \theta)^2 - g^{jl}(x)\partial_j\theta\partial_l\theta \asymp 0, \tag{79}$$

where $\asymp$ means the coincidence of the Taylor coefficients of both sides considered as functions of x depending on t as a parameter at the points $\gamma(t), t \in [t_-, t_+]$. The amplitude functions u_n, $n = 0, \dots, N$ can be constructed as solutions of the transport equations

$$\mathcal{L}_\theta u_n \asymp (\partial_t^2 - \Delta_g + q)u_{n-1}, \quad \text{with } u_{-1} = 0. \tag{80}$$

Here $\mathcal{L}_\theta$ is the transport operator

$$\mathcal{L}_\theta u = 2\partial_t\theta\partial_t u - 2\langle \nabla\theta, \nabla u\rangle_g + (\partial_t^2 - \Delta_g)\theta \cdot u. \tag{81}$$

The following existence result is proven e.g. in [KKL], [KK]:

THEOREM 7. *Let $\gamma(t)$, $t \in [t_-, t_+]$ be a geodesic lying in int(M) when $t \in (t_-, t_+)$. Let $\theta_0(x)$, $u_n^0(x)$, $n = 0, 1, \dots$ be functions with θ_0 satisfying (78) for $t = t_0 \in [t_-, t_+]$. Assume that $\partial_t\gamma(t_0) = c\, Grad\, \theta_0(\gamma(t_0))$, $c > 0$.*

Then there are functions $\theta(x,t)$ and $u_n(x,t)$ satisfying (78)–(80) such that $\theta(x, t_0) = \theta_0(x)$, $u_n(x, t_0) = u_n^0(x,)$. Moreover, there is a solution $u_\varepsilon(x,t)$ of equation

$$(\partial_t^2 - \Delta_g + q)u_\epsilon(x,t) = 0, \quad (x,t) \in M \times [t_-, t_+], \tag{82}$$

such that

$$|u_\epsilon(x,t) - \chi(x,t)U_\epsilon(x,t)| \le C_{\widetilde{N}}\epsilon^{\widetilde{N}}, \tag{83}$$

where $\widetilde{N} \to \infty$ when $N \to \infty$. Here χ is the cut-off function, $\chi = 1$ near the trajectory $(\gamma(t), t)$, $t \in [t_-, t_+]$.

In the other words, for an arbitrary geodesic and matching initial data there is a Gaussian beam that propagates along this geodesic.

Next we consider a class of boundary sources in (61) which generate Gaussian beams. Let $z_0 \in \partial M$, $t_0 > 0$, and let $z = (z^1, \dots z^{m-1})$ be a local system of coordinates on ∂M near z_0. Consider a class of functions $f_\epsilon = f_{\epsilon, z_0, t_0}(z,t)$ on the boundary cylinder $\partial M \times \mathbb{R}$, where

$$f_\epsilon(z,t) = (\pi\epsilon)^{-m/4}\chi(z,t)\exp\{i\epsilon^{-1}\Theta(z,t)\}V(z,t). \tag{84}$$

Here χ is a smooth cut-off function near (z_0, t_0) and

$$\Theta(z,t) = -(t - t_0) + \frac{1}{2}\langle H_0(z - z_0), (z - z_0)\rangle + \frac{i}{2}(t - t_0)^2, \tag{85}$$

where $\langle \cdot\,, \cdot\rangle$ is the complexified Euclidean inner product, $\langle a, b\rangle = \sum a_j b_j$, and H_0 is a symmetric matrix with a positive definite imaginary part, i.e.,

$H_0 = H_0^t$, $\operatorname{Im} H_0 > 0$. At last $V(z,t)$ is a smooth function having non-zero value at (z_0, t_0).

The following result in valid (see e.g. [KKL]).

LEMMA 8. *For any function V and $t < t_0 + \tau(z_0)$ the solution $u_{\varepsilon,V}$ of problem (61) is a Gaussian beam propagating along the normal geodesic $\gamma_{z_0,\nu}$. Here $\tau(z_0)$ is the first time when the geodesic $\gamma_{z_0,\nu}$ hits ∂M.*

4.2 Reconstruction. Using the Gaussian beams described in Lemma 8 we can give an alternative method of the reconstruction of M, g, q and η of the Schrödinger operator in the orbit on an unknown operator A.

Indeed, by Corollary 2, the hyperbolic form $\mathcal{B}^{2T}$ uniquely determines $||P_{y,\tau} u^{\kappa f}(t)||$. Let $f = f_\varepsilon$ be of form (84)–(85) with $V = 1$. Then due to Lemma 8 $u^{\kappa f}(t)$, $f = f_\varepsilon$ is a Gaussian beam $u_{\varepsilon,\kappa}$ propagating along $\gamma_{z_0,\nu}$. The asymptotic expansion (76) of a Gaussian beam implies that for $s < \tau_{S,T}(z_0)$, $s + t_0 \le T$,

$$\lim_{\varepsilon\to 0} ||P_{y,\tau} u_{\varepsilon,\kappa}(s+t_0)|| = \begin{cases} h(t), & d(\gamma_{z_0,\nu}(s), y) < \tau, \\ 0, & d(\gamma_{z_0,\nu}(s), y) > \tau, \end{cases}$$

where $h(t)$ is a strictly positive function. Thus we can find $\min(d(\gamma_{z_0,\nu}(s), y), T)$. □

To find q, η and $\kappa|_{\partial M}$ we need to consider the Gaussian beam $u_{\varepsilon,\kappa}(x,t)$ more carefully. In particular, it follows from [KKL], Ch. 2.4 that

$$\text{(86)} \qquad \lim_{\epsilon\to 0} ||u_{\varepsilon,\kappa}(t)||^2 = h^2(t) = \frac{|\kappa(z_0)|^2 \sqrt{g(z_0,0))}}{\sqrt{\det(\operatorname{Im} H(t))}\, |\det(Y(t))|}.$$

Here the complex valued matrices $H(t)$ and $Y(t)$ explicitly depend only on the metric tensor g which is already in our disposal and the initial matrix $H_0(t_0)$. More precisely, these matrices are solutions of some Cauchy problems for the Hamilton system of equations and a linear system of ordinary differential equations along the geodesic $\gamma_{z_0,\nu}(s)$. Using (86) together with Corollary 2 this gives us $\kappa(z_0) > 0$.

To find potential q we can use the second term of the asymptotic expansion of $||P_{y,\tau} u_{\varepsilon,\kappa}(s+t_0)||^2$. Using results of section 2.4.19 of [KKL] we can easily obtain that for $d(\gamma_{z_0,\nu}(s), y) < \tau$

$$\text{(87)} \quad \lim_{\varepsilon\to 0} \varepsilon^{-1}(||P_{y,\tau} u_{\varepsilon,\kappa}(s+t_0)||^2 - h^2(t)) = a(s)\int_0^s q(\gamma_{z_0,\nu}(t))dt + b(s).$$

Here functions $a(s)$ and $b(s)$ again depend only on the metric g and initial data (84), (85). These functions can be found by solving some Cauchy problems for a system of ordinary differential equations along $\gamma_{z_0,\nu}$. Moreover, $a(s) \neq 0$ for any s. Therefore, by using (87) we can find the integral of the potential q along the normal geodesic $\gamma_{z_0,\nu}[0,s]$ and thus q itself. □

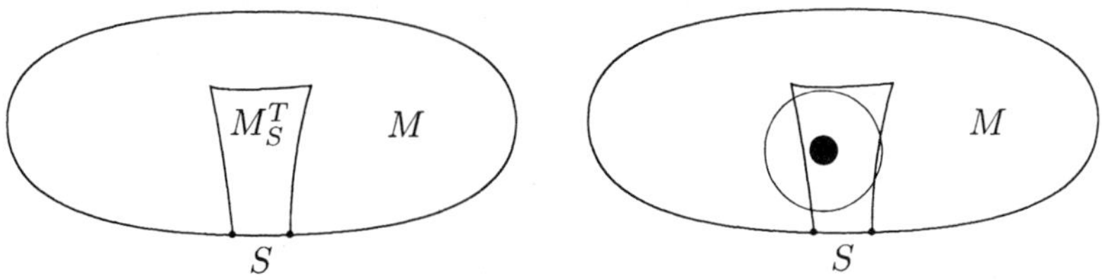

FIG. 2. *Left:* Set M_S^T. *Right:* M^2 *is union of domain of influences of balls* $B_r(z) \subset M_S^T$.

5. Reconstruction with data given on a part of the boundary.

5.1. Local constructions. In this section we will generalize constructions of Section 3 to the case when $\mathcal{B}_S^{2T}$ is given for open subset $S \subset \partial M$, i.e., the form $f \mapsto \widetilde{\mathcal{B}}^{2T}[\kappa f, \kappa f] = \mathcal{B}^{2T}[f,f]$, $f \in C_0^\infty(S \times \mathbb{R}_+)$ is given.

The construction of $M(S,T)$ will be given by iterating local constructions. First, we will construct a subset of M adjacent to S.

We use the function $\tau_{S,T} : S \to \mathbb{R}$,

$$\tau_{S,T}(z) = \sup\{s \in [0,T] : d(\gamma_{z,\nu}(s), S) = s\}.$$

We note that the function $\tau_{S,T}$ is, in general, not continuous but only upper semicontinuous. Let

$$\Omega_{S,T} = \{(z,s) \in S \times \mathbb{R}_+ : \quad s < \tau_{S,T}(z)\}$$

be the open set that lies under the graph of $\tau_{S,T}$. The exponential map,

$$\exp_{\partial M} : \Omega_{S,T} \to M, \quad (z,s) \mapsto \gamma_{z,\nu}(s), \tag{88}$$

is a diffeomorphism between $\Omega_{S,T}$ and M_S^T,

$$M_S^T = \exp_{\partial M}(\Omega_{S,T}) \subset M,$$

(see Fig 2).

Let $\widetilde{g} = (\exp_{\partial M})^* g$ be a metric on $\Omega_{S,T}$ which makes $\exp_{\partial M}$ an isometry. We will first construct the function $\tau_{S,T}$ and, therefore, $\Omega_{S,T}$ and then the metric tensor $\widetilde{g}$ in this set.

We want to apply the results of Section 3. We first observe that the main tools, Lemma 3 and Lemma 4 remain valid with $\mathcal{B}_S^{2T}$ instead of $\mathcal{B}^{2T}$ if we take $f, h \in C_0^\infty(S \times \mathbb{R}_+)$, $\Gamma_1, \Gamma_2 \subset S$.

To construct $\tau_{S,T}$ we observe that for $s < T$ we have $s \leq \tau_{S,T}(y)$ if and only if for any $t < s$ and any neighborhood $\Gamma \subset S$ of y, $M(\Gamma, s) \not\subset M(S,t)$. On the other hand, by Theorem 5, $M(\Gamma, s) \subset M(S,t)$ if and only if

$$\|P_{S,t} u^{\kappa f}(s)\| = \|u^{\kappa f}(s)\| \tag{89}$$

for all $f \in C_0^\infty(\Gamma \times (0,s))$. As both sides of (89) can be computed when $\mathcal{B}_S^{2T}$ is given, we see that $\mathcal{B}_S^{2T}$ determines $\tau_{S,T}$.

Using some local coordinates $y = (y^1, \dots, y^{(m-1)})$ on S, we obtain local coordinates $(y^1, \dots, y^{(m-1)}, s)$ on $\Omega_{S,T}$.

Next we construct the metric $\widetilde{g}$ in these local coordinates. Using part *ii.* of the proof of Lemma 5 when $\Gamma \subset S$, and $z, y \in S$, we see that $\mathcal{B}_S^{2T}$ determines $\min(d(\gamma_{y,\nu}(s), z), T)$ for $z, y \in S$ and $s < T$.

Again, we use evaluation functions

$$E_z : \Omega_{S,T} \to \mathbb{R}, \quad E_z(y, s) = \min(d(z, \gamma_{(y,\nu)}(s)), T) = r^T_{(y,s)}(z),$$

where $z \in S$. Then $E_z(y, s)$, considered as a function of (y, s) with fixed z is a truncated distance function. Thus using them as in section 3.4 we find $\widetilde{g}$. Further steps to find $q|_{M_S^T}$ and η, $\kappa|_S$ are also the same.

5.2. Global constructions.

LEMMA 9. *Assume that we are given (M_S^T, g) and the gauge-equivalence class $\sigma_S a(x, D)|_{M_S^T}$ as well as the form $\mathcal{B}_S^{2T}$. Let $B_r(z) \subset M_S^T$ be a ball and consider the manifold $M \setminus B_r(z)$ with the boundary $\partial M \cup S_1$, $S_1 = \partial B_r(z)$. Then we can find the form $\mathcal{B}_{S_1}^{2T_1}$, $T_1 < T - (r + d(z, S))$ upto a gauge transformation.*

Proof. As we know the orbit $\sigma_S a(x, D)|_{M_S^T}$ we can work with the restriction to M_S^T of the corresponding Schrödinger operator $-\Delta_g + q$.
(i) We start with the reconstruction of the values of waves $u^f(x, t)$, $f \in \mathring{C}^\infty(S \times \mathbb{R}_+)$ in M_S^T. For at any point $x = \gamma_{y,\nu}(s) \in M_S^T$ it is possible to construct a wave u^{f_0} with $u^{f_0}(x, s) > 0$. Indeed, we can take, for instance, the real part of a Gaussian beam, $f_0 = f_{\varepsilon,y,0}$ (84) with sufficiently small ε. By remark after Lemma 4 we can find the inner products $\big((P_{y,s} - P_{S,s-\delta})u^f(t), u^{f_0}(s)\big)$ for $t + s < 2T$, $s < T$. As the metric in M_S^T is already found similar considerations to that in Lemma 5 make possible to find $u^f(\gamma_{y,\nu}(s), t)$. Thus we obtain $u^f(x, t)$ for $t < 2T - d(x, S)$ when $x \in M_S^T$, $f \in \mathring{C}^\infty(S \times \mathbb{R}_+)$.

Let now $G(x, y, t)$ be the Green function

$$(90) \qquad \begin{aligned} &(\partial_t^2 - \Delta_g + q)G(x, y, t) = \delta_y(x)\delta(t) \text{ in } M \times \mathbb{R}, \quad y \in M \\ &BG(\cdot\,, y, \cdot)|_{\partial M \times \mathbb{R}} = 0; \quad G(x, y, t)|_{t<0} = 0. \end{aligned}$$

The distribution G has a limit when $y \to \partial M$, $G(\cdot\,, y, \cdot) \in \mathcal{D}'(M \times \mathbb{R})$. It is then the solution of problem (61) with $\kappa f = \delta_{\partial M,y}(x)\delta(t)$ where $\delta_{\partial M,y}$ is the delta-function on ∂M. Choosing a sequence $f_j \in C_0^\infty(S \times \mathbb{R}_+)$, $f_j \to \delta_{\partial M,y}(x)\delta(t)$ in $\mathcal{D}'(\partial M \times \mathbb{R})$ we find $G(x, y, t)$ as a limit of $u^{f_j}(x, t)$. Thus we can find $G(x, y, t)$ for $y \in S$, $x \in M_S^T$, $t + d(x, S) < 2T$.

Let us now fix $x \in M_S^T$. Because $G(x, y, t) = G(y, x, t)$ we know $G(x, y, t)$ when $x \in S$, $y \in M_S^T$, and $t + d(y, S) < 2T$.

Let $B_r(z) \subset M_S^T$ be a ball and consider the initial boundary value problem

$$(91) \qquad \begin{aligned} &\partial_t^2 u - \Delta_g u^F + q u^F = F, \quad \text{in} \quad M \times \mathbb{R}, \\ &Bu^F\big|_{\partial M \times \mathbb{R}} = 0, \quad u^F\big|_{t=0} = 0, \quad \partial_t u^F\big|_{t=0} = 0, \end{aligned}$$

where $F(x,t) \in \mathring{C}^\infty(B_r(z) \times \mathbb{R}_+)$. Since

$$u^F(x,t) = \int_{\mathbb{R}} \int_{B_r(z)} G(x,y,t-s)F(y,s)\, dV_g(y) ds, \tag{92}$$

we can find $u^F(x,t)$ for $x \in S$, $t < 2T - (d(z,S)+r)$.

Next we consider the inner product $w(t,s) = (u^F(t), u^h(s))$ where $h \in \mathring{C}^\infty(S \times \mathbb{R}_+)$. The same considerations as in the proof of Lemma 3 show that

$$(\partial_t^2 - \partial_s^2) w(t,s) = -\int_{\partial M} u^F(t)|_S \cdot \overline{h(s)}\, dS_g + \int_M F(t)\overline{u^h(s)} dV_g. \tag{93}$$

Our previous results imply that we can evaluate the rhs of (93) for $t, s < 2T - (d(z,S)+r)$. Therefore we can find $(u^F(t), u^h(s))$ for any $F \in \mathring{C}^\infty(B_r(z) \times \mathbb{R}_+)$ and $h \in \mathring{C}^\infty(S \times \mathbb{R}_+)$ for $t + s < 2T - (d(z,S)+r)$.

Using an approximation to $(P_{y,s} - P_{S,s-\delta})u^h(s)$ by the waves $u^{\eta_j}, \eta_j \in C_0^\infty(S \times (0,s))$ as described in the proof of Lemma 4 we find the inner products $((P_{y,s} - P_{S,s-\delta})u^h(s), u^F(t))$. Because $u^h(\gamma_{y,\nu}(s), s)$ is known this gives us $u^F(x,t)$ for any $F(x,t) \in \mathring{C}^\infty(B_r(z) \times \mathbb{R}_+)$, $t + d(x,S) < 2T - (d(z,S)+r)$, $x \in M_S^T$. Approximating $\delta_z(y)\delta(t)$ by smooth functions $F(t,y)$ it is thus possible to determine $G(x,z,t)$ in $X = \{(x,z,t) : d(x,S) + d(z,S) + t < 2T,\ x, z \in M_S^T\}$.

(ii) Consider the manifold $M \backslash B_r(z)$ and the initial boundary value problem

$$\begin{aligned} &\partial_t^2 e^f - \Delta_g e^f + q e^f = 0 \quad \text{in} \quad M\backslash B_r(z) \times \mathbb{R}_+ \\ &B e^f|_{\partial M \times \mathbb{R}_+} = 0,\ \partial_\nu e^f|_{S_1 \times \mathbb{R}_+} = f; \quad e^f|_{t=0} = e^f_t|_{t=0} = 0, \end{aligned} \tag{94}$$

where $S_1 = \partial B_r(z)$ and $f \in C_0^\infty(S_1 \times \mathbb{R}_+)$. If $\widetilde{e}^f$ is a smooth continuation of e^f inside $B_r(z) \times \mathbb{R}_+$ then $\widetilde{e}^f$ is the solution to (91) with $F = (\partial_t^2 - \Delta_g + q)\widetilde{e}^f$.

Because we know $u^F(x,t)$ in $B_r(z) \times (0, 2T(z,r))$, $T(z,r) = T - (d(z,S)+r)$ we use the above considerations to find $e^f|_{S_1 \times (0,2T(z,r))}$ for any $f \in C_0^\infty(S_1 \times \mathbb{R}_+)$, i.e., to construct the response operator $\Lambda_{S_1}^{2T(z,r)}$ for the manifold $M \setminus B_r(z)$. As the metric near $B_r(z)$ is known this provides us with the hyperbolic form $\mathcal{B}_{S_1}^{2T(z,r)}$ of the Schrödinger operator. □

After this it is possible to iterate the previous construction: We construct the manifold and the Schrödinger operator in the domain (see Fig. 2)

$$M^2 = \bigcup_{z \in M^1, r > 0} M_{\partial B_r(z)}^{T(z,r)} \cup M^1, \quad \text{where } M^1 = M_S^T.$$

To glue together $M_{\partial B_r(z)}^{T(z,r)}$ with different z and r we use the fact that if $G(x,y,t) = G(\widetilde{x}, y, t)$ for some $t > 0$ and $y \in \Omega$ where $\Omega \subset M$ is open then $x = \widetilde{x}$.

In M^2 we can again take balls $B_r(z)$, and iterate the construction. Then it takes only a finite number of iterations to reconstruct any compact subset in $(M(S,T),g)$ and at most numerable many to construct the whole manifold $(M(S,T),g)$ and the Schrödinger operator on it and, therefore, the orbit $\sigma_S a(x,D)|_{M(S,T)}$.

We note that the extensive use of the Green functions for a step by step reconstruction of the manifold and the operator is somewhat similar to that in [LaU] which, however, deals with a fixed-frequency inverse problem.

6. Appendix: Elliptic operators on manifolds of bounded geometry. Here we consider a class of second-order elliptic self-adjoint differential operators on non-compact manifolds with boundary. Manifolds and coefficients of the operators are C^∞-smooth. Moreover, all Riemannian manifolds (M,g) are assumed to be complete as metric spaces, i.e., Cauchy sequences converge.

We denote by $B_r(x)$ the balls of (M,g) and by $B_{\partial,r}(y)$ the balls of $(\partial M, g_{\partial M})$, where $g_{\partial M}$ is the metric inherited from (M,g).

Assume that there is $r_0 > 0$ such that

a. For any $x_0 \in M$ and $r \leq \min(r_0, d_g(x_0, \partial M))$ the ball $B_r(x_0)$ is the domain of Riemannian normal coordinates $x = (x^1, \cdots, x^m)$ centered in x_0;

b. For any $y_0 \in \partial M$ and $r \leq r_0$ the ball $B_{\partial,r}(y_0)$ is the domain of Riemannian normal coordinates (on ∂M) $y = (y^1, \cdots, y^{m-1})$ centered in y_0;

c. For any $x_0 \in M$ with $d_g(x_0, \partial M) < r_0$ there exists a unique nearest boundary point $y_0 = y(x_0) \in \partial M$ and, when $r < r_0$, the cylinder $C_r(x_0) = \{x \in M : d_g(x, \partial M) < r,\ y(x) \in B_{\partial,r}(y_0)\}$, is the domain of boundary normal coordinates $x = (y^1, \cdots, y^{m-1}, n)$ where $(y^1, \cdots, y^{m-1})$ are Riemannian coordinates on $B_{\partial,r}(y(x_0))$ and $n = d_g(x, \partial M)$.

We note that the supremum of all such r_0 is called the injectivity radius of $(M, \partial M)$. In the future we denote by $\widehat{B_r}(x_0)$, $x_0 \in M$, $r \leq r_0$ either a ball $B_r(x_0)$ if $r \geq d_g(x_0, \partial M)$ or a cylinder $C_r(x_0)$ if $r < d_g(x_0, \partial M)$. Moreover, by normal coordinates in $\widehat{B_r}(x_0)$ we mean Riemannian normal coordinates if $\widehat{B_r}(x_0) = B_r(x_0)$ and boundary normal coordinates if $\widehat{B_r}(x_0) = C_r(x_0)$.

Remark 6. By the Hopf-Rinow theorem any two points on a complete Riemannian manifold without boundary can be connected by a shortest geodesic. Moreover, for any x there is $r(x) > 0$ such that $B_{r(x)}(x)$ is a domain of Riemannian normal coordinates. Using the Hopf's double of M and a locally finite partition of unity, a manifold $(M, g, \partial M)$ can be considered as a subset of a manifold without boundary. Then the Hopf-Rinow theorem implies that any $x \in M$ can be connected with ∂M by a shortest geodesic which is normal to ∂M. Moreover, if $y \in \partial M$ there is $r(y) > 0$ such that $C_{r(y)}(y)$ is a domain of boundary normal coordinates (for results on geometry see e.g. [Cv]). What is essential in conditions a–c is that $r(x)$ is uniformly bounded from below.

DEFINITION 4. *A complete connected smooth non-compact Riemannian manifold $(M, g, \partial M)$ is called a k-finite, $k \in \mathbb{Z}_+$, manifold of bounded geometry if there is r_0 such that the injectivity radius of Riemannian normal coordinates and injectivity radius of boundary normal coordinates are*

larger than r_0 and for some $C > 0$
i. In any generalized geodesic ball $\widehat{B_r}(x_0)$ of radius $r < r_0$ there exist normal coordinates $x = (x^1, \cdots, x^m)$ such that in these coordinates the metric tensor satisfies

$$C^{-1} I \le g \le CI, \quad |\partial^\alpha g_{ij}| \le C \quad \text{when } |\alpha| \le k. \tag{95}$$

ii. If $(\widehat{B_r}(x_0), x)$ and $(\widehat{B_r}(\widetilde{x}_0), \widetilde{x})$ are above coordinates, then transition functions satisfy $||\widetilde{x} \circ x^{-1}||_{C^{k+1}} \le C$.

Remark 7. Conditions of this definition are satisfied if, for example, the injectivity radii of the Riemannian normal coordinates in M, the boundary normal coordinates in M and Riemannian normal coordinates in ∂M are uniformly bounded, and in addition the Riemannian curvature tensor R and the second fundamental form of the boundary S have bounded covariant derivatives upto order $k + 1$. We note that above geometric requirements can be made weaker by using harmonic coordinates, see e.g., [HH], [dTK] and [KaKuLT].

Examples of manifolds of bounded geometry are given by manifolds which asymptotically at infinity look like as a finite number of cones and waveguides.

Manifolds of bounded geometry enjoy a number of important properties:
(i) For any $r < r_0$ the manifold M has a covering of a countable set of generalized balls $\widehat{B_{r/2}}(x_i)$ which have a finite intersection index I, i.e., if we take more than I balls $\widehat{B_r}(x_i)$ they will have empty intersection. In addition balls $\widehat{B_{r/4}}(x_i)$ do not intersect. This result is a direct generalization to the manifolds with boundary of the result by Gromov [Gr1] using uniform estimate of the volume of balls $B_R(x)$ (see e.g. [KaKuLa]).
(ii) There is a partition of unity ϕ_i, $\text{supp}(\phi_i) \subset \widehat{B_{r/2}}(x_i)$ which satisfy $||\phi_i||_{C^{k+1}} \le C_k$ in normal coordinates satisfying (95).
(iii) There is an invariant definition of spaces $C^q(M)$ for $q \le k+1$. Also the Sobolev spaces $H^s(M)$ have an invariant meaning for $0 \le s \le k+1$. Furthermore, for such s

$$H^s(M) = \text{cl}(C_c^\infty(M)), \tag{96}$$

where $C_c^\infty(M)$ consists of C^∞-functions which are equal to 0 outside some ball $B_R(x)$. The space of Sobolev functions having compact support is denoted by $H_c^s(M)$. For these results we refer to [Sh1].
(iv) Similar Sobolev spaces can be defined on ∂M and the usual embedding and extension theorems remain valid for $H^s(M)$, $s \le k+1$.

For k-finite manifold of bounded geometry we can generalize the basic results for second order elliptic operators given in [Sh1] for non-compact manifolds without boundary.

Let us consider elliptic operators with smooth coefficients of the form

$$a(x,D)v(x) = -\Delta_g\, v(x) + (V(x), \nabla v(x))_g + c(x)v(x), \tag{97}$$

V is a real vector field (1st order operator) and c is a real valued function (cf. (1)). If $a(x,D)$ is symmetric on $C_0^\infty(\mathrm{int}(M))$ with respect to a smooth measure $d\mu = \rho dV_g$, $\rho > 0$, $dV_g = g^{1/2}\, dx^1 \cdots dx^m$, then

$$a(x,D)v(x) = -\rho^{-1} g^{-1/2}(\partial_i g^{1/2} g^{ij} \rho \partial_j v(x)) + q(x)v(x) \tag{98}$$

with a smooth real q (cf. (5)).

LEMMA 10. *Let $(M,g,\partial M)$ be a k-finite manifold of bounded geometry and $a(x,D)$ be an operator of form (98) with $\rho, \rho^{-1} \in C^\infty(M) \cap C_b^2(M)$ $q \in C^\infty(M) \cap C_b^0(M)$.*
Then for any real valued $\eta \in C^\infty(\partial M) \cap C_b^1(\partial M)$ the operator A,

$$Av = a(x,D)v; \quad D(A) = \{v \in H^2(M) : \partial_\nu v + \eta v|_{\partial M} = 0\} \tag{99}$$

is self-adjoint and bounded from below, i.e. $A \geq -c_0$. Moreover, $D((A + c)^{1/2}) = H^1(M)$ when $c > c_0$.

We remind the readers that C_b^k stands for the class of functions having k uniformly bounded derivatives.

Lemma 10 can be proved by following arguments of [Sh1]. The idea of the proof is to consider a minimal operator A_0 defined by (98) on $C_\eta^\infty(M)$, i.e., the set of smooth functions satisfying the Robin boundary condition (99). Then $A = \mathrm{cl}(A_0)$ is a self-adjoint operator with $D(A) = \{v \in H^2(M) : \partial_\nu v + \eta v|_{\partial M} = 0\}$. This may be shown by the same technique of minimal-maximal operators and methods based on finite propagation speed as in [Sh1] (also [Ch]).

Remark 8. For readers convenience we note that Green's formula is valid on the manifolds of bounded geometry. Indeed, by using definition (96) we see for instance that for $u, v \in \mathcal{D}(A) \cap H_c^1(M)$

$$\int_M (v\, a(x,D)u - u\, a(x,D)v)\, \rho dV_g(x) = \int_{\partial M} (v\, Bu - u\, Bv))\, \rho dS_g(x).$$

This fact is used many times in our considerations.

REFERENCES

[B1] Belishev, M. An approach to multidimensional inverse problems for the wave equation. (Russian) *Dokl. Akad. Nauk SSSR* **297** (1987), no. 3, 524–527; translated in *Soviet Math. Dokl.* **36** (1988), no. 3, 481–484.

[B2] Belishev, M. Boundary control in reconstruction of manifolds and metrics (the BC method). *Inverse Problems* **13** (1997), **5**, R1–R45.

[BKa] Belishev, M. I., Kachalov, A. P. Boundary control and quasiphotons in a problem of the reconstruction of a Riemannian manifold from dynamic data. (Russian) Zap. Nauchn. Sem. POMI **203** (1992), 21–50; transl. *J. Math. Sci.* **79** (1996), 1172–1190.

[BKu1] Belishev, M., Kurylev, Y. Inverse spectral problem for the plane waves scattering in a half-space with local inhomogeneities. (Russian) *Zurn. Vichisl. Matem. i Matem. Phys.* **29** (1989), 1045-1056.

[Cl] Calderón, A.P. On an inverse boundary value problem. *Seminar on Numerical Analysis and its Applications to Continuum Physics (Rio de Janeiro, 1980)*, pp. 65–73, Soc. Brasil. Mat., Rio de Janeiro, 1980.

[Cv] Chavel I. Riemannian Geometry - a Modern Introduction. Camb. Tracts in Math. **108**. Camb. Univ. Press, Cambridge, 1993, pp. 386.

[CIN] Cheney, M., Isaacson, D. and Newell, J. C. Electrical impedance tomography. *SIAM Rev.* **41** (1999), 85–101.

[Ch] Chernoff P. Essential self-adjointness of powers of generators of hyperbolic equations. *J. Funct. Anal.*, **12** (1973), 401-414.

[dTK] DeTurck D., Kazdan J. Some regularity theorems in Riemannian geometry. *Ann. Sci. cole Norm. Sup. (4)* **14** (1981), 249–260.

[Gr1] Gromov M. Curvature, diameter and Betti numbers. *Comment. Math. Helv.*, **56** (1986), 179-195.

[HH] Hebey E., Herzlich M. Harmonic coordinates, harmonic radius and convergence of Riemannian manifolds, *Rend. di Matem., ser. VII*, **17**, Roma (1997), 569-605.

[Ho] Hórmander L. Remarks on Holmgren's uniqueness theorem. *Ann. Inst. Fourier* **43** (1993), 1223–1251.

[Is] Isakov, V. Inverse Problems for Partial Differential Equations. Applied Mathematical Sciences, **127**. Springer-Verlag, New York, 1998, pp. 284.

[KK] Katchalov, A., Kurylev, Y. Multidimensional inverse problem with incomplete boundary spectral data. *Comm. Part. Diff. Equations* **23** (1998), no. 1-2, 55–95.

[KKL] Katchalov, A., Kurylev, Y., Lassas, M. *Inverse boundary spectral problems*, Chapman&Hall/CRC, 2001, pp. 290.

[KKLM] Katchalov, A., Kurylev, Y., Lassas, M., Mandache N. equivalence of time-domain inverse problems and boundary spectral problems. Submitted to *Inv. Probl.*

[KaL] Katchalov A., Lassas M. *Gaussian beams and inverse boundary spectral problems*, in: New Geom. and Anal. Meth. in Inv. Probl. (Eds. Y. Kurylev and E. Somersalo), *Springer Lect. Notes*, to appear.

[KaKuLa] Katsuda A., Kurylev Y., Lassas M. Stability on inverse boundary spectral problem, in: New Geom. and Anal. Meth. in Inv. Probl. (Eds. Y. Kurylev and E. Somersalo), *Springer Lect. Notes*, to appear.

[KaKuLT] Katsuda A, Kurylev Y, Lassas M. and Taylor M. Geometric convergence for manifolds with boundary. In preparation.

[Ko] Koosis P. Introduction to H_p Spaces, Cambr. Univ. Press, 1998, pp. 287.

[K1] Kurylev, Y. Multi-dimensional inverse boundary problems by BC-method: groups of transformations and uniqueness results. *Math. Comput. Modelling* **18** (1993), no. 1, 33–45.

[K2] Kurylev, Y. Multidimensional Gel'fand inverse problem and boundary distance map. *Inv. Probl. related Geom.*, Mito (1997), 1–15.

[KL1] Kurylev, Y., Lassas, M. Hyperbolic inverse problem with data on a part of the boundary. *Differential Equations and Mathematical Physics (Birmingham, AL, 1999)*, 259–272, AMS/IP Stud. Adv. Math., **16**, Amer. Math. Soc., Providence, 2000.

[KL2] Kurylev, Y., Lassas, M. Hyperbolic inverse problem and unique continuation of Cauchy data of solutions along the boundary, *Proc. Roy. Soc. Edinburgh, Ser. A*, to appear.

[LaU] Lassas M., Uhlmann G. On determining a Riemannian manifold from the Dirichlet-to-Neumann map. *Ann. Sci. Ecole Norm. Sup.* (5) **34** (2001), 771–787.

[LU] Lee, J., Uhlmann, G. Determining anisotropic real-analytic conductivities by boundary measurements. *Comm. Pure Appl. Math.* **42** (1989), no. 8, 1097–1112.

[NSU] Nachman, A., Sylvester, J. Uhlmann, G. An n-dimensional Borg-Levinson theorem. *Comm. Math. Phys.* **115** (1988), no. 4, 595–605.

[RS] Reed M., Simon B. *Methods of Modern Mathematical Physics. v.1*, Acad. Press, New York-London, 1972.

[SM] Schotland, John C.; Markel, Vadim A. Inverse problem in optical diffusion tomography. I. Fourier-Laplace inversion formulas. J. Opt. Soc. Amer. A 18 (2001), no. 6, 1336–1347.

[Sh1] Shubin M. Spectral theory of elliptic operators on non-compact manifolds. *Asterisque* **207** (1992), 37–108.

[Sy] Sylvester, J. An anisotropic inverse boundary value problem. *Comm. Pure Appl. Math.*, **43** (1990), 201–232.

[SU] Sylvester, J., Uhlmann, G. A global uniqueness theorem for an inverse boundary value problem. *Ann. of Math.* (2) **125** (1987), no. 1, 153–169.

[Ta1] Tataru, D. Unique continuation for solutions to PDEs; between Hörmander's theorem and Holmgren's theorem. *Comm. Part. Diff. Equations* **20** (1995), no. 5–6, 855–884.

[Ta2] Tataru D. Unique continuation for operators with partially analytic coefficients. *J. Math. Pures Appl.* **78** (1999), 505–521.

RAY TRANSFORM AND SOME RIGIDITY PROBLEMS FOR RIEMANNIAN METRICS*

VLADIMIR SHARAFUTDINOV†

This is a survey of the ray transform of symmetric tensor fields on Riemannian manifolds. In the case of second rank tensor fields, the ray transform arises in the linearization of the boundary rigidity problem which is discussed in Section 1. In Section 2 we introduce a class of Riemannian manifolds, convex non-trapping manifolds (CNTM), for which the ray transform can be defined in a very natural way. In the case of positive rank tensor fields, the ray transform has a non-trivial kernel containing the space of potential fields. The principal question is: for which CNTM's does the kernel of the ray transform coincide with the space of potential fields? For such a manifold, we can go further and ask: is there a stability estimate in the problem of recovering the solenoidal part of a tensor field from its ray transform? Some results on these questions are listed. Integral geometry is closely related to inverse problems for kinetic and linear transport equations that are discussed in Section 3. In Section 4 we present some results on the nonlinear boundary rigidity problem whose derivation is based on stability estimates for the ray transform. Section 5 is devoted to the periodic version of the ray transform, i.e., to the question: to what extent is a tensor field on a closed Riemannian manifold determined by its integrals over all closed geodesics? Anosov manifolds, i.e., closed Riemannian manifolds with geodesic flow of Anosov type constitute the most natural class for investigating the latter question. The question is closely related to the classical spectral rigidity problem: to what extent is the geometry of a Riemannian manifold determined by the eigenvalue spectrum of the Laplace — Beltrami operator?

We give no proofs here, most of the proofs can be found in [47]. Some applications of the ray transform remain out of the present survey. We now mention the most important such applications. The detailed theory of the ray transform on Euclidean space, including explicit inversion formulas of Radon type, is developed in Chapter 2 of [43]. Applications of the ray transform to tomography problems of the propagation of electromagnetic and elastic waves in slightly anisotropic media are presented in Chapters 5–7 of [43]. Some new applications of the ray transform to inverse problems of elasticity have been found in [38].

1. The boundary rigidity problem and linearization of the problem. The general boundary rigidity problem reads: to which extent is a Riemannian metric on a compact manifold with boundary determined from the distances between boundary points?

*Supported by CRDF, Grant RM2-2242; and by NSF, Grant DMS-9765792.

†Sobolev Institute of Mathematics, 4 Koptyug Avenue, Novisibirsk, 630090, Russia.

For the case in which M is a bounded domain of Euclidean space and the metric is conformal to the Euclidean one, this problem is called the inverse kinematic problem which arose in Geophysics and has a long history starting at least in the early part of the 20th century with Herglotz [22]. He considered the case where M is a ball $\{x \in \mathbf{R}^3 \mid r = |x| \leq R\}$ equipped with a spherically symmetric metric $ds^2 = dx^2/c^2(r)$ where $c(r)$ is a positive function depending only on the radius $r = |x|$. Herglotz found a formula to determine $c(r)$ from the boundary distance function. Physically this corresponds to the case of a spherically symmetric Earth model with an index of refraction depending only on the radius. The boundary distance function corresponds to the travel times of e.g. acoustic waves going through the Earth and measured at the surface. The general problem for the case that the sound speed depends on all variables has been extensively studied (see for instance [40] and the references given there). Also, this problem has a close connection to other inverse problems related to determining the sound speed from boundary measurements, see [51].

More precisely, the boundary rigidity problem can be formulated as follows. Let (M, g) be a compact Riemannian manifold with boundary ∂M. Let g' be another Riemannian metric on M. We say that g and g' have the same *boundary distance-function* if $d_g(x, y) = d_{g'}(x, y)$ for arbitrary boundary points $x, y \in \partial M$, where d_g (resp. $d_{g'}$) represents distance in M with respect to g (resp. g').

It is easy to give examples of pairs of metrics with the same boundary distance-function. Indeed, if $\varphi : M \to M$ is an arbitrary diffeomorphism of M onto itself which is the identity on the boundary, then the metrics g and $g' = \varphi^* g$ have the same boundary distance-function. Here $g' = \varphi^* g$ is the pull-back of g under φ; (i.e., for arbitrary vectors $\xi, \eta \in T_x M$ we have $\langle \xi, \eta \rangle'_x = \langle \varphi_* \xi, \varphi_* \eta \rangle_{\varphi(x)}$, where $\varphi_* : T_x M \to T_{\varphi(x)} M$ is the differential of φ at x and $\langle\ ,\ \rangle$ (resp. $\langle\ ,\ \rangle'$) is the inner product with respect to the metric g (resp. g')).

We say that a compact Riemannian manifold is *boundary rigid* if this is the only type of nonuniqueness. More precisely, (M, g) is *boundary rigid* if for every Riemannian metric g' on M with the same boundary distance-function as g, there is a diffeomorphism $\varphi : M \to M$ which is the identity on the boundary and for which $g' = \varphi^* g$.

The following question of stability in this problem seems to be important as well: are two metrics close (in some sense) to each other when their boundary distance functions are close?

There are examples of manifolds that are obviously not boundary rigid. Indeed, one can construct a metric g with a point $x_0 \in M$ such that $dist(x_0, \partial M) > \sup_{x,y \in \partial M} d_g(x, y)$. For such a metric, d_g is independent of a change of g in a neighborhood of x_0. Therefore it is necessary to impose some a-priori restrictions on the metric. One such restriction is to assume that the Riemannian manifold is *simple*, i.e., given two points there is a unique geodesic joining the points and ∂M is strictly convex. ∂M is strictly convex if the second fundamental form of the boundary is positive definite at every boundary point.

There is another class of manifolds, SGM-manifolds (strong geodesic minimizing), which is quite natural for considering the boundary rigidity problem. Roughly speaking, a compact Riemannian manifold is an SGM-manifold if the length of every geodesic segment is equal to the distance between its endpoints. (See the chapter by C.B. Croke in this book for further discussion of this condition, and [10] for the precise definition.) SGM-manifolds constitute a more general class than simple manifolds.

Although the boundary rigidity problem has been extensively studied in the last two decades, there are very few global results for this problem. There is a conjecture that every simple Riemannian manifold is boundary rigid. It has been proved that a simple metric is uniquely determined in a prescribed conformal class by the boundary distance function [34, 6, 10]. In the two-dimensional case, boundary rigidity has been proved for metrics of constant Gaussian curvature [30] and of nonpositive curvature [9, 36]. Boundary rigidity of flat metrics has been proved in the multidimensional case [17]. Only recently some local results were obtained in [12, 50] and [27] in which one assumes that the metric is a-priori close to a given metric. We will discuss these results in Section 4.

Let us linearize the boundary rigidity problem. To this end we suppose g^τ to be a family, of simple metrics on M, smoothly depending on $\tau \in (-\varepsilon, \varepsilon)$. Let us fix $p, q \in \partial M$, $p \neq q$, and put $a = d_{g^0}(p, q)$. Let $\gamma^\tau : [0, a] \to M$ be the geodesic, of the metric g^τ, for which $\gamma^\tau(0) = p$ and $\gamma^\tau(a) = q$. Let $\gamma^\tau = (\gamma^1(t, \tau), \dots, \gamma^n(t, \tau))$ be the coordinate representation of γ^τ in a local coordinate system, $g^\tau = (g^\tau_{ij})$. Simplicity of g^τ implies smoothness for the functions $\gamma^i(t, \tau)$. The equation

$$\frac{1}{a}[d_{g^\tau}(p, q)]^2 = \int_0^a g^\tau_{ij}(\gamma^\tau(t))\dot{\gamma}^i(t, \tau)\dot{\gamma}^j(t, \tau)dt \tag{1.1}$$

holds (where the dot denotes differentiation with respect to t). Differentiating (1.1) with respect to τ and putting then $\tau = 0$, we get

$$\begin{aligned} \frac{1}{a}\frac{\partial}{\partial \tau}\bigg|_{\tau=0}[d_{g^\tau}(p, q)]^2 &= \int_0^a f_{ij}(\gamma^0(t))\dot{\gamma}^i(t, 0)\dot{\gamma}^j(t, 0)\,dt \\ &+ \int_0^a \frac{\partial}{\partial \tau}\bigg|_{\tau=0}\left[g^0_{ij}(\gamma^\tau(t))\dot{\gamma}^i(t, \tau)\dot{\gamma}^j(t, \tau)\right]dt \end{aligned} \tag{1.2}$$

where $f_{ij} = \frac{\partial}{\partial\tau}\big|_{\tau=0} g^\tau_{ij}$. The second integral on the right-hand side of (1.2) is equal to zero since the geodesic γ^0 is an extremal of the functional $E_0(\gamma) = \int_0^a g^0_{ij}(\gamma(t))\dot\gamma^i(t)\dot\gamma^j(t)\,dt$. Thus we come to the equation

$$\frac{1}{a}\frac{\partial}{\partial\tau}\bigg|_{\tau=0}[d_{g^\tau}(p,q)]^2 = If(\gamma_{pq}) \equiv \int\limits_{\gamma_{pq}} f_{ij}(x)\dot x^i\dot x^j\,dt \tag{1.3}$$

in which γ_{pq} is a geodesic of the metric g^0 and t is the arc length of this geodesic in the metric g^0.

If the boundary distance function d_{g^τ} does not depend on τ, then the left-hand side of (1.3) is equal to zero. On the other hand, if each of the metrics g^τ is boundary rigid, then there exists a one-parameter family of diffeomorphisms $\varphi^\tau : M \to M$ such that $\varphi^\tau|_{\partial M} = \mathrm{Id}$ and $g^\tau = (\varphi^\tau)^*g^0$. Written in coordinate form, the last equation gives

$$g^\tau_{ij} = (g^0_{kl}\circ\varphi^\tau)\frac{\partial\varphi^k(x,\tau)}{\partial x^i}\frac{\partial\varphi^l(x,\tau)}{\partial x^j}$$

where $\varphi^\tau(x) = (\varphi^1(x,\tau),\dots,\varphi^n(x,\tau))$. Differentiating this relation with respect to τ and putting $\tau = 0$, we get the equation

$$(dv)_{ij} \equiv \frac{1}{2}(\nabla_i v_j + \nabla_j v_i) = \frac{1}{2}f_{ij}, \tag{1.4}$$

for the vector field $v = \frac{d}{d\tau}|_{\tau=0}\varphi^\tau$ where $\nabla_i v_j$ are covariant derivatives of the field v in the metric g^0. The condition $\varphi^\tau|_{\partial M} = \mathrm{Id}$ implies that $v|_{\partial M} = 0$. We thus come to the following question which is a linearization of the boundary rigidity problem: to what extent is a symmetric tensor field $f = (f_{ij})$ on a simple Riemannian manifold (M, g^0) determined by the family of integrals (1.3) which are known for all $p, q \in \partial M$? In particular, is it true that the equality $If(\gamma_{pq}) = 0$ for all $p, q \in \partial M$ implies existence of a covector field v such that $v|_{\partial M} = 0$ and $dv = f$? In the latter case (M, g^0) is called a *deformation boundary rigid* manifold.

Let us generalize this linear problem to tensor fields of arbitrary degree. Given a Riemannian manifold (M, g), by $C^\infty(S^m\tau'_M)$ we denote the space of smooth covariant tensor fields of rank m on M. The first order differential operator

$$d = \sigma\nabla : C^\infty(S^{m-1}\tau'_M) \to C^\infty(S^m\tau'_M) \tag{1.5}$$

is called the *inner derivative*. Here ∇ is the covariant derivative with respect to the metric g and σ is the symmetrization operator. In the case of $m = 2$, (1.5) coincides with (1.4). A tensor field $f \in C^\infty(S^m\tau'_M)$ is said to be a *potential field* if there exists a field $v \in C^\infty(S^{m-1}\tau'_M)$ vanishing on the boundary, $v|_{\partial M} = 0$, such that $f = dv$. In full analogy with the above considerations, we pose the following

PROBLEM 1.1 (the integral geometry problem for tensor fields). *Let (M,g) be a simple Riemannian manifold, and $m \geq 0$ be an integer. To what extent is a symmetric tensor field $f \in C^\infty(S^m\tau'_M)$ determined by the set of the integrals*

$$If(\gamma_{pq}) = \int\limits_{\gamma_{pq}} f_{i_1\dots i_m}(x)\dot{x}^{i_1}\dots\dot{x}^{i_m}\,dt \tag{1.6}$$

that are known for all $p,q \in \partial M$? Here γ_{pq} is the geodesic with endpoints p,q and t is the arc length along this geodesic. In particular, does the equality $If(\gamma_{pq}) = 0$ for all $p,q \in \partial M$ imply existence of a field $v \in C^\infty(S^{m-1}\tau'_M)$, such that $v|_{\partial M} = 0$ and $dv = f$?

By the *ray transform* of the field f we will mean the function If that is determined by formula (1.6) on the set of geodesics joining boundary points. In Section 2 this problem will be generalized to a wider class of metrics and to tensor fields of less regularity.

2. Ray transform on a CNTM. A compact Riemannian manifold (M,g) with boundary is called a *convex non-trapping manifold* (CNTM briefly), if it satisfies two conditions: 1) the boundary ∂M is strictly convex; 2) for every point $x \in M$ and every vector $0 \neq \xi \in T_xM$, the maximal geodesic $\gamma_{x,\xi}(t)$ satisfying the initial conditions $\gamma_{x,\xi}(0) = x$ and $\dot\gamma_{x,\xi}(0) = \xi$ is defined on a finite segment $[\tau_-(x,\xi), \tau_+(x,\xi)]$. We recall that a geodesic $\gamma : [a,b] \to M$ is maximal if it cannot be extended to a segment $[a-\varepsilon_1, b+\varepsilon_2]$, where $\varepsilon_i \geq 0$ and $\varepsilon_1 + \varepsilon_2 > 0$. The second condition is equivalent to the absence of a geodesic of infinite length in M.

Remark. In [43] and [47], the term CDRM (compact dissipative Riemannian manifold) was used instead of CNTM.

By $TM = \{(x,\xi) \mid x \in M,\ \xi \in T_xM\}$ we denote the tangent bundle of the manifold M, and by $\Omega M = \{(x,\xi) \in TM \mid |\xi| = 1\}$ we denote its submanifold of unit vectors. We introduce the following submanifolds of TM:

$$T^0M = \{(x,\xi) \in TM \mid \xi \neq 0\};$$
$$\partial_\pm\Omega M = \{(x,\xi) \in \Omega M \mid x \in \partial M,\ \pm\langle\xi,\nu(x)\rangle \geq 0\},$$

where ν is the unit vector of the outer normal to the boundary. Note that $\partial_+\Omega M$ and $\partial_-\Omega M$ are compact manifolds with common boundary $\Omega(\partial M)$, and $\partial\Omega M = \partial_+\Omega M \bigcup \partial_-\Omega M$.

While defining a CNTM, we have determined two functions $\tau_\pm : T^0M \to \mathbf{R}$. One can easily see that $\tau_\pm(x,\xi)$ is smooth near any point (x,ξ) such that the geodesic $\gamma_{x,\xi}(t)$ intersects ∂M transversely for $t = \tau_\pm(x,\xi)$. By strict convexity of ∂M, this condition holds for all $(x,\xi) \in T^0M$ except for the points of the set $T^0(\partial M)$. We thus conclude that $\tau_\pm$ are smooth on $T^0M \setminus T^0(\partial M)$. All points of the set $T^0(\partial M)$ are singular points for $\tau_\pm$, since one can easily see that some derivatives of these functions are unbounded in a neighborhood of such a point. Nevertheless, the following claim is valid:

LEMMA 2.1. *Let (M, g) be a CNTM. The function $\tau : \partial\Omega M \to \mathbf{R}$ defined by the equality*

$$\tau(x,\xi) = \begin{cases} \tau_+(x,\xi), & \text{if } (x,\xi) \in \partial_-\Omega M, \\ \tau_-(x,\xi), & \text{if } (x,\xi) \in \partial_+\Omega M \end{cases}$$

is smooth. In particular, $\tau_- : \partial_+\Omega M \to \mathbf{R}$ is a smooth function.

In definition (1.6) of the ray transform on a simple manifold, we parameterized the set of maximal geodesics by endpoints. Dealing with a CNTM, it is more comfortable to parameterize the set of maximal oriented geodesics by points of the manifold $\partial_+\Omega M$.

Let $C^\infty(\partial_+\Omega M)$ be the space of smooth functions on the manifold $\partial_+\Omega M$. The *ray transform* on a CNTM M is the linear operator

$$I : C^\infty(S^m\tau'_M) \to C^\infty(\partial_+\Omega M) \tag{2.1}$$

defined by the equality

$$\begin{aligned} If(x,\xi) &= \int\limits_{\tau_-(x,\xi)}^{0} \langle f(\gamma_{x,\xi}(t)), \dot\gamma^m_{x,\xi}(t)\rangle\, dt \\ &= \int\limits_{\tau_-(x,\xi)}^{0} f_{i_1\dots i_m}(\gamma_{x,\xi}(t))\dot\gamma^{i_1}_{x,\xi}(t)\dots\dot\gamma^{i_m}_{x,\xi}(t)\, dt, \end{aligned} \tag{2.2}$$

where $\gamma_{x,\xi} : [\tau_-(x,\xi), 0] \to M$ is the maximal geodesic satisfying the initial conditions $\gamma_{x,\xi}(0) = x$ and $\dot\gamma_{x,\xi}(0) = \xi$. By Lemma 2.1, the right-hand side of equation (2.2) is a smooth function on $\partial_+\Omega M$.

For a compact manifold M, the topological Hilbert space $H^k(S^m\tau'_M)$ consists of rank m symmetric tensor fields whose coordinates in a local coordinate system are locally square integrable together with all partial derivatives of order $\le k$. By $\|\cdot\|_k$ we denote one of equivalent norms on the space which is defined by choosing a finite atlas. In a similar way the topological Hilbert space $H^k(\partial_+\Omega M)$ of functions on $\partial_+\Omega M$ is defined.

THEOREM 2.2. *The ray transform on a CNTM is extendible to a bounded operator*

$$I : H^k(S^m\tau'_M) \to H^k(\partial_+\Omega M) \tag{2.3}$$

for every integer $k \ge 0$.

For a field $v \in C^\infty(S^{m-1}\tau'_M)$ and a geodesic $\gamma : (a, b) \to M$, the following equality is evidently valid:

$$\frac{d}{dt}\big[v_{i_1\dots i_{m-1}}(\gamma(t))\dot\gamma^{i_1}(t)\dots\dot\gamma^{i_{m-1}}(t)\big] = (dv)_{i_1\dots i_m}(\gamma(t))\dot\gamma^{i_1}(t)\dots\dot\gamma^{i_m}(t). \tag{2.4}$$

Let M be a CNTM. Given a field $v \in C^\infty(S^{m-1}\tau'_M)$ satisfying the boundary condition $v|_{\partial M} = 0$, equality (2.4) and definition (2.2) of the ray transform imply immediately that $I(dv) = 0$. From this, using Theorem 2.2 and boundedness of the trace operator $H^{k+1}(S^m\tau'_M) \to H^k(S^m\tau'_M|_{\partial M})$, $v \mapsto v|_{\partial M}$, we obtain the following

LEMMA 2.3. *Let M be a CNTM, $k \geq 0$ and $m \geq 0$ be integers. If a field $v \in H^{k+1}(S^m\tau'_M)$ satisfies the boundary condition $v|_{\partial M} = 0$, then $Idv = 0$.*

We want to distinguish a subspace in $H^k(S^m\tau'_M)$ which is a complement to the space of potential fields. The most natural contender for such a complement is the kernel of the dual to the d operator.

Given a Riemannian manifold (M,g), the *divergence operator* $\delta : C^\infty(S^m\tau'_M) \to C^\infty(S^{m-1}\tau'_M)$ is defined in coordinate form by the formula $(\delta u)_{i_1\dots i_{m-1}} = g^{jk}\nabla_k u_{ji_1\dots i_{m-1}}$. The operators d and $-\delta$ are dual to each other with respect to the L_2-product

$$(u,v)_{L_2} = \int\limits_M \langle u(x), v(x)\rangle\, dV^n(x)$$

on the space $L_2(S^m\tau'_M) = H^0(S^m\tau'_M)$. Here $dV^n(x) = [\det(g_{ij})]^{1/2}dx^1 \wedge \dots \wedge dx^n$ is the Riemannian volume form and $\langle u,v\rangle = g^{i_1j_1}\dots g^{i_mj_m}u_{i_1\dots i_m} \times \bar v_{j_1\dots j_m}$ is the point-wise dot-product of tensors.

The next theorem generalizes the well-known fact about decomposition of a vector field ($m=1$) into potential and solenoidal parts to symmetric tensor fields of arbitrary degree.

THEOREM 2.4. *Let M be a compact Riemannian manifold with boundary; let $k \geq 1$ and $m \geq 0$ be integers. For every field $f \in H^k(S^m\tau'_M)$, there exist uniquely determined ${}^sf \in H^k(S^m\tau'_M)$ and $v \in H^{k+1}(S^{m-1}\tau'_M)$ such that*

$$f = {}^sf + dv, \quad \delta\, {}^sf = 0, \quad v|_{\partial M} = 0. \tag{2.5}$$

The estimates

$$\|{}^sf\|_k \leq C\|f\|_k, \quad \|v\|_{k+1} \leq C\|\delta f\|_{k-1}$$

are valid where the constant C is independent of f. In particular, ${}^sf$ and v are smooth if f is smooth.

We call the fields ${}^sf$ and dv the *solenoidal* and *potential* parts of the field f.

We return to considering the ray transform on a CNTM. By Lemma 2.1, the ray transform pays no heed to the potential part of (2.5): $Idv = 0$. Consequently, given the ray transform If, we can hope to recover only the solenoidal part of the field f. We thus come to the following equivalent form of Problem 1.1.

Problem 2.5 (problem of inverting the ray transform). *For which CNTM can the solenoidal part of any field $f \in H^k(S^m\tau'_M)$ be recovered from the ray transform If?*

We will now list some known results on Problem 2.5. The first theorem we are going to present includes some smallness assumption on the curvature. To formulate the assumption, we need some preliminary definitions.

Let M be a Riemannian manifold. For a point $x \in M$ and a two-dimensional subspace $\sigma \subset T_xM$, by $K(x,\sigma)$ we denote the sectional curvature at the point x and in the two-dimensional direction σ. For $(x,\xi) \in T^0M$ we put $K(x,\xi) = \sup_{\sigma\ni\xi} K(x,\sigma)$, $\quad K^+(x,\xi) = \max\{0, K(x,\xi)\}$. For a CNTM (M,g), we introduce the following invariant:

$$k^+(M,g) = \sup_{(x,\xi)\in\partial_-\Omega M} \int\limits_0^{\tau_+(x,\xi)} tK^+(\gamma_{x,\xi}(t), \dot\gamma_{x,\xi}(t))\,dt. \tag{2.6}$$

We recall that here $\gamma_{x,\xi} : [0, \tau_+(x,\xi)] \to M$ is the maximal geodesic satisfying the initial conditions $\gamma_{x,\xi}(0) = x$ and $\dot\gamma_{x,\xi}(0) = \xi$. Note that $k^+(M,g)$ is a dimensionless quantity, i.e., it does not vary under multiplication of the metric g by a positive constant.

For $x \in \partial M$, we denote by $j_\nu : C^\infty(S^m\tau'_M|_{\partial M}) \to C^\infty(S^{m-1}\tau'_M|_{\partial M})$ the operator of contraction with the unit outer normal vector, ν, to the boundary. In local coordinates this operator is defined by the equality $(j_\nu f)_{i_1\dots i_{m-1}} = \nu^k f_{ki_1\dots i_{m-1}}$.

Theorem 2.6. *Let $n \ge 2$, $m \ge 0$ be integers, and (M,g) be an n-dimensional CNTM satisfying the condition*

$$\begin{aligned} &k^+(M,g) < (n+2m-1)/m(m+n) \qquad &&\textit{for}\quad m > 0,\\ &k^+(M,g) < 1 &&\textit{for}\quad m = 0. \end{aligned} \tag{2.7}$$

For every tensor field $f \in H^1(S^m\tau'_M)$, the solenoidal part ${}^sf$ is uniquely determined by the ray transform If and the following conditional stability estimate is valid:

$$\begin{aligned} \|{}^sf\|_0^2 &\le C\left(m\|j_\nu\, {}^sf|_{\partial M}\|_0 \cdot \|If\|_0 + \|If\|_1^2\right)\\ &\le C_1\left(m\|f\|_1 \cdot \|If\|_0 + \|If\|_1^2\right) \end{aligned} \tag{2.8}$$

where constants C and C_1 are independent of f.

This theorem was first proved [37] in the case of a non-positively curved manifold. In [42] the theorem was proved under the condition $k^+(M,g) < 1/(m+1)$ which is slightly stronger than (2.7). In the case of condition (2.7) the proof was presented in [47]. We will make a few remarks about the theorem.

The first summand on the right-hand side of estimate (2.8) shows that the problem of recovering ${}^sf$ from If is perhaps only a conditionally well-posed problem: for stably determining ${}^sf$, we need to have an a priori estimate for $\|f\|_1$. Note that this summand has appeared due to the method applied in our proof; the author knows no example demonstrating that the problem is in fact only conditionally well-posed. The factor m before the first summand emphasizes that in the case $m = 0$ the problem is well-posed.

We emphasize that (2.7) is a restriction only on the positive values of the sectional curvature, which is of an integral nature, moreover. Condition (2.7) is satisfied for a non-positively curved manifold and for any sufficiently small convex piece of an arbitrary Riemannian manifold.

The right-hand side of inequality (2.7) takes its maximal value for $m = 0$. If a CNTM (M, g) satisfies the condition $k^+(M, g) < 1$, then the following hold: (1) M is diffeomorphic to the ball, and (2) the metric g is simple.

In [26], there are some generalizations of Theorem 2.6 concerning estimating ${}^sf$ in $\|\cdot\|_k$-norms.

We denote the kernel of operator (2.3) by $Z^k(S^m\tau'_M)$. Let us recall that a tensor field $f \in H^k(S^m\tau'_M)$ is called potential if it can be represented in the form $f = dv$ with some $v \in H^{k+1}(S^{m-1}\tau'_M)$ satisfying the boundary condition $v|_{\partial M} = 0$. Let $P^k(S^m\tau'_M)$ be the subspace, of $H^k(S^m\tau'_m)$, consisting of all potential fields. By Lemma 2.3, there is the inclusion

$$P^k(S^m\tau'_M) \subset Z^k(S^m\tau'_M). \tag{2.9}$$

Problem 2.5 of inverting the ray transform is equivalent to the following question: For what classes of CNTMs and for what values of k and m can the inclusion in (2.9) be replaced with equality? As can be easily shown, if the answer is positive for $k = k_0$, then it is positive for $k \geq k_0$.

THEOREM 2.7. *Given a simple compact Riemannian manifold* (M, g), *inclusion* (2.9) *is of a finite codimension for all* m *and* $k \geq 1$.

THEOREM 2.8. *If* (M, g) *is a simple compact Riemannian manifold, then the inclusion* (2.9) *is an equality for* $m = 0$ *or* $m = 1$ *and for all* $k \geq 1$.

Theorem 2.7 is proved in [45], and the method of the proof gives Theorem 2.8 as a simple corollary of some intermediate estimate. However, Theorem 2.8 was known before; for $m = 0$ it was proved in [34, 6]; and for $m = 1$, in [2]. There is still no answer to the following question:

PROBLEM 2.9. *Given a simple Riemannian manifold, is the codimension* $c_{k,m}(M, g)$ *of the inclusion* (2.9) *independent of* k? *In other words, does there exist a complement of* $P^k(S^m\tau'_M)$, *in* $Z^k(S^m\tau'_M)$, *consisting of smooth tensor fields?*

Before citing the next result, we recall some definitions concerning conjugate and focal points. A Riemannian manifold (M, g) *has no conjugate points* if, for every geodesic γ, a non-zero Jacobi vector field along γ can not vanish at two different points of γ. (M, g) *has no focal points* if, for every geodesic $\gamma : [a, b] \to M$ and every non-zero Jacobi field $Y(t)$ along

γ satisfying the initial condition $Y(a) = 0$, the modulus $|Y(t)|$ is a strictly increasing function on $[a, b]$, i.e., $d|Y(t)|^2/dt > 0$ for $t \in (a, b]$.

The following result has been obtained in the two dimension case only [49].

THEOREM 2.10. *A compact simply connected two-dimensional Riemannian manifold (M, g) with strictly convex boundary and with no focal points is deformation boundary rigid, i.e., for a field $f \in C^\infty(S^2\tau'_M)$, the equality $If = 0$ implies existence of a covector field $v \in C^\infty(\tau'_M)$ such that $v|_{\partial M} = 0$ and $f = dv$.*

If a Riemannian manifold has no focal points, then it has no conjugate points. This implies that a manifold in Theorem 2.10 is simple and, in particular, is diffeomorphic to the disk D^2.

In all above-listed results, Theorems 2.6–2.8 and 2.10, the manifold under consideration turns out to be a simple manifold. This is not the case of the next two theorems. To formulate the first of the theorems, we need a preliminary definition.

Let (M, g) be a Riemannian manifold and $G \subset \partial M$ be a part of the boundary. Denote by Γ_G the set of all geodesics $\gamma : [a, b] \to M$ with endpoints in G. Assume Γ_G to have a natural structure of a smooth manifold. Then we can define the ray transform $I_G : C^\infty(S^m\tau'_M) \to C^\infty(\Gamma_G)$ by equation (2.2).

THEOREM 2.11. *Let g be a Riemannian metric on the spherical shell*

$$D = \{x \in \mathbf{R}^n \mid \rho_0 \le |x| \le \rho_1\} \qquad (0 < \rho_0 < \rho_1,\ n \ge 2).$$

Assume g to be invariant under all orthogonal transformations of $\mathbf{R}^n$ and such that the sphere $S_\rho = \{x \mid |x| = \rho\}$ is strictly convex for every $\rho \in [\rho_0, \rho_1]$. Let $G = S_{\rho_1}$. If a symmetric tensor field $f \in C^\infty(S^m\tau'_D)$ is in the kernel of the ray transform I_G, then there exists a tensor field $v \in C^\infty(S^{m-1}\tau'_D)$ meeting the boundary condition $v|_G = 0$ and such that $dv = f$.

This theorem was proved in [39] in the case of $m = 0$, and in [46] in the general case.

Finally, we present a recent result concerning the case of manifolds with non-trivial topology and nonconvex boundary [48].

THEOREM 2.12. *Under hypotheses of Theorem 2.6, let D be an open subset of M whose closure $\bar{D}$ is contained in $M \backslash \partial M$. Assume the boundary ∂D to be a smooth submanifold of M. If a tensor field $f \in C^\infty(S^m\tau'_{\bar{D}})$ satisfies*

$$\int\limits_a^b f_{i_1\dots i_m}(\gamma(t))\dot\gamma^{i_1}(t)\dots\dot\gamma^{i_m}(t)\,dt = 0$$

for every geodesic $\gamma : [a, b] \to \bar{D}$ with endpoints on ∂D: $\gamma(a), \gamma(b) \in \partial D$; then f is a potential field, i.e., there exists a field $v \in C^\infty(S^{m-1}\tau'_{\bar{D}})$ vanishing on ∂D and such that $f = dv$.

The set of geodesics $\gamma : [a, b] \to \bar{D}$, $\gamma(a), \gamma(b) \in \partial D$ participating in the latter theorem can have a bad topology and can do not constitute a smooth manifold. Therefore we avoid using the ray transform in the statement of the theorem.

Note that Theorem 2.6 gives the stability estimate for the solenoidal part of a field f, while Theorems 2.7–2.8 and 2.10–2.12 are purely uniqueness statements. Most probably, some stability estimates can be obtained in Theorems 2.7–2.8 and 2.10–2.11 too, but with respect to some stronger norms of If than the norm used in (2.8). This is not the case with Theorem 2.12 because all arguments in the proof of this theorem are very unstable.

3. The kinetic equation. The proofs of Theorems 2.6–2.8, 2.10 have much in common. At least all the proofs start in the same way, namely, with reducing the question under consideration to the corresponding inverse problem for some differential equation on ΩM. The latter equation, called the kinetic equation, is worth discussing here because it has a simple physical interpretation and has a number of important generalizations.

Given a field $f \in C^\infty(S^m \tau'_M)$ on a CNTM M, we define the function $u \in C(T^0 M)$ by the equation

$$u(x, \xi) = \int\limits_{\tau_-(x,\xi)}^{0} \langle f(\gamma_{x,\xi}(t)), \dot{\gamma}^m_{x,\xi}(t) \rangle \, dt \tag{3.1}$$

using the same notation as in definition (2.2) of the ray transform. The difference between (2.2) and (3.1) is the fact that the first of them is considered only for $(x, \xi) \in \partial_+ \Omega M$ while the second one, for all $(x, \xi) \in T^0 M$. In particular, we have the boundary condition

$$u|_{\partial_+ \Omega M} = If. \tag{3.2}$$

Since $\tau_-(x, \xi) = 0$ for $(x, \xi) \in \partial_- \Omega M$, we have the second boundary condition

$$u|_{\partial_- \Omega M} = 0. \tag{3.3}$$

In particular, the homogeneous boundary condition

$$u|_{\partial \Omega M} = 0 \tag{3.4}$$

is satisfied if f is in the kernel of the ray transform.

The function $u(x, \xi)$ is smooth at the same points at which $\tau_-(x, \xi)$ is smooth. This holds, as we know, at all points of the open subset $T^0 M \setminus T(\partial M)$ of the manifold $T^0 M$.

The function $u(x, \xi)$ is positively homogeneous in its second argument:

$$u(x, \lambda\xi) = \lambda^{m-1} u(x, \xi) \qquad (\lambda > 0) \tag{3.5}$$

and satisfies the equation

$$Hu = f_{i_1 \dots i_m}(x)\xi^{i_1} \dots \xi^{i_m} \tag{3.6}$$

on $T^0M \setminus T(\partial M)$, where $H \in C^\infty(\tau_{TM})$ is the geodesic vector field on TM defined in local coordinates by the formula

$$H = \xi^i \frac{\partial}{\partial x^i} - \Gamma^i_{jk}(x)\xi^j\xi^k \frac{\partial}{\partial \xi^i}. \tag{3.7}$$

Γ^i_{jk} are the Christoffel symbols of the metric g. One can easily derive (3.6) from definition (3.1). In fact (3.6) is nothing more than the Newton — Leibnitz formula for integral (3.1).

The manifold ΩM is invariant with respect to the geodesic flow. This means that the field H is tangent to ΩM at all points of the manifold ΩM and, consequently, equation (3.6) can be considered on ΩM. Sometimes this is more convenient because of the compactness of ΩM.

The operator H is related to the inner derivative d by the following equation:

$$H\left(v_{i_1 \dots i_{m-1}}(x)\xi^{i_1} \dots \xi^{i_{m-1}}\right) = (dv)_{i_1 \dots i_m}(x)\xi^{i_1} \dots \xi^{i_m}, \tag{3.8}$$

which can be proved by an easy calculation in coordinates.

If $u(x,\xi)$ is a homogeneous polynomial of degree $m-1$ in ξ, then Hu is a homogeneous polynomial of degree m, as is seen from (3.7). Problem 2.5 is equivalent to the converse statement: for which CNTM's is any solution to the boundary value problem (3.6), (3.4) a homogeneous polynomial in ξ? Indeed, in such the case f should be a potential field, as is seen from (3.8).

The equation

$$Hu = F(x,\xi) \tag{3.9}$$

on ΩM, with the right-hand side depending arbitrarily on ξ, is called the (*stationary, unit-velocity*) *kinetic equation of the metric* g. It has a simple physical interpretation. Let us imagine a stationary distribution of particles moving in M. Every particle moves along a geodesic of the metric g with unit speed, the particles do not influence one another nor the medium. Assume that there are also sources of particles in M. By $u(x,\xi)$ and $F(x,\xi)$ we mean the densities of particles and sources with respect to the volume form $dV^n(x) \wedge d\omega_x^{n-1}(\xi)$ on ΩM, where $dV^n(x)$ is the Riemannian volume form on M and $d\omega_x^{n-1}(\xi)$ is the corresponding volume form on the unit sphere $\Omega_x M$. Then equation (3.9) holds.

If the source $F(x,\xi)$ is known then, to get a unique solution u to equation (3.9), one has to set the incoming flow $u|_{\partial_-\Omega M}$. In particular, the boundary conditions (3.3) mean the absence of incoming flow. The second boundary conditions (3.2), i.e., the outgoing flow $u_{out} = u|_{\partial_+\Omega M}$, must be used for the inverse problem of determining the source. This inverse problem has the very essential (although not quite physical) requirement that the source depends polynomially on the direction ξ. The operator d gives us the following means of constructing sources which are invisible from outside and polynomial in ξ: if $v \in C^\infty(S^{m-1}\tau'_M)$ and $v|_{\partial M} = 0$, then the source $F(x,\xi) = (dv)_{i_1\dots i_m}\xi^{i_1}\dots\xi^{i_m}$ is invisible from outside. Does this construction exhaust all sources that are invisible from outside and polynomial in ξ? That is is the physical interpretation of Problem 2.5.

Let us consider the kinetic equation with an isotropic source

$$Hu(x,\xi) = f(x). \tag{3.10}$$

It describes the distribution of particles (or radiation) moving along geodesics of a given Riemannian metric with unit speed and not interacting with each other nor with the medium. If we wish to take account of interaction of particles with the medium, then we have to insert extra summands into the equation. The simplest such summand describes attenuation of particles by the medium. In the case of isotropic attenuation $\alpha(x)$ the kinetic equation (3.10) is replaced with the following:

$$(H + \alpha(x))u(x,\xi) = f(x). \tag{3.11}$$

The boundary value problem (3.11),(3.3) has an explicit solution. In particular, the outgoing flow is expressed by the integral

$$\begin{aligned} u_{out}(x,\xi) &= I^\alpha f(x,\xi) \\ &\equiv \int\limits_{\tau_-(x,\xi)}^{0} f(\gamma_{x,\xi}(t)) \exp\left[-\int\limits_{t}^{0} \alpha(\gamma_{x,\xi}(s))\,ds\right] dt \quad ((x,\xi)\in\partial_+\Omega M), \end{aligned} \tag{3.12}$$

where $\gamma_{x,\xi} : [\tau_-(x,\xi),0] \to M$ is the maximal geodesic satisfying the initial conditions $\gamma_{x,\xi}(0) = x$ and $\dot\gamma_{x,\xi}(0) = \xi$. The operator

$$I^\alpha : C^\infty(M) \to C^\infty(\partial_+\Omega M) \tag{3.13}$$

defined by (3.12) is called the *attenuated ray transform corresponding to the attenuation* α. It plays a key role in problems of emission tomography. Statements of problems of emission tomography can vary considerably. For instance, the problem of simultaneously determining the source f and the attenuation α is of great practical import. We will here deal with a more modest problem of determining the source f on condition that the attenuation α is known. We will restrict ourselves to considering the attenuated ray transform of scalar functions. In the case of symmetric tensor fields, investigation of the attenuated ray transform comes across the following fundamental question: does there exist, for I^α, an analog of the operator d of inner differentiation?

Increasing the complexity, the next summand usually included in the kinetic equation is the scattering integral describing the effects of collision of particles with motionless atoms of the medium. The kinetic equation with the scattering integral is conventionally called the linear transport equation. Here we restrict ourselves to considering the *stationary unit-velocity transport equation with isotropic scattering diagram*:

$$(H + \alpha(x))\, u(x,\xi) = \frac{1}{\omega_{n-1}} \int\limits_{\Omega_x M} \sigma(x; \langle \xi, \xi' \rangle) u(x,\xi')\, d\omega_x(\xi') + f(x). \tag{3.14}$$

Here $\Omega_x M = \Omega M \cap T_x M$ is the unit sphere at the point x; $d\omega_x$ is the volume form on the sphere $\Omega_x M$ induced by the metric g; ω_{n-1} is the volume of the unit sphere in $\mathbf{R}^n$. The coefficient $\sigma \in C^\infty(M \times [-1,1])$ is called the *scattering diagram*. We supply (3.14) with the homogeneous boundary condition (absence of incoming flow)

$$u|_{\partial_- \Omega M} = 0, \tag{3.15}$$

and consider the inverse problem of recovering the source $f(x)$ from the outgoing flow

$$u|_{\partial_+ \Omega M} = u_{out}(x,\xi). \tag{3.16}$$

Before formulating the result on the inverse problem, we introduce some notation.

Given functions $\alpha \in C^\infty(M)$ and $\sigma \in C^\infty(M \times [-1,1])$, we define the function $\kappa = \kappa[\alpha,\sigma] \in C(M)$ as follows. For $n = \dim M \geq 3$, we expand $\sigma(x;\mu)$ in a Fourier series in Gegenbauer's polynomials:

$$\sigma(x;\mu) = \sum_{k=0}^{\infty} \sigma_k(x) C_k^{(n/2-1)}(\mu), \tag{3.17}$$

and put

$$\kappa(x) = \max_{k \geq 1} \left| \frac{n-2}{n+2k-2} \sigma_k(x) - \alpha(x) \right|. \tag{3.18}$$

For $n = 2$, formulas (3.17) and (3.18) are replaced with the following:

$$\begin{aligned} \sigma(x;\cos\theta) &= \sum_{k=-\infty}^{\infty} \sigma_k(x) e^{ik\theta}, \\ \kappa(x) &= \max_{|k| \geq 1} |\sigma_k(x) - \alpha(x)|. \end{aligned} \tag{3.19}$$

Note that $\kappa(x)$ is independent of $\sigma_0(x)$. In particular, $\kappa(x) = |\alpha(x)|$ if the scattering diagram $\sigma(x;\mu) = \sigma(x)$ does not depend on μ.

We say that a linear system $d^2y/dt^2 + A(t)y = 0$ $(y = (y_1 \ldots, y_n))$ of second order differential equations has no conjugate points on a segment

$[a, b]$ if there is no nontrivial solution to the system which vanishes at two different points of the segment.

THEOREM 3.1. *Let (M, g) be a CNTM of dimension $n \geq 2$ and let $\alpha \in C^\infty(M)$ and $\sigma \in C^\infty(M \times [-1, 1])$ be two functions. Assume that, for every $(x, \xi) \in \Omega M$, the equation*

$$\frac{D^2\eta}{dt^2} + \overset{\kappa}{R}(t)\eta = 0 \tag{3.20}$$

lacks conjugate points on the geodesic $\gamma = \gamma_{x,\xi} : [\tau_-(x, \xi), \tau_+(x, \xi)] \to M$. Here $D/dt = \dot\gamma^i \nabla_i$ is the covariant derivative along γ, and $\overset{\kappa}{R}(t) : T_{\gamma(t)}M \to T_{\gamma(t)}M$ is the linear operator whose matrix is defined in local coordinates by the equation

$$\overset{\kappa}{R}{}^p_k(t) = \left[g^{pi}\left(R_{ijkl} + \kappa^2(g_{ik}g_{jl} - g_{il}g_{jk})\right)\right]_{x=\gamma(t)} \dot\gamma^j(t)\dot\gamma^l(t), \tag{3.21}$$

where (R_{ijkl}) is the curvature tensor and the function $\kappa(x)$ is defined by (3.17)–(3.19). *Then every function $f \in H^1(M)$ can be uniquely recovered from the trace* (3.16) *of a solution to the boundary value problem* (3.14)–(3.15), *and further, the stability estimate*

$$\|f\|_{L_2(M)} \leq C\|u_{out}\|_{H^1(\partial_+\Omega M)} \tag{3.22}$$

holds with some constant C independent of f.

This theorem is proved in [44]. We now formulate some corollaries of the theorem which are related to the cases in which either the scattering integral is absent or the metric g is Euclidean. Both cases are significant for applications.

COROLLARY 3.2. *Let (M, g) be a* CNTM *and $\alpha \in C^\infty(M)$. Assume that equation* (3.20) *with*

$$\overset{\kappa}{R}{}^p_k(t) = \overset{\alpha}{R}{}^p_k(t) = [g^{pi}(R_{ijkl} + |\alpha|^2(g_{ik}g_{jl} - g_{il}g_{jk}))]_{x=\gamma(t)}\dot\gamma^j(t)\dot\gamma^l(t)$$

lacks conjugate points on the geodesic $\gamma = \gamma_{x,\xi} : [\tau_-(x, \xi), \tau_+(x, \xi)] \to M$ for every $(x, \xi) \in \Omega M$. Then the attenuated ray transform

$$I^\alpha : H^1(M) \to H^1(\partial_+\Omega M)$$

is injective and the stability estimate

$$\|f\|_{L_2(M)} \leq C\|I^\alpha f\|_{H^1(\partial_+\Omega M)}$$

holds with some constant C independent of f.

In the case of $\alpha \equiv \sigma \equiv 0$ equation (3.20) transforms into the classical Jacobi equation

$$\frac{D^2\eta}{dt^2} + R(\dot\gamma, \eta)\dot\gamma = 0,$$

and the operator (3.13) coincides with the ray transform (2.1) for $m = 0$. In this case Corollary 3.2 coincides with the claim of Theorem 2.8 for $m = 0$.

We now discuss in brief the role of the curvature tensor in Theorem 3.1 and Corollary 3.2. It is well known [16] that, if all sectional curvatures are nonpositive, then the Jacobi equation lacks conjugate points on a geodesic segment of any length. Of course, this property may fail when we add the summand with the factor κ^2 to the right-hand side of (3.21). Nevertheless, the general tendency is preserved: the more negative the sectional curvature is, the larger the values that κ^2 may assume without violating the assumptions of Theorem 3.1.

We now consider the case in which M is a bounded domain in $\mathbf{R}^n$ and the metric g coincides with the Euclidean metric. In this case equation (3.14) becomes the classical transport equation

$$\xi^i \frac{\partial u(x,\xi)}{\partial x^i} + \alpha(x)u(x,\xi) = \frac{1}{\omega_{n-1}} \int\limits_{|\xi'|=1} \sigma(x;\langle\xi,\xi'\rangle)u(x,\xi')\,d\xi' + f(x), \tag{3.23}$$

and system (3.20) reduces to the single scalar equation

$$\frac{d^2\eta}{dt^2} + \kappa^2\eta = 0. \tag{3.24}$$

We thus obtain

COROLLARY 3.3. *Let M be a closed bounded domain in $\mathbf{R}^n$ with smooth strictly convex boundary. Let functions $\alpha \in C^\infty(M)$ and $\sigma \in C^\infty(M \times [-1,1])$ be such that equation* (3.24) *lacks conjugate points on any straight line segment $\gamma : [a,b] \to M$; here $\kappa = \kappa[\alpha,\sigma]$ is defined by formulas* (3.17)–(3.19). *Then every function $f \in H^1(M)$ is uniquely recoverable from the trace* (3.16) *of the solution to the boundary value problem* (3.23), (3.15) *and the stability estimate* (3.22) *is valid.*

A number of conditions are known which ensure the absence of conjugate points for a scalar equation. Some of them are based on the Sturm comparison theorems, and the others, on Lyapunov's integral estimates [21]. The simplest of them guarantees the absence of conjugate points for equation (3.24) if the inequality $\kappa_0 \operatorname{diam} M < \pi$ is valid with $\kappa_0 = \sup\limits_{x\in M} \kappa(x)$.

Finally, if $\sigma \equiv 0$, we get the attenuated ray transform on $\mathbf{R}^n$ which is conveniently written down as

$$I^\alpha f(x,\xi) = \int\limits_{-\infty}^{\infty} f(x+t\xi) \exp\left[-\int\limits_{t}^{\infty} \alpha(x+s\xi)\,ds\right] dt \quad (x\in\mathbf{R}^n,\ 0\neq\xi\in\mathbf{R}^n),$$

on assuming that the functions f and α are extended to be zero outside M. Invertibility of the attenuated ray transform on $\mathbf{R}^n$ was recently proved [5], and an explicit inversion procedure was found [35].

4. Local boundary rigidity. The boundary C^∞-jet of a metric is determined by the boundary distance function under a weaker convexity of the boundary than the one used in the definition of a CNTM.

Given a connected Riemannian manifold with boundary (M, g), we say that the boundary ∂M is weakly convex if the following holds: for every two points $p_0, p_1 \in \partial M$, $p_0 \neq p_1$, there exists a geodesic $\gamma : [0,1] \to M$ joining these points, $\gamma(0) = p_0$, $\gamma(1) = p_1$, such that the length of γ is equal to $d_g(p_0, p_1)$, and all interior points of γ belong to $M \setminus \partial M$.

THEOREM 4.1. *Let (M, g) be a connected Riemannian manifold with weakly convex boundary. Then the C^∞-jet of the metric g at the boundary is uniquely determined by the boundary distance function d_g in the following sense. If ∂M is weakly convex with respect to another metric g' on M, then the equality $d_g = d_{g'}$ implies the existence of a diffeomorphism $\varphi : M \to M$ which is the identity on the boundary, $\varphi|_{\partial M} = Id$, and such that the metrics g and $g'' = \varphi^* g'$ satisfy the following: In any local coordinate system $(x^1, \dots, x^n)$ defined in a neighborhood of a boundary point, we have $D^\alpha g|_{\partial M} = D^\alpha g''|_{\partial M}$ for every multi-index α.*

This result was proved in two dimensions in [31] and for $|\alpha| \leq 2$ in [30]. In the general case it is proved in [27].

We will present two results on local boundary rigidity which are very similar to each other but are obtained in different ways.

THEOREM 4.2. *There is a C^{12}-neighborhood W of the standard Euclidean metric on the closed ball $B = \{x \in \mathbf{R}^n \mid |x| \leq 1\}$ such that, for every two metrics $g, g' \in W$, coincidence of the boundary distance functions of the metrics implies that they are isometric via an isometry that is the identity on the boundary ∂B.*

THEOREM 4.3. *Let an n-dimensional CNTM (M, g) satisfy the inequality*

$$k^+(M, g) < (n+3)/2(n+2) \tag{4.1}$$

with $k^+(M, g)$ defined by (2.6). Then there is a neighborhood $W \subset C^{3,\alpha} \times (S^2\tau'_M)$ of g, with any $0 < \alpha < 1$, such that if a metric $g' \in W$ has the same boundary distance-function as g, then there exists a $C^{3,\alpha}$-diffeomorphism $\varphi : M \to M$, the identity on ∂M, such that $g' = \varphi^* g$; moreover, φ tends to the identity as g' tends to g (both in $C^{3,\alpha}$-topology).

Theorem 4.2 is proved in [50], and Theorem 4.3, in [12]. By Theorem 2.6, condition (4.1) guarantees invertibility of the ray transform for second rank tensor fields. This fact is explicitly used in the proof of Theorem 4.3. In fact, the proof of Theorem 4.2 also uses in an implicit way a version of the weighted ray transform of the tensor field $g - g'$. At first sight, Theorem 4.2 seems to be a special case of Theorem 4.3. However, there is an essential distinction in the hypotheses of these theorems: one of the metrics is fixed in Theorem 4.3, while both the metrics can vary in W in Theorem 4.2. Nevertheless, it was proved recently that Theorem 4.3 can be generalized in spirit of Theorem 4.2 [Matti Lassas. Private communication].

Local uniqueness theorems can be considered part of a program for proving finiteness theorems. The other part being compactness results: one has to prove compactness, in the $C^{3,\alpha}$-topology, of the family of metrics with the same boundary distance function as a given metric, assuming the family to be factored by the group of diffeomorphisms that fix the boundary. This part of the program is not realized yet. Gromov's method for proving compactness of families of metrics has a version for manifolds with boundary [25], but it gives compactness in weaker topologies than $C^{3,\alpha}$.

We conclude the section with presenting a semilocal result. Before we state the result we introduce some notation.

Let (M, g) be a compact Riemannian manifold. If T is a real covariant tensor field of degree m on M, then its modulus, which is defined in local coordinates by

$$|T|_g^2 = g^{i_1 j_1} \dots g^{i_m j_m} T_{i_1 \dots i_m} T_{j_1 \dots j_m},$$

is independent of the choice of coordinates. By

$$\|T\|_{C^k(M,g)} = \sum_{l=0}^{k} \sup_{x \in M} |\underbrace{\nabla \dots \nabla}_{l} T(x)|_g$$

we denote the C^k-norm of the tensor field T. Here ∇ is the covariant derivative. We denote by R_g the curvature tensor of the metric g, and by e the Euclidean metric.

THEOREM 4.4. *Let $D \subset \mathbf{R}^n$ be a closed bounded domain with a smooth strictly convex (with respect to the Euclidean metric) boundary ∂D. Let $K > 0$ and g be a Riemannian metric on D satisfying the conditions $\|R_g\|_{C^k(D,g)} \le K$, $k^+(D, g) < 1/4$, where $k = [n/2] + 18$ and $[n/2]$ denotes the integer part of $n/2$. Let g' be another Riemannian metric on D satisfying $d_g = d_{g'}$. Then there exists $\varepsilon = \varepsilon(K, D, n) > 0$ such that if $\|g'_{ij} - \delta_{ij}\|_{C^l(D,e)} < \varepsilon$ with $l = [n/2] + 20$, then the metrics g and g' are isometric via an isometry which is the identity on the boundary.*

The proof presented in [27] is as follows: first, by using Theorem 4.1, the claim is reduced to the special case when D is the unit ball; then we prove that g is C^{12}-close to the Euclidean metric in appropriate coordinates; and then we apply Theorem 4.2.

5. Spectral rigidity. Let (M, g) be a closed (= compact without boundary) Riemannian manifold. For a symmetric tensor field $f \in C^\infty \times (S^m \tau'_M)$ and a closed geodesic $\gamma : [a, b] \to M$, we may consider the integral

$$If(\gamma) = \oint_\gamma \langle f, \dot\gamma^m \rangle dt = \int_a^b f_{i_1 \dots i_m}(\gamma(t)) \dot\gamma^{i_1}(t) \dots \dot\gamma^{i_m}(t)\, dt. \tag{5.1}$$

The integrand of (5.1) is written with respect to local coordinates. Nevertheless, it is evidently invariant, i.e., independent of the choice of coordinates. Here we do not use the term ray transform because the set of closed geodesics does not constitute a manifold, and use $If(\gamma)$ as the notation for the right-hand side integral on (5.1). Let $Z^\infty(S^m\tau'_M)$ denote the subspace of $C^\infty(S^m\tau'_M)$ consisting of all fields f such that $If(\gamma) = 0$ for every closed geodesic γ. For $m > 0$ this subspace is not zero as is seen from the following argument. A tensor field f is called a *potential field* if it can be represented in the form $f = dv$ for some $v \in C^\infty(S^{m-1}\tau'_M)$. Let $P^\infty(S^m\tau'_M)$ denote the space of all potential fields. If $f = dv$, then the integrand of (5.1) equals to $d(v_{i_1\dots i_{m-1}}(\gamma(t))\dot\gamma^{i_1}(t)\dots\dot\gamma^{i_{m-1}}(t))/dt$. Therefore there is an inclusion

$$P^\infty(S^m\tau'_M) \subset Z^\infty(S^m\tau'_M). \tag{5.2}$$

The principal question is: for what classes of closed Riemannian manifolds and for what values of m is the inclusion (5.2) in fact an equality? Of course, the question is reasonable only for manifolds that have sufficiently many closed geodesics. It turns out that Anosov manifolds constitute the most natural class for investigating the question.

We recall the definition of an Anosov flow. Let $H \in C^\infty(\tau_N)$ be a vector field, on a closed manifold N, not vanishing at any point, and $G^t : N \to N$ be the flow generated by the vector field. G^t is called an *Anosov flow* if, for every point $x \in N$, the tangent space T_xN splits into the direct sum of three subspaces $T_xN = \{H(x)\} \oplus X^s(x) \oplus X^u(x)$, where $\{H(x)\}$ is the one-dimensional subspace spanned by the vector $H(x)$, and the two other subspaces are such that for $\xi \in X^s(x)$, $\eta \in X^u(x)$ the differential d_xG^t satisfies the estimates

$$\begin{aligned} |(d_xG^t)\xi| &\le ae^{-ct}|\xi| \quad \text{for } t>0, & |(d_xG^t)\xi| &\ge be^{-ct}|\xi| \quad \text{for } t<0;\\ |(d_xG^t)\eta| &\le ae^{ct}|\eta| \quad \text{for } t<0, & |(d_xG^t)\eta| &\ge be^{ct}|\eta| \quad \text{for } t>0, \end{aligned}$$

where a, b, c are positive constants independent of x, ξ, η. If such a splitting exists, then it is unique, and $\dim X^s(x)$ is independent of x. The subspaces X^s and X^u are called the *stable* and *unstable subspaces* respectively.

An *Anosov manifold* is a closed Riemannian manifold whose geodesic flow $G^t : \Omega M \to \Omega M$ is of Anosov type. The following two claims are valid for such a manifold: (1) the orbit of a point (x, ξ) with respect to the geodesic flow is dense in ΩM for almost all $(x, \xi) \in \Omega M$; (2) the set of $(x, \xi) \in \Omega M$, such that the geodesic $\gamma_{x,\xi}$ is closed, is dense in ΩM. See [3] for proofs. A closed Riemannian manifold of negative sectional curvature is an Anosov manifold, and the class of Anosov manifolds is wider than the class of closed negatively curved manifolds.

For an Anosov manifold, the question of equality in (5.2) can be reduced to the inverse problem for the kinetic equation. This is done with the help of the following:

THEOREM 5.1 (the Livčic theorem). *Let $H \in C^\infty(\tau_N)$ be a vector field on a closed manifold N which generates an Anosov flow. If a function $F \in C^\infty(N)$ integrates to zero over every closed orbit of the flow, then there exists a function $u \in C^\infty(N)$ such that $Hu = F$.*

A. N. Livčic [28] constructed the function u and proved that it is Hölder-continuous. Smoothness of the function was proved later [29].

For a field $f \in Z^\infty(S^m\tau'_M)$ on an Anosov manifold, Livčic's theorem gives a function $u \in C^\infty(\Omega M)$ satisfying the kinetic equation

$$Hu(x,\xi) = f_{i_1\dots i_m}(x)\xi^{i_1}\dots\xi^{i_m}. \tag{5.3}$$

The question of equality in (5.2) is thus equivalent to the following one: is any solution to (5.3) a homogeneous polynomial of degree $m-1$ in ξ?

We list some results on the problem.

THEOREM 5.2. *For an Anosov manifold of nonpositive sectional curvature, the equality $P^\infty(S^m\tau'_M) = Z^\infty(S^m\tau'_M)$ holds for all m.*

This theorem is proved in [11]. It can be considered as a periodic analog of Theorem 2.6. In [11], the theorem is formulated for negatively curved manifolds, but only nonpositivity of the curvature and the Anosov type of the geodesic flow are used in the proof.

THEOREM 5.3. *For an Anosov manifold, inclusion* (5.2) *has a finite codimension for every m.*

This theorem is proved in [13]. It is the periodic version of Theorem 2.7. [13] contains also the following two theorems that are the periodic analog of Theorem 2.8.

THEOREM 5.4. *Let (M,g) be an Anosov manifold. If a function $f \in C^\infty(M)$ integrates to zero over every closed geodesic then f must itself be zero.*

THEOREM 5.5. *Let (M,g) be an Anosov manifold, and f be a smooth 1-form on M. If f integrates to zero around every closed geodesic, then f is an exact form.*

The following periodic version of Theorem 2.10 is proved in [49]:

THEOREM 5.6. *For an Anosov surface without focal points, inclusion* (5.2) *is equality for $m = 2$.*

In the famous lecture by M. Kac [23], the following question was posed: can one hear the shape and size of a drum? The question is posed more precisely as follows.

Let (M,g) be a closed Riemannian manifold, and $\Delta : C^\infty(M) \to C^\infty(M)$ be the corresponding Laplace — Beltrami operator. Being an elliptic operator, $-\Delta$ has an infinite discrete eigenvalue spectrum $\operatorname{Spec}(M,g) = \{0 = \lambda_0 < \lambda_1 \le \lambda_2 \le \dots\}$. Two closed Riemannian manifolds are called *isospectral* if their eigenvalue spectra coincide. Kac's question can be formulated as follows: do there exist isospectral but not isometric manifolds?

The first example of isospectral manifolds was found by J. Milnor in the dimension 16 [32]. Later M. Vignéras [52] showed that even in the class of closed manifolds of constant negative curvature there are isospectral but not homeomorphic manifolds of any dimension. In order to avoid these examples and linearize the problem, V. Guillemin and D. Kazhdan introduced in [18] the following definition of spectral rigidity.

A smooth one-parameter family g^τ $(-\varepsilon < \tau < \varepsilon)$ of metrics on a closed manifold M is called a *deformation* of a metric g if $g^0 = g$. Such a family is called an *isospectral deformation* if the spectrum of the Laplace — Beltrami operator Δ^τ of the metric g^τ is independent of τ. A deformation g^τ is called a *trivial deformation* if there exists a family φ^τ of diffeomorphisms of M such that $g^\tau = (\varphi^\tau)^* g$. A manifold (M, g) is called *spectrally rigid* if it does not admit a nontrivial isospectral deformation.

Since [18] was published, a number of examples of isospectral deformations of compact manifolds have been given [15, 41]. Hence to rule out isospectral deformations there must be some extra assumption.

For Anosov manifolds, the spectral rigidity problem relates closely to integral geometry. In particular, the following claim is stated in [18].

THEOREM 5.7. *An Anosov manifold (M, g) is spectrally rigid if the inclusion* (5.2) *is equality for $m = 2$.*

This theorem is formulated in [18] in the case of negatively curved manifolds. Nevertheless, it is valid for Anosov manifolds too because the proof uses only the fact that the index of any closed geodesic is zero. In fact, Theorem 5.7 is a simple corollary of a deep relationship between the eigenvalue spectrum of an elliptic self-dual differential operator and the singular support of the trace of the wave kernel, established by J. J. Duistermaat and V. Guillemin in [14]. Being applied to the Laplace — Beltrami operator, this gives: if two Anosov manifolds have the same eigenvalue spectrum, then they have the same length spectrum. The length spectrum is the set of lengths of closed geodesics. The latter statement implies Theorem 5.7 with the help of a linearization like we have used in Section 1.

Comparing Theorem 5.7 with Theorems 5.2 and 5.6, we obtain the following results.

THEOREM 5.8. *An Anosov manifold of nonpositive sectional curvature is spectrally rigid.*

THEOREM 5.9. *An Anosov surface without focal points is spectrally rigid.*

Theorem 5.3 says us that, for an Anosov manifold, the space of infinitesimal isospectral deformations has a finite dimension modulo trivial deformations.

For two-dimensional manifolds of negative curvature, Theorem 5.8 was proved by V. Guillemin and D. Kazhdan in [18]. The same authors proved this fact for n-dimensional manifolds [19] under a pointwise curvature pinching assumption. That result was later extended by Min-Oo [33] to the case where the curvature operator is negative definite.

A closed Riemannian manifold is said to have a simple length spectrum if there do not exist two different closed geodesics such that the ratio of their lengths is a rational number. This is a generic condition.

THEOREM 5.10. *Let (M, g) be an Anosov manifold with simple length spectrum, and $\Delta : C^\infty(M) \to C^\infty(M)$ be the corresponding Laplace — Beltrami operator. If real functions $q_1, q_2 \in C^\infty(M)$ are such that the operators $\Delta + q_1$ and $\Delta + q_2$ have coincident eigenvalue spectra, then $q_1 \equiv q_2$.*

This result follows from Theorem 5.4 because, under hypotheses of Theorem 5.10, eigenvalue spectrum of the operator $\Delta + q$ determines integrals of the potential q over closed geodesics, as is shown in [18].

Finally, we will discuss two open questions.

(1) Does there exist an Anosov surface that is not spectrally rigid? It is known [24, 4] that an Anosov manifold has no conjugate points but can have focal points. In view of Theorem 5.9, if such a surface exists, it must have focal points. An example of an Anosov surface with focal points is constructed in [20]. This example is obtained from a negatively curved surface by a spherical perturbation of the metric in a geodesic ball. By Theorem 2.11, spherically symmetric metrics are deformation boundary rigid. Therefore the surface constructed in [20] does not seem to be a contender for a non spectrally rigid surface. On the other hand, no other example of an Anosov surface with focal points is known.

(2) The following conjecture was stated by C.B. Croke: Given a closed negatively curved manifold (M, g), there are at most finitely many, modulo isometries, negatively curved metrics on M which are isospectral to g. To prove the conjecture, we have to start with proving the following periodic analog of Theorem 4.3: there exists a neighborhood of g containing no metric that is isospectral but not isometric to g. This statement is still an open question. The main difficulty here lies in the non-existence of an appropriate "approximate Livčic theorem". The second part of the proof of the conjecture should be some statement on the compactness of the family of metrics isospectral to a given one. A number of such compactness results already exists — see for example [1] and [7].

REFERENCES

[1] ANDERSON M. (1991). Remarks on the compactness of isospectral sets in low dimensions. *Duke Math. J.*, **63**, 699–711.

[2] ANIKONOV YU. E. AND ROMANOV V.G. (1976). On uniqueness of definition of first-order form by its integrals over geodesics. *Ill-posed Math. and Geophys. Problems.* Novosibirsk, 22–27 (in Russian).

[3] ANOSOV D. (1967). *Geodesic Flows on Closed Riemannian Manifolds with Negative Curvature*, Proc. Steclov Inst. of Math., Vol. 90.

[4] ANOSOV D.V. (1985). On geodesic flows satisfying the condition (Y). *Proc. Steclov Inst. of Math.*, **167**, 3–24.

[5] ARBUZOV E.V., BUKHGEIM A.L., AND KAZANTSEV S.G. (1998). Two-dimensional tomography problems and the theory of A-analytic functions. *Siberian Adv. Math.*, **8**, 1–20.

[6] BERNSTEIN I.N. AND GERVER M.L. (1980). Conditions of distinguishability of metrics by godographs. *Methods and Algorithms of Interpretation of Seismological Information.* Computerized Seismology, **13**, Nauka, Moscow, 50–73 (in Russian).

[7] BROOKS R., PERRY P., AND PETERSEN P. (1992). Compactness and finiteness theorems for isospectral manifolds. *J. Reine Angew. Math.*, **426**, 67–89.

[8] CHEEGER J. (1970). Finiteness theorems for Riemannian manifolds. *Am. J. Math.*, **92**, 61–75.

[9] CROKE C.B. (1990). Rigidity for surfaces of non-positive curvature. *Comment. Math. Helv.*, **65**, 150–169.

[10] CROKE C.B. (1991). Rigidity and the distance between boundary points. *J. Diff. Geometry,* **33**, 445–464.

[11] CROKE C.B. AND SHARAFUTDINOV V.A. (1998). Spectral rigidity of a compact negatively curved manifold *Topology*, **37**(6), 1265–1273.

[12] CROKE C.B., DAIRBEKOV N.S., AND SHARAFUTDINOV V.A. (2000). Local boundary rigidity of a compact Riemannian manifold with curvature bounded above. *Trans. AMS*, **352**, 3937–3956.

[13] DAIRBEKOV N.S. AND SHARAFUTDINOV V.A. Some questions of integral geometry on Anosov manifolds. *Dynamical Systems and Ergodic Theory*, to appear.

[14] DUISTERMAAT J.J. AND GUILLEMIN V. (1975). The spectrum of positive elliptic operators and periodic bicharacteristics. *Inventions,* **24**, 39–80.

[15] GORDON C. AND WILSON E. (1984). Isospectral deformations of compact solvmanifolds. *J. Diff. Geom.*, **19**, 241–256.

[16] GROMOLL D., KLINQENBERG W., AND MEYER W. (1968). *Riemannsche Geometrie im Grossen.* Springer-Verlag, Berlin - Geidelberg - New York.

[17] GROMOV M. (1983). Filling Riemannian manifolds. *J. Differential Geometry*, **18**, 1–148.

[18] GUILLEMIN V. AND KAZHDAN D. (1980). Some inverse spectral results for negatively curved 2-manifolds. *Topology*, **19**, 301–312.

[19] GUILLEMIN V. AND KAZHDAN D. (1980). Some inverse spectral results for negatively curved n-manifolds. *Proceedings of Symposia in Pure Math.*, **36**, 153–180.

[20] GULLIVER R. (1975). On the variety of manifilds without conjugate points. *Trans. Amer. Math. Soc.*, **210**, 185–201.

[21] HARTMAN P. (1964). *Ordinary Differential Equations.* Wiley, New York.

[22] HERGLOTZ, G. (1905). Uber die Elastizitaet der Erde bei Beruecksichtigung ihrer variablen Dichte. *Zeitschr. fur Math. Phys.*, **52**, 275–299.

[23] KAC M. (1966). Can one hear the shape of a drum? *Amer. Math. Monthly*, **73**.

[24] KLINGENBERG W. (1974). Riemannian manifolds with geodesic flow of Anosov type. *Ann. of Math.,* **99**(1), 1–13.

[25] KODANI S. (1990). Convergence theorem for Riemannian manifolds with boundary. *Compositio. Math.*, **75**(2), 171–192.

[26] KOMAROV P.L. (2000). Stability estimates for a solution to an integral geometry problem in Sobolev norms. *Siberian Math. J.*, **41**(3), 602–614.

[27] LASSAS M., SHARAFUTDINOV V., AND UHLMANN G. Semiglobal boundary rigidity for Riemannian metrics. *Mathematischen Annalen.* To appear.

[28] LIVĆIC A.N. (1971). Some homological properties of U-systems. *Mat. Zametki*, **10**, 555–564.

[29] DE LA LLAVE R., MARCO J.M., AND MORIYON R. (1986). Canonical perturbation theory of Anosov Systems and regularity results for the Livsic cohomology equation. *Annals of Math.*, **123**, 537–611.

[30] MICHEL, R. (1981). Sur la rigidité imposee par la longuer des géodesiques. *Invent. Math.*, **65**, 71–84.

[31] MICHEL R. (1994). Restriction de la distance géodésique a un arc et rigidité. *Bull. Soc. Math. France*, **122**, 435–442.

[32] MILNOR J. (1964). Eigenvalues of the Laplace operator on certain manifolds. *Proc. Acad. Nat. Sci. U.S.A.*, **51**, 542–542.

[33] MIN-OO M. (1986) Spectral rigidity for manifolds with negative curvature operator. *Contemp. Math.*, **51**, 99–103.

[34] MUKHOMETOV R.G. (1982). On a problem of reconstructing Riemannian metrics. *Siberian Math. J.*, **22**(3), 420–433.

[35] NOVIKOV R.G. An inversion formula for the attenuated X-ray transform. *Ark. Mat.*, to appear.

[36] OTAL J.-P. (1990). Sur les longuer des géodesiques d'une métrique a courbure négative dans le disque. *Comm. Math. Helv.*, **65**, 334–347.

[37] PESTOV L.N. AND SHARAFUTDINOV V.A. (1988). Integral geometry of tensor fields on a manifold of negative curvative. *Siberian Math. J.*, **29**(3), 427–441.

[38] RACHELE L. Uniqueness of the density in an inverse problem for isotropic elastodynamics. *Communications in PDE*, to appear.

[39] ROMANOV V.G. (1967). Reconstructing a function by means of integrals along a family of curves. *Sibirsk. Mat. Zh.*, **8**(5), 1206–1208.

[40] ROMANOV V.G. (1987). *Inverse Problems of Mathematical Physics.* VNU-SCIENCEPRESS, Utreht, The Netherlands.

[41] SCHUETH D. (1999). Continuous families of isospectral metrics on simply connected manifolds. *Ann. of Math.*, **149**(1), 287–308.

[42] SHARAFUTDINOV V.A. (1992). Integral geometry of a tensor field on a manifold whose curvative is bounded above. *Siberian Math. J.*, **33**(3), 524–533.

[43] SHARAFUTDINOV V.A. (1994). *Integral Geometry of Tensor Fields.* VSP, Utrecht, the Netherlands.

[44] SHARAFUTDINOV V.A. (1995). Modified horizontal derivative and some of its applications. *Siberian Math. J.*, **36**(3), 664–700.

[45] SHARAFUTDINOV V.A. (1995). Finiteness theorem for the ray transform on a Riemannian manifold. *Inverse Problems,* **11**, 1039–1050.

[46] SHARAFUTDINOV V.A. (1997). Integral geometry of a tensor field on a surface of revolution. *Siberian Math. J.* **38**(3), 603–620.

[47] SHARAFUTDINOV V.A. (1999). *Ray Transform on Riemannian Manifolds. Eight Lectures on Integral Geometry.* http://www.math.washington.edu/~sharafut/Ray_transform.dvi.

[48] SHARAFUTDINOV V.A. Integral geometry of tensor fields on a nonconvex domain. To appear.

[49] SHARAFUTDINOV V. AND UHLMANN G. (2001). On deformation boundary rigidity and spectral rigidity of Riemannian surfaces with no focal points. *J. of Differential Geometry,* **56**, 93–110.

[50] STEFANOV P. AND UHLMANN G. (1998). Rigidity for metrics with the same lengths of geodesics. *Math. Research Letters*, **5**, 83–96.

[51] UHLMANN G. (1998). Inverse boundary value problems for partial differential equations. *Proceedings of the International Congress of Mathematicians.* Vol. III (Berlin, 1998). Doc. Math. Vol. III, 77–86

[52] VIGNÉRAS M. (1980). Variétés riemanniennes isospectrales et non isométriques. *Ann. of Math.*, **110**, 21–32.

UNIQUE CONTINUATION PROBLEMS FOR PARTIAL DIFFERENTIAL EQUATIONS*

DANIEL TATARU†

1. Introduction. The aim of this article is to give an overview of the main problems and results in unique continuation. Broadly speaking, an unique continuation result is any statement of the following type:

Given a linear partial differential operator P and two regions $A \subset B$, a solution u to $Pu = 0$ is uniquely determined in the larger set B by its values (behavior) in the smaller set A.

A quantitative version of such a statement is called a stability result:

If two solutions u_1 and u_2 to $Pu = 0$ are close in A then they must be close in B.

To this setup one can also associate a corresponding existence result,

A solution u to $Pu = 0$ in A extends to a solution in B.

If both the uniqueness and the existence results hold then we say that the solvability problem for P (in this setup) is well-posed. However, most interesting unique continuation problems arise in contexts where the corresponding existence result does not hold. For this reason unique continuation problems are often referred to as "uniqueness in ill-posed problems".

This presentation is focused more on the geometrical aspects of the problems and less on the many relevant technical issues. Consequently, we choose to limit our scope in two ways:

(i) As a method for proving unique continuation results we only discuss the L^2 Carleman estimates.

(ii) We do not aim for sharp statements concerning the regularity of the coefficients and of the solution.

2. The classical unique continuation problem. Consider a partial differential operator $P(x, D)$ of order m in $\mathbb{R}^n$. Let Σ be an oriented hypersurface in $\mathbb{R}^n$, which we represent as a nondegenerate level set of a smooth function, $\Sigma = \{\phi = 0\}$. The sign of ϕ away from Σ is determined by the orientation of Σ. Denote the two sides of Σ by $\Sigma^+ = \{\phi > 0\}$ and $\Sigma^- = \{\phi < 0\}$. Then we define the unique continuation property across Σ for solutions to $P(x, D)u = 0$ as follows:

DEFINITION 1. *We say that unique continuation property across Σ holds for the operator $P(x, D)$ if for each $x_0 \in \Sigma$ and V a neighborhood of x_0 the following holds: Let u be a solution for $P(x, D)u = 0$ in V so that $u = 0$ in $\Sigma^+ \cap V$. Then $u = 0$ near x_0.*

*Research partially supported by NSF grant DMS-9622942.

†Department of Mathematics, University of California at Berkeley, Berkeley, CA 94720.

In other words, the values of a solution u to $Pu = 0$ on one side of Σ (i.e. in Σ^+) uniquely determine its values on the other side (i.e. in Σ^-) near Σ. Note that the choice of the orientation is important; in effect, as we shall see later, in most cases if unique continuation holds in one direction then it does not hold in the opposite direction. If the surface Σ is noncharacteristic then we can also introduce the Cauchy problem:

DEFINITION 2 (Cauchy problem). *Given functions $(u_0, u_1, \cdots, u_{m-1})$ on Σ, find u in Σ^- so that:*

$$\begin{cases} P(x,D)u = 0 & \text{in } \Sigma^- \\ u = u_0, \dfrac{\partial u}{\partial \nu} = u_1, \cdots, \dfrac{\partial^{m-1} u}{\partial \nu^{m-1}} = u_{m-1} & \text{on } \Sigma \end{cases}$$

It is easy to see that the unique continuation property is equivalent to (local) uniqueness in the Cauchy problem.

Roughly speaking the unique continuation property says that the information about a solution u to $Pu = 0$ in Σ^- can be retrieved from information about u in Σ^+. Thus one may hope to gain some intuition about the unique continuation problem by looking at special solutions u to $Pu = 0$.

For instance, if P has a real symbol and is not elliptic, then one can consider high frequency solutions which are localized near null bicharacteristic rays of P, given by

$$\dot{x} = p_\xi, \qquad \dot{\xi} = -p_x$$

on $\{p(x,\xi) = 0\}$ where $p(x,\xi)$ is the principal symbol of P. Then it is natural to ask that all such rays which intersect a small neighborhood of Σ in Σ^- must also intersect Σ^+. A strong form of this condition can be expressed in the form

$$\{p,\phi\} = p = 0 \implies \{p,\{p,\phi\}\} > 0 \qquad \text{on } T^*_\Sigma\mathbb{R}^n \tag{1}$$

where $\{\cdot,\cdot\}$ is the Poisson bracket of two symbols, $\{p,q\} = p_\xi q_x - p_x q_\xi$. The Poisson bracket of p and q is the derivative of q along the bicharacteristic (Hamilton) flow of p. Then (1) says that ϕ is strictly convex on null bicharacteristics of P near its critical points. In geometrical terms, this means that all null bicharacteristics which are tangent to Σ must curve away from Σ toward Σ^+. One can think of (1) as some sort of convexity condition for Σ^-, not with respect to all straight lines but instead with respect to the null bicharacteristic flow of P.

Even in the case when P is elliptic, a second obstruction for unique continuation could arise if we were able to construct highly localized approximate solutions u to $Pu = 0$ in Σ^- which decay exponentially toward Σ. This corresponds to replacing real phase functions with complex phase functions, where the imaginary part of the phase is a positive multiple of the function ϕ. To study such solutions one needs to complexify the symbol of p in the direction of the inner normal, i.e. to look at $p_\phi = p(x, \xi + i\tau\nabla\phi)$ where $\tau > 0$. It is not possible to simply look at the geometry of the bicharacteristics for this new symbol, because it is complex valued. Instead this is superseded by the (necessary) local solvability condition for p_ϕ, which has the form

$$\{\Re p_\phi, \Im p_\phi\} \leq 0 \qquad \text{on } \{p_\phi = 0\}$$

If this is violated then we get a sufficient condition which roughly guarantees the absence of such exponentially decreasing solutions, namely

$$\{\Re p_\phi, \Im p_\phi\} > 0 \quad \text{on } T^*_\Sigma \mathbb{R}^n \cap \{p_\phi = 0\}, \ \tau > 0 \tag{2}$$

In addition observe that ϕ is only determined up to multiplication by a smooth positive function, while the above inequality is not invariant with respect to such a transformation. As we need (2) to hold only for some selection of ϕ, a short algebraic computation leads to the relaxed condition

$$\{\Re p_\phi, \Im p_\phi\} > 0 \quad \text{on } T^*_\Sigma \mathbb{R}^n \cap \{p_\phi = \{p_\phi, \phi\} = 0\}, \ \tau > 0 \tag{3}$$

It also interesting to note that as a limit of (3) as $\tau \to 0$ one retrieves (1) (with "$>$" replaced by "$\geq$", though).

Based on this analysis we introduce the pseudoconvexity condition for surfaces,

DEFINITION 3. *We say that the surface Σ is strongly pseudoconvex with respect to P if either*

a) P is elliptic and (3) holds.

b) P has real principal symbol and (1), (3) hold.

Now we are ready to state the main result (see [5] and references therein):

THEOREM 4 (Hörmander). *Suppose that the oriented surface Σ is strongly pseudoconvex with respect to P. Then unique continuation across Σ holds for P.*

This is nicely complemented by the following negative result, proved in [1]:

THEOREM 5 (Alihnac). *Suppose that the condition (1) for the oriented surface Σ strictly fails at some $(x_0, \xi) \in T^*_\Sigma \mathbb{R}^n$. Then there is a function u supported in Σ^- and a smooth function a so that $x_0 \in supp\ u$ but $x_0 \notin supp\ (P + a)u$.*

These two results show that the notion of pseudoconvexity is the right concept to use for unique continuation problems.

We now sketch the proof of the positive result. This is based on a one parameter family of weighted estimates, called Carleman estimates. The proper set of weights to use is tied to the notion of pseudoconvex functions.

DEFINITION 6. *We say that the function ϕ is strongly pseudoconvex with respect to P if either*

a) P is elliptic and (2) holds.

b) P has real principal symbol and (1), (2) hold.

Obviously the nondegenerate level sets of pseudoconvex functions are pseudoconvex surfaces. There is a weaker converse to this statement, namely

REMARK 7. *Suppose that Σ is strongly pseudoconvex with respect to P, and let ϕ be a defining function for Σ. Then for sufficiently large λ the function $e^{\lambda\phi}$ is strongly pseudoconvex with respect to P near Σ.*

Now we can state the estimates:

THEOREM 8 (Carleman estimates). *Suppose that ϕ is strongly pseudoconvex with respect to P in some compact region $\Omega \subset \mathbb{R}^n$. Then there are $c, \tau_0 > 0$ so that for all functions u supported in Ω we have:*

(a) P is elliptic:

$$\tau^{-1}\|e^{\tau\phi}u\|^2_{H^m_\tau} \leq c\|e^{\tau\phi}P(x,D)u\|^2_{L^2}, \quad \tau \geq \tau_0 \tag{4}$$

(b) P has real principal symbol:

$$\tau\|e^{\tau\phi}u\|^2_{H^{m-1}_\tau} \leq c\|e^{\tau\phi}P(x,D)u\|^2_{L^2}, \quad \tau \geq \tau_0 \tag{5}$$

Here the H^m_τ norms are weighted Sobolev norms,

$$\|v\|_{H^m_\tau} = \|(D^2+\tau^2)^{\frac{m}{2}}v\|_{L^2}$$

Proof of Theorem 8. We first eliminate the exponential weight from (4). If we set $v = e^{\tau\phi}u$ then

$$e^{\tau\phi}P(x,D)u = e^{\tau\phi}P(x,D)e^{-\tau\phi}v = P_\phi(x,D,\tau)v$$

where the conjugated operator P_ϕ has the form

$$P_\phi(x,D,\tau) = e^{\tau\phi}P(x,D)e^{-\tau\phi} = P(x, D + i\tau\nabla\phi)$$

Hence (4) for instance reduces to

$$\tau^{-1}\|v\|^2_{H^m_\tau} \leq c\|P_\phi(x,D,\tau)v\|^2_{L^2}, \quad \tau \geq \tau_0 \tag{6}$$

Given the definition of the H^m_τ norms, it follows that all the lower order terms in P are controlled by the left hand side for large τ, therefore they are negligible. We decompose P_ϕ into a part with real symbol and one with imaginary symbol,

$$P_\phi = P^r_\phi + iP^i_\phi$$

and compute

$$\begin{aligned}\|P_\phi v\|^2_{L^2} &= \|P^r_\phi v\|^2_{L^2} + \|P^i_\phi v\|^2_{L^2} + 2\Im\langle P^r_\phi v, P^i_\phi v\rangle \\ &= \|P^r_\phi v\|^2_{L^2} + \|P^i_\phi v\|^2_{L^2} + \langle Qv, v\rangle\end{aligned}$$

where

$$Q = i((P^r_\phi)^* P^i_\phi - (P^i_\phi)^* P^r_\phi)$$
$$= i(((P^r_\phi)^* - P^r_\phi)P^i_\phi - ((P^i_\phi)^* - P^i_\phi)P^r_\phi + [P^r_\phi, P^i_\phi])$$

and the notation $[\cdot,\cdot]$ stands for the commutator of two operators. Since both P^r_ϕ and P^i_ϕ have real principal symbols $\Re P_\phi$, respectively $\Im P_\phi$, it follows that the principal symbol of Q is a homogeneous polynomial of degree $2m-1$ in ξ,τ which has the form

$$q(x,\xi,\tau) = (\Re p_\phi)_{x\xi}\Im p_\phi - (\Im p^i_\phi)_{x\xi}\Re p_\phi + \{\Re p_\phi, \Im p_\phi\}$$

Now we consider separately the two cases:

a) If P is elliptic then

$$p_\phi = 0 \implies |\xi| \leq c\tau$$

From (2) and the above relation we get

$$\tau q(x,\xi,\tau) > 0 \qquad \text{on } \{p_\phi = 0\}$$

therefore for large enough C we have

$$\tau q(x,\xi,\tau) + C|p_\phi|^2 > c(\tau^2 + \xi^2)^m$$

Hence Garding's inequality implies that

$$\tau\langle Qv, v\rangle + C(\|P^r_\phi v\|^2 + \|P^i_\phi v\|^2) \geq c\|v\|^2_{H^m_\tau} - d\|v\|_{H^{m-1}_\tau}$$

The second term on the right is negligible for large enough τ, therefore (6) follows.

b) If P has real principal symbol then we want to argue as above, but in addition we need to consider what happens when τ is small compared to $|\xi|$. In this case for small τ we have

$$\Re p_\phi = p + O(\tau^2), \qquad \Im p_\phi = \tau\{p,\phi\} + O(\tau^3)$$

Then we use (1) for $\tau \ll |\xi|$ and (2) otherwise to conclude that for large enough C we have

$$\tau^{-1}q(x,\xi,\tau) + C(\tau^{-2}|\Im p_\phi|^2 + (\tau^2+\xi^2)^{-1}|\Re p_\phi|^2) \geq c(\xi^2+\tau^2)^{m-1}$$

Therefore Garding's inequality shows that

$$\tau^{-1}\langle Qv, v\rangle + C(\|P^r_\phi v\|^2_{H^{-1}_\tau} + \tau^{-2}\|P^i_\phi v\|^2_{L^2}) \geq c\|v\|^2_{H^{m-1}_\tau} - d\|v\|_{H^{m-2}_\tau}$$

which in turn yields the counterpart of (5) after the conjugation, namely

$$\tau\|v\|^2_{H^{m-1}_\tau} \leq c\|P_\phi(x,D,\tau)v\|^2_{L^2}, \quad \tau \geq \tau_0 \tag{7}$$

□

Proof of Theorem 4. Here we show how to use the Carleman estimates in order to obtain the unique continuation result. By Remark 7, we can represent Σ as a level set of a strongly pseudoconvex function ϕ, $\Sigma = \{\phi = 0\}$. We do not use the Carleman estimates with respect to the function ϕ but with respect to a modified function, namely

$$\psi(x) = \phi(x) - 2\delta^3 + \delta|x - x_0|^2, \qquad \delta > 0$$

For small δ this is a small C^2 perturbation of ϕ, therefore it remains strongly pseudoconvex with respect to P. Observe also that for small δ

$$\{\psi > -\delta^3\} \cap \Sigma^- \subset B(x_0, \delta) \subset V$$

We choose a smooth cutoff function χ supported in $\{\psi > -\delta^3\}$ which equals 1 in $\{\psi > 0\}$. We apply the Carleman estimate (say (4)) to χu to get

$$\tau^{-1}\|e^{\tau\psi}\chi u\|^2_{H^m_\tau} \leq c\|e^{\tau\psi}P(x,D)\chi u\|^2_{L^2}, \quad \tau \geq \tau_0$$

According to our hypothesis on u, it follows that $P(\chi u)$ is supported in $\{\psi \leq 0\}$. Then we further get

$$\tau^{-1}\|e^{\tau\psi}\chi u\|^2_{H^m_\tau} \leq c\|P(x,D)\chi u\|^2_{L^2}, \quad \tau \geq \tau_0$$

Letting $\tau \to \infty$ in this relation we get $\chi u = 0$ in $\{\psi > 0\}$, i.e. $u = 0$ in $\{\psi > 0\}$. But this region contains a neighborhood of x_0. □

Example 1. Consider a second order elliptic operator $P = a^{ij}(x)\partial_i\partial_j$. Then

$$p_\phi(x,\xi,\tau) = -a^{ij}(x)(\xi_i + i\tau\partial_i\phi)(\xi_j + i\tau\partial_j\phi)$$

Hence $\Im p_\phi = -2a^{ij}(x)\xi_i\partial_j\phi$ therefore $\{\Im p_\phi, \phi\} = -2a^{ij}\partial_i\phi\partial_j\phi \neq 0$. Then (3) is always satisfied, which implies that any smooth surface is pseudoconvex with respect to P.

Example 2. Consider the wave operator, $P = \partial_t^2 - \Delta_x$. Arguing as above we see that the condition (3) is always satisfied, at least as long as Σ is noncharacteristic. However, now (1) is nontrivial. Since P has constant coefficients, its bicharacteristics are straight therefore any convex surface is also pseudoconvex. However, there are also nonconvex surfaces which are still pseudoconvex, for instance the positive level sets of the function $\phi = x^2 - t^2$.

3. Strong unique continuation. As seen before, in the case of second elliptic equations any surface is pseudoconvex. Hence, if a solution vanishes in an open set then it must vanish everywhere. Shrinking the open set to a point one can deduce that the same conclusion follows provided the solution decays rapidly enough near that point. What is the precise meaning of "rapidly enough"? The answer to this was provided in dimension 2 by Carleman [3] in 1939 (this is where the Carleman estimates were first introduced), and in higher dimension by Aronszajn [2]. See also [7] for up to date references.

DEFINITION 9. *Let $x_0 \in \mathbb{R}^n$. We say that a function $u \in L^2_{loc}$ vanishes of infinite order at x_0 if for all integers N we have*

$$\lim_{r \to 0} r^{-N} \int_{B(x_0,r)} |u|^2 dx = 0$$

DEFINITION 10. *We say that strong unique continuation holds for an operator P if for each x_0 and each solution u to $Pu = 0$ near x_0 the following statement is true. If u vanishes of infinite order at x_0 then u is zero near x_0.*

THEOREM 11. *Strong unique continuation holds for second order elliptic operators whose principal part has real coefficients.*

The proof of this theorem uses a family of Carleman estimates with weights which blow up polynomially at x_0. To simplify the statement, set $x_0 = 0$. If $P = \Delta$ then one can verify directly that the function $\phi(x) = -\ln|x|$ satisfies a degenerate pseudoconvexity condition, which easily leads to weaker Carleman estimates of the form

$$r^{-1}(\tau \| |x|^{-\tau-1} u\|_{L^2} + \| |x|^{-\tau} \nabla u\|_{L^2}) \leq \| |x|^{-\tau} \Delta u\|_{L^2} \quad \text{supp } u \subset B(0,r)$$

In order to work with a general second order elliptic operator P or to get stronger estimates one has to add more convexity to this weight. A sample result follows.

THEOREM 12. *Let $P(x, D)$ be a second order elliptic operator with real coefficients so that $P(0, D) = \Delta$. Then for sufficiently large λ the function $\phi(x) = -\ln(|x| + \lambda|x|^2)$ is strongly pseudoconvex with respect to P near 0, and the following estimate holds for all u supported near 0 which vanish of infinite order at 0.*

$$\begin{aligned} \tau^3 \|(|x| + \lambda|x|^2)^{-\tau-1} u\|^2_{L^2} + \tau \|(|x| + \lambda|x|^2)^{-\tau} \nabla u\|_{L^2} \\ \leq c\|(|x| + \lambda|x|^2)^{-\tau} \Delta u\|^2_{L^2} \end{aligned} \tag{8}$$

4. Operators with analytic coefficients. The unique continuation results described in the previous section can be considerably improved in the special case when one has some information about the analyticity of the coefficients. The classical example in this direction is Holmgren's theorem for noncharacteristic surfaces. Recall that a surface $\Sigma = \{\phi = 0\}$ is called noncharacteristic if $p(x, \nabla\phi) \neq 0$ for $x \in \Sigma$.

THEOREM 13 (Holmgren). *Assume that P has analytic coefficients and that Σ is noncharacteristic. Then unique continuation across Σ holds for the operator P.*

However, it is equally interesting to consider also the case when the coefficients are analytic only with respect to some of the variables. A good example is the case of autonomous evolution equations, i.e. with time independent coefficients. A natural setup to consider is as follows. We split the coordinates $x = (x_a, x_b)$ and consider partial differential operators P whose coefficients are analytic as functions of x_a, but possibly nonanalytic as functions of x_b. We define the subset of the cotangent bundle $\Gamma = \{\xi_a = 0\}$, which one can think of as the conormal bundle of the foliation of $\mathbb{R}^n$ by the hyperplanes $x_b = \text{const}$, along which the coefficients are analytic.

THEOREM 14. *Assume that the coefficients of P are analytic with respect to x_a. Let Σ be an oriented hypersurface so that the pseudoconvexity condition (3) for Σ with respect to P holds in Γ. Assume that either of the following two conditions holds:*

a) P is elliptic in Γ.

b) P has real symbol on Γ and Γ is invariant with respect to the null bicharacteristic flow of P (i.e. $p_{x_a} = 0$ on $\{p = 0, \xi_a = 0\}$) .

Then unique continuation across Σ holds for the operator P.

This result was first proved in Tataru [15] for a more restrictive class of operators, which was subsequently relaxed in [12], [4] and finally, in the present form, in [17].

Comparing this with Theorem 4, one can see that the assumptions are almost identical, with the only difference that now we are asking for these assumptions to be satisfied on a smaller subset of the cotangent bundle, namely on Γ.

Suppose we consider the case of operators whose coefficients are analytic with respect to all variables. Then $x = x_a$, $\Gamma = \{\xi = 0\}$ so we only need to ask for the pseudoconvexity condition (3) at a single point. There it reads

$$\Re p_x(x, \nabla\phi)\bar{p}_\xi(x, \nabla\phi) + p_\xi(x, \nabla\phi)\nabla^2\phi\, \bar{p}_\xi(x, \nabla\phi) > 0 \quad \text{on } \{p(x, \nabla\phi) = 0\}$$

If Σ is noncharacteristic then $p(x, \nabla\phi) \neq 0$ on Σ so this is satisfied by default and we retrieve Holmgren's theorem. But even if Σ is characteristic at some points, the above condition may still hold, so we get a slight improvement over Holmgren's theorem.

The key step in the proof of Theorem 14 is still a Carleman estimate, but this time the weight is a pseudodifferential weight.

THEOREM 15. *Let P be as in Theorem 14. Let ϕ be a smooth function which is real analytic in x_a and is strongly pseudoconvex with respect to P on Γ within some compact region $\Omega \subset \mathbb{R}^n$. Then there are positive constants c, d so that for all functions u supported in Ω all $\epsilon > 0$ sufficiently small and all τ sufficiently large we have:*

(a) P is elliptic in Γ:

$$\tau^{-1}\|e^{-\frac{\epsilon D_a^2}{2\tau}} e^{\tau\phi} u\|^2_{H^m_\tau} \leq c(\|e^{-\frac{\epsilon D_a^2}{2\tau}} e^{\tau\phi} P(x, D)u\|^2_{L^2} + e^{-d\epsilon\tau}\|e^{\tau\phi}u\|^2_{H^m_\tau}) \tag{9}$$

(b) P has real symbol in Γ:

$$\tau\|e^{-\frac{\epsilon D_a^2}{2\tau}} e^{\tau\phi} u\|^2_{H^{m-1}_\tau} \leq c(\|e^{-\frac{\epsilon D_a^2}{2\tau}} e^{\tau\phi} P(x, D)u\|^2_{L^2} + e^{-d\epsilon\tau}\|e^{\tau\phi}u\|^2_{H^{m-1}_\tau}). \tag{10}$$

Proof. (Sketch) We first conjugate with respect to the exponential weight as before. Then (9) for instance reduces to

$$(11) \qquad \tau^{-1}\|e^{-\frac{\epsilon D_a^2}{2\tau}} v\|_{H_\tau^m}^2 \leq c(\|e^{-\frac{\epsilon D_a^2}{2\tau}} P_\phi v\|_{L^2}^2 + e^{-d\epsilon\tau}\|v\|_{H_\tau^m}^2).$$

However, this time we also need to conjugate with respect to the Gaussian. This interacts with the coefficients of P_ϕ. A simple computation gives the conjugate of x_a,

$$e^{-\frac{\epsilon D_a^2}{2\tau}} x_a = (x_a + i\frac{\epsilon D_a}{\tau})e^{-\frac{\epsilon D_a^2}{2\tau}}.$$

It is slightly more difficult to do the same for powers of x_a. The result has a simple form if one uses the Weyl calculus:

$$e^{-\frac{\epsilon D_a^2}{2\tau}} x_a^\alpha = OP^w(x_a + i\frac{\epsilon \xi_a}{\tau})e^{-\frac{\epsilon D_a^2}{2\tau}}.$$

To conjugate the coefficients of P_ϕ we first extend them to smooth compactly supported functions in $\mathbb{C}^{n_a} \times \mathbb{R}^{n_b}$ which are holomorphic in x_a in a small (complex) neighborhood of Ω. Then a somewhat more involved computation shows that for v supported in Ω we have

$$e^{-\frac{\epsilon D_a^2}{2\tau}} P_\phi = P_{\phi,\epsilon} e^{-\frac{\epsilon D_a^2}{2\tau}} + R_{\phi,\epsilon}$$

where, assuming that P_ϕ is written as a Weyl operator, we have

$$P_{\phi,\epsilon} = OP^w(p_\phi(x_a + i\frac{\epsilon\xi_a}{\tau}, x_b, \xi, \tau))$$

and the remainder R is exponentially small,

$$\|Rv\|_{H_\tau^m} \leq e^{-c\tau}\|v\|_{H_\tau^m} \qquad \text{supp } v \subset \Omega.$$

Hence the estimate (11) reduces to

$$(12) \qquad \tau^{-1}\|w\|_{H_\tau^m}^2 \leq c(\|P_{\phi,\epsilon} w\|_{L^2}^2 + e^{-d\epsilon\tau}\|v\|_{H_\tau^m}^2) \qquad w = e^{-\frac{\epsilon D_a^2}{2\tau}}.$$

The second right hand side term accounts for the part of w whose Fourier transform is supported in $|\xi_a| > d\tau$. On the other hand, if d is small enough then the region $|\xi_a| \leq d\tau$ is close to Γ therefore the pseudoconvexity condition holds there. If in addition ϵ is sufficiently small then $P_{\epsilon,\phi}$ is a small perturbation of P_ϕ therefore the proof of (11) is a straightforward modification of the proof of (6).

The case (b) raises some additional difficulties because when $\tau = 0$ the estimate (10) is degenerate while on the other hand $P_{\phi,\epsilon}$ no longer has real symbol (like P_ϕ). This is where the invariance of Γ with respect to the null bicharacteristic flow comes in, as it insures that $P_{\epsilon,\phi}$ satisfies a principal normality condition on Γ,

$$\{\Re p_{\epsilon,\phi}, \Im p_{\epsilon,\phi}\} \leq c|\xi|^{m-1}|p_{\epsilon,\phi}| \quad \text{on } \{\xi_a = 0, \ \tau = 0\}.$$

This suffices in order to carry out the proof of the counterpart of (12), though it is a bit more difficult. □

Proof of Theorem 14. This is done in the same manner as the proof of Theorem 4. The δ there must be chosen much smaller than ϵ, so that the second term on the right in the estimates (9), (10) can be neglected. The nontrivial part of the argument is summarized in the following

LEMMA 16. *Assume that*

$$\|e^{-\frac{\epsilon D_a^2}{2\tau}} e^{\tau\psi} u\|_{L^2} \leq c \qquad \tau > \tau_0.$$

Then $u = 0$ *in* $\{\psi > 0\}$.

To prove this take a function v with compactly supported Fourier transform and define the entire function

$$g(s) = \langle v, e^{is\psi} u \rangle$$

g is trivially bounded on the real axis, and by hypothesis it follows that it is also bounded on the positive imaginary axis. But g is of exponential type, therefore the Phragmen-Lindelof theorem implies that g is bounded in the upper half-plane. Then its Fourier transform,

$$\hat{g}(\theta) = \langle v, \delta_{\psi=\theta} u \rangle$$

is supported in $\mathbb{R}^-$. Hence

$$\langle v, 1_{\psi>0} u \rangle = 0.$$

By density this must hold for all $v \in L^2$ therefore $u = 0$ in $\{\psi > 0\}$. □

Example 3. Consider a wave operator with time independent coefficients, $P = \partial_t^2 - g^{ij}(x)\partial_i\partial_j$ with positive definite g^{ij}. Its symbol is $p(x, x, \xi) = -s^2 + g^{ij}(x)\xi_i\xi_j$ and $\Gamma = \{s = 0\}$. Then P is elliptic in Γ. On the other hand, as in Example 1, we can argue that (3) is satisfied by default for a surface Σ provided that Σ is noncharacteristic. Hence we conclude that for P we have unique continuation across any noncharacteristic surface. This was the first problem to be considered in this direction, and a partial result for it by Robbiano [11] opened the way for further developments.

5. Boundary value problems.

Let Ω be an open subset of $\mathbb{R}^n$ with smooth boundary $\partial\Omega$. Consider a boundary value problem in Ω,

$$\left\{ \begin{array}{ll} P(x, D)u = 0 & \text{in } \Omega \\ B(x, D)u = 0 & \text{in } \partial\Omega. \end{array} \right. \tag{13}$$

Here $B = \{B_k\}_{k=1,k_0}$ with $k_0 \le m-1$ are defined on the boundary $\partial\Omega$. We denote their orders by $m_k \le m-1$.

Given a smooth oriented surface $\Sigma = \{\phi = 0\}$ one can still ask whether unique continuation across Σ holds for solutions to this equation. At points on Σ which are inside Ω one can disregard the boundary condition and look at this problem as before. However, at points on $\Sigma \cap \partial\Omega$ the boundary condition plays an essential role. Some additional condition must be imposed on the boundary operators in order for such a result to hold. By analogy with the similar condition arising in the study of elliptic or hyperbolic boundary value problems, such a condition is called a Lopatinskii condition. Then the main result is

THEOREM 17. *Assume that Σ is strongly pseudoconvex with respect to P and that the pair (B, P) satisfies the Lopatinskii condition with respect to $d\phi$. Then unique continuation across Σ holds for solutions to (13).*

Of course what is missing in this statement is a definition of what the Lopatinskii condition is, and this is where things get a bit tricky.

On one hand, there is a strong form of the Lopatinskii condition, called the strong Lopatinskii condition. This is an algebraic condition connecting the symbols of P, B, the conormal direction ν to $\partial\Omega$ and the conormal direction $d\phi$ to Σ. The precise definition is somewhat lengthy, so we refer the reader to [14] for details. It is an analogue in this set-up of Sakamoto's strong Lopatinskii condition which guarantees the strong L^2 well-posedness of hyperbolic initial/boundary value problems.

This is necessary and sufficient in order for a strong form of the Carleman estimates to hold:

THEOREM 18. *Assume that ϕ is strongly pseudoconvex with respect to P and that the pair (B, P) satisfies the strong Lopatinskii condition with respect to $d\phi$ in a compact subset K of Ω. Then for all u supported in K we have:*

(a) P is elliptic:

$$\begin{aligned}(14)\qquad &\tau^{-1}\|e^{\tau\phi}u\|^2_{H^m_\tau(\Omega)} + \tau\sum_{j=0}^{m-1}\|e^{\tau\phi}\partial^j u\|^2_{H^{m-k-1}_\tau(\partial\Omega)}\\ &\le c(\|e^{\tau\phi}P(x,D)u\|^2_{L^2(\Omega)} + \tau\sum_k\|e^{\tau\phi}B_k u\|^2_{H^{m_k}_\tau(\partial\Omega)}),\quad \tau \ge \tau_0\end{aligned}$$

(b) P has real principal symbol:

$$\begin{aligned}(15)\qquad &\tau\|e^{\tau\phi}u\|^2_{H^{m-1}_\tau(\Omega)} + \tau\sum_{j=0}^{m-1}\|e^{\tau\phi}\partial^j u\|^2_{H^{m-k-1}_\tau(\partial\Omega)}\\ &\le c(\|e^{\tau\phi}P(x,D)u\|^2_{L^2(\Omega)} + \tau\sum_k\|e^{\tau\phi}B_k u\|^2_{H^{m_k}_\tau(\partial\Omega)}),\quad \tau \ge \tau_0.\end{aligned}$$

In both cases the constant c can be chosen not to depend on τ.

For the proof we refer the reader to [14].

On the other hand one also expects that in certain cases only a weaker form of these estimates may hold, which nevertheless imply the unique continuation property.

THEOREM 19. *Assume that ϕ is strongly pseudoconvex with respect to P and that the pair (B, P) satisfies the weak Lopatinskii condition with respect to $d\phi$ in a compact subset K of Ω. Then for all u supported in K with $Bu = 0$ on $\partial\Omega$ we have:*

(a) P is elliptic:

$$\tau^{-1}\|e^{\tau\phi}u\|^2_{H^m_\tau(\Omega)} \leq c\|e^{\tau\phi}P(x,D)u\|^2_{L^2(\Omega)}, \quad \tau \geq \tau_0 \tag{16}$$

(b) P has real principal symbol:

$$\tau\|e^{\tau\phi}u\|^2_{H^{m-1}_\tau(\Omega)} \leq c\|e^{\tau\phi}P(x,D)u\|^2_{L^2(\Omega)}, \quad \tau \geq \tau_0. \tag{17}$$

In both cases the constant c can be chosen not to depend on τ.

This is only a quasi-theorem in that it is missing the proof and a clear definition of what the weak Lopatinskii condition is. One could conceivably take the estimates as a definition of the weak Lopatinskii condition and attempt to produce an algebraic characterization of it. However, this is a highly nontrivial endeavor even in the case of the L^2 well-posedness for second order hyperbolic initial boundary value problems. At the very least, one may keep this in mind as a general principle and prove it in special cases.

Example 4. Let $P = g^{ij}(x)\partial_i\partial_j$ be a second order elliptic operator. Let ν be the outer normal to the boundary $\partial\Omega$. N be the Neumann operator, $Nu = \nu_i g^{ij}\partial_j$. Then the strong Lopatinskii condition is satisfied if B is either Dirichlet, $Bu = u$ or Neumann, $Bu = Nu$, if and only if $N\phi < 0$ on $\partial\Omega$.

Example 5. Let $P = \partial_t^2 + g^{ij}(t,x)\partial_i\partial_j$ be a second order wave operator. Suppose that the boundary is time-like, and set $N = -\nu_0\partial_0 + \nu_i g^{ij}\partial_j$. Then

(i) The strong Lopatinskii condition is satisfied if B is Dirichlet, $Bu = u$, if and only if $N\phi < 0$ on $\partial\Omega$.

(ii) The weak Lopatinskii condition is satisfied if B is Neumann, $Bu = Nu$, if and only if $N\phi = 0$ on $\partial\Omega$.

6. Anisotropic unique continuation. In the classical unique continuation results one works with partial differential operators whose principal symbol is homogeneous of a certain order, and all lower order terms play no role in the geometry of the problem. Such a set-up fits some operators but does not fit others. Consider for instance the heat operator $\partial_t - \Delta$, the Schroedinger operator $\partial_t - i\Delta$ or the KdV operator $\partial_t - \partial_x^3$. In all these examples the time derivative is of lower order than the spatial derivatives, however one can only neglect it at the expense of completely changing the type of the operator. The natural conclusion is that one has to adapt the setup to the operator being studied. We call such operators, which require a different set-up, anisotropic. For the heat and Schroedinger operators one needs a calculus where one time derivative carries the same weight as two space derivatives. For the KdV operator, one time derivative must carry the same weight as three space derivatives. Such a calculus can be constructed in a very similar framework. However, to keep things simple we confine ourselves to the case which corresponds to both the heat and Schroedinger operators. To fix the notations suppose we work in $\mathbb{R} \times \mathbb{R}^n$, (t, x) are the physical variables and (s, ξ) are the Fourier variables. For a multiindex $\alpha = (\alpha_0, \alpha_1, \cdots, \alpha_n)$ we set $\partial^\alpha = \partial_t^{\alpha_0} \partial_{x_1}^{\alpha_1} \cdots \partial_{x_n}^{\alpha_n}$ and we define the order of this operator by

$$|\alpha| = 2\alpha_0 + \alpha_1 + \cdots + \alpha_n.$$

For a partial differential operator we let the principal part be the part containing the highest order terms according to this method of counting.

A straightforward observation in this case is that one has to modify the definition of the Poisson bracket, and set

$$\{p(t,x,s,\xi), q(t,s,x,\xi)\} = p_\xi(t,x,s,\xi) q_x(t,x,s,\xi) - p_x(t,x,s,\xi) q_\xi(t,x,s,\xi).$$

Missing on the right are the terms which correspond to t and s derivatives, which in this context become lower order. This is because differentiation with respect to s reduces the order by 1 in s, which means by 2 with respect to the anisotropic definition of the order.

Unique continuation problems for operators of this type were considered as early as 1958, see Mizohata [9]. Using the modified Poisson bracket in the definition of the pseudoconvexity condition Isakov [6] proved that all the classical results in Section 2 can be transferred to the anisotropic case without any change. However, such a result is not entirely satisfactory as it is not properly adapted to this context.

One way to think about this is that equations such as the Schroedinger or the heat equation have an infinite speed of propagation in time. In other words, high frequency information propagate essentially along time slices (at least locally in space). This may lead one to believe that the unique continuation results should also apply along time slices, rather than in space-time domains.

A naive attempt would be to show that if a solution vanishes on some part of a time slice then it must also vanish on some other part. But this cannot work in general; consider for instance for the Schroedinger equation, where one can prescribe the initial data arbitrarily on a given time slice. The same example shows that even assuming that the solution vanishes of infinite order is insufficient. As it turns out, one needs an exponential rate of decay in order for such an argument to work.

To state the result we need to redefine the notion of pseudoconvexity. For this we consider oriented surfaces Σ within a fixed time-slice $\{t = t_0\}$.

DEFINITION 20. *An oriented surface* $\Sigma = \{\phi(x) = 0\} \subset \{t = t_0\}$ *is strongly pseudoconvex with respect to* P *if either*

a) P *is elliptic and (3) holds.*

b) P *has real principal symbol and (1), (3) hold.*

This definition is seemingly identical with Definition 3 but one has to keep in mind that the principal part of P is now determined with respect to the anisotropic scaling. Also we are using the anisotropic version of the Poisson bracket, which only requires x derivatives of ϕ.

DEFINITION 21. *We say that anisotropic unique continuation across* $\Sigma \subset \{t = t_0\}$ *holds for the operator* P *if for each* $\epsilon \neq 0$, $x_0 \in \Sigma$ *and* V *a neighborhood of* x_0 *in* $\mathbb{R}^n$ *the following holds: Let* u *be a solution for* $P(x, D)u = 0$ *in* $V \times [t_0, t_0 + \epsilon]$ *so that*

$$\lim_{t \to t_0,\ t \in (t_0, t_0+\epsilon]} u(t,x) e^{\frac{M}{t}} = 0 \qquad \text{for all } M > 0 \tag{18}$$

uniformly for $x \in \Sigma^+ \cap V$. *Then the same holds for all* x *in a neighborhood of* x_0.

Note that we only ask for one-sided information on the solution u, either for $t > t_0$ (when $\epsilon > 0$) or for $t < t_0$ (when $\epsilon < 0$). Then the following result is proved in Tataru [16].

THEOREM 22. *Assume that the oriented surface* $\Sigma \subset \{t = t_0\}$ *is strongly pseudoconvex with respect to* P. *Then anisotropic unique continuation across* Σ *holds for the operator* P.

If we define strongly pseudoconvex functions in the same manner as strongly pseudoconvex surfaces, then the corresponding Carleman estimates have the form (see [16]):

THEOREM 23. *Assume that the function* ϕ *is strongly pseudoconvex with respect to* P *in a compact region* $\{t_0\} \times K$. *Let* g *be a smooth positive function so that*

$$|g^{(j)}(t)| \leq c_j g(t)^j \qquad j \geq 1 \tag{19}$$

Then for small enough ϵ *and for all functions* u *supported in* $[t_0 - \epsilon, t_0 + \epsilon] \times K$ *we have*

(a) P *is elliptic:*

$$\tau^{-1} \| g^{-\frac{1}{2}} e^{\tau\phi(x)g(t)} u \|^2_{H^m_{\tau g}} \leq c \| e^{\tau\phi(x)g(t)} P(x,D) u \|^2_{L^2}, \quad \tau \geq \tau_0 \tag{20}$$

(b) P *has real principal symbol:*

$$\tau \| g^{\frac{1}{2}} e^{\tau\phi(x)g(t)} u \|^2_{H^{m-1}_{\tau g}} \leq c \| e^{\tau\phi(x)g(t)} P(x,D) u \|^2_{L^2}, \quad \tau \geq \tau_0. \tag{21}$$

In both cases the constant c *can be chosen not to depend on* τ *and on* g *subject to (19).*

Here the spaces $H^m_{\tau g}$ are defined as before but with τ replaced by τg. The function g depends on t but this does not cause problems in the definition of the norm because, by (19), the function g varies slowly enough.

As a limit of functions g which satisfy (19) uniformly we obtain functions g which blow up, for instance

$$g(t) = |t - t_0|^{-1} = \lim_{\delta \to 0} (\delta^2 + (t - t_0)^2)^{-\frac{1}{2}}.$$

Using this weight the (weaker) two sided version of the unique continuation result immediately follows as in the proof of Theorem 4. The only difference is that here it suffices to use the estimate for fixed τ, as opposed to letting $\tau \to \infty$.

It is somewhat more difficult, but one can also show that the corresponding one sided estimates with the weight $|t - t_0|^{-1}$ are also valid. The price one needs to pay for this is to increase the lower limit for the values of τ for which the estimate holds.

Example 6. Let P be the parabolic operator $P = \partial_t - g^{jk}\partial_j\partial_k$. Then P is elliptic, and its principal symbol is $p = is + g^{jk}\xi_j\xi_k$. Hence the symbol of the conjugated operator is $p_\phi = is + g^{ij}(\xi_i + i\partial_i\phi)(\xi_j + i\partial_j\phi)$. A short computation yields $\{\Im p_\phi, \phi\} = 2g^{ij}\partial_i\phi\partial_j\phi \neq 0$, therefore (3) always holds. Hence all surfaces Σ are strongly pseudoconvex with respect to P. This example also shows that Theorem 22 is sharp, in the sense that one cannot relax the decay requirement at t_0. Indeed, consider the parabolic initial value problem with initial data $u = u_0$ at time 0. If u_0 vanishes in some open set ω, then on compact subsets of ω the solution u decays like $e^{-\frac{c}{t}}$ as t approaches 0.

Example 7. Let P be the Schroedinger operator $P = i\partial_t - \Delta$. Then $p = -s + \xi^2$ and $p_\phi = -s + (\xi + i\tau\nabla\phi)^2$. Since $\{\Im p_\phi, \phi\} = 2|\nabla\phi|^2 \neq 0$ it follow that (3) always holds. On the other hand, the bicharacteristic flow for P contains all straight lines on time slices, therefore (1) says that a surface Σ is strongly pseudoconvex if and only if it is convex.

7. Parabolic strong unique continuation. As we have seen in Example 6, the anisotropic unique continuation result is sharp for the forward heat equation, i.e. if we have information only for $t > t_0$. If, on the other hand, we also know that the function u is also a solution to the heat equation before the time t_0, then we can get a much stronger result, which is the parabolic counterpart of the elliptic strong unique continuation result.

A second order parabolic operator is an operator of the form

$$P = \partial_t - a^{ij}(t,x)\partial_i\partial_j, \qquad \{a^{ij}\} \text{ real, positive definite}$$

possibly with arbitrary lower order terms. Given $(t_0, x_0) \in \mathbb{R} \times \mathbb{R}^n$, a backward parabolic neighborhood of (t_0, x_0) is a set containing a cylinder $[t_0 - r^2, t_0] \times B(x_0, r)$.

DEFINITION 24. *We say that a function $u \in L^2_{loc}$ vanishes of infinite order at (t_0, x_0) if for all integers N we have*

$$\lim_{r \to 0} r^{-N} \int_{t_0 - r^2}^{t_0} \int_{B(x_0, r)} |u|^2 dt dx = 0.$$

DEFINITION 25. *We say that parabolic strong unique continuation holds for an operator P if for each $(t_0, x_0) \in \mathbb{R} \times \mathbb{R}^n$ and each solution u to $Pu = 0$ in a backward parabolic neighbourhood of (t_0, x_0) the following statement is true. If u vanishes of infinite order at (t_0, x_0) then $u(t_0, x)$ is zero for x near x_0.*

THEOREM 26. *Parabolic strong unique continuation holds for second order parabolic operators with real coefficients.*

A result of this type was first proved by Lin [8] in a more restrictive setting, but a slightly weaker formulation has been known for much longer, see for instance ([13]). As stated the result was proved in [10].

One can prove this result using a class of Carleman estimates with weights which blow up polynomially at (t_0, x_0). Set $(t_0, x_0) = 0$. If $P = \partial_t - \Delta$ then we have a nice family of weights for which a degenerate estimate holds.

$$r^{-2}(\| |t|^{-\tau} e^{-\frac{x^2}{4t}} u\|_{L^2} \leq \| |t|^{-\tau} e^{-\frac{x^2}{4t}} (\partial_t - \Delta u)\|_{L^2} \quad \text{supp } u \subset [-r^2, 0] \times B(0, r).$$

In order to work with a general second order parabolic operator P or to get stronger estimates one needs to add more convexity to the weight.

THEOREM 27. *Let P be a second order parabolic operator with real coefficients so that $P(0, D) = \partial_t - \Delta$. Then for sufficiently large λ the function $\phi(x, t) = (|t| + \sqrt{\tau |t|^3})^{-\tau} e^{-\frac{x^2}{4t}}$ is strongly pseudoconvex with respect to P near 0, and the following estimate holds for all u supported in $[-c\tau^{-1}, 0]$ which vanish of infinite order at 0.*

$$\begin{aligned} \tau^{\frac{3}{2}} \|(|t| + \sqrt{\tau |t|^3})^{-\tau - \frac{3}{2}} e^{-\frac{x^2}{4t}} u\|_{L^2} + \tau^{\frac{1}{2}} \|(|t| + \sqrt{\tau |t|^3})^{-\tau - \frac{1}{2}} e^{-\frac{x^2}{4t}} \nabla_x u\|_{L^2} \\ \leq \|(|t| + \sqrt{\tau |t|^3})^{-\tau} e^{-\frac{x^2}{4t}} u\|_{L^2}. \end{aligned} \tag{22}$$

REFERENCES

[1] S. ALINHAC AND M.S. BAOUENDI. A nonuniqueness result for operators of principal type. *Math. Z.*, **220**(4): 561–568, 1995.

[2] N. ARONSZAJN. A unique continuation theorem for solutions of elliptic partial differential equations or inequalities of second order. *J. Math. Pures Appl. (9)*, **36**: 235–249, 1957.

[3] T. CARLEMAN. Sur un problème d'unicité pur les systèmes d'équations aux dérivées partielles à deux variables indépendantes. *Ark. Mat., Astr. Fys.*, **26**(17): 9, 1939.

[4] LARS HÖRMANDER. On the uniqueness of the Cauchy problem under partial analyticity assumptions. unpublished.

[5] LARS HÖRMANDER. *The analysis of linear partial differential operators. IV.* Springer-Verlag, Berlin, 1985. Fourier integral operators.

[6] VICTOR ISAKOV. Carleman type estimates in an anisotropic case and applications. *J. Differential Equations*, **105**(2): 217–238, 1993.

[7] HERBERT KOCH AND DANIEL TATARU. Carleman estimates and unique continuation for second-order elliptic equations with nonsmooth coefficients. *Comm. Pure Appl. Math.*, **54**(3): 339–360, 2001.

[8] FANG-HUA LIN. A uniqueness theorem for parabolic equations. *Comm. Pure Appl. Math.*, **43**(1): 127–136, 1990.

[9] SIGERU MIZOHATA. Unicité du prolongement des solutions pour quelques opérateurs différentiels paraboliques. *Mem. Coll. Sci. Univ. Kyoto. Ser. A. Math.*, **31**: 219–239, 1958.

[10] CHI-CHEUNG POON. Unique continuation for parabolic equations. *Comm. Partial Differential Equations*, **21**(3–4): 521–539, 1996.

[11] LUC ROBBIANO. Théorème d'unicité adapté au contrôle des solutions des problèmes hyperboliques. *Comm. Partial Differential Equations*, **16**(4–5): 789–800, 1991.

[12] LUC ROBBIANO AND CLAUDE ZUILY. Uniqueness in the Cauchy problem for operators with partially holomorphic coefficients. *Invent. Math.*, **131**(3): 493–539, 1998.

[13] TAIRA SHIROTA. A theorem with respect to the unique continuation for a parabolic differential equation. *Osaka Math. J.*, **12**: 377–386, 1960.

[14] D. TATARU. Carleman estimates and unique continuation for solutions to boundary value problems. *J. Math. Pures Appl. (9)*, **75**(4): 367–408, 1996.

[15] DANIEL TATARU. Unique continuation for solutions to PDE's; between Hörmander's theorem and Holmgren's theorem. *Comm. Partial Differential Equations*, **20**(5–6): 855–884, 1995.

[16] DANIEL TATARU. Carleman estimates, unique continuation and controllability for anizotropic PDEs. In *Optimization methods in partial differential equations (South Hadley, MA, 1996)*, pp. 267–279. Amer. Math. Soc., Providence, RI, 1997.

[17] DANIEL TATARU. Unique continuation for operators with partially analytic coefficients. *J. Math. Pures Appl. (9)*, **78**(5): 505–521, 1999.

REMARKS ON FOURIER INTEGRAL OPERATORS*

MICHAEL TAYLOR†

Abstract. We prove some estimates on several types of Fourier integral operators, emphasizing $\mathfrak{h}^1 \to \mathfrak{h}^1$ estimates and $\mathfrak{h}^1 \to$ bmo estimates. The results are mostly special cases of more general known results, but the proofs of the special cases presented here are simpler than the usual proofs. Furthermore, the special cases treated here arise quite commonly.

Introduction. In addition to the L^2-Sobolev estimates on Fourier integral operators, whose significance for the theory of linear PDE is well known, there has been a fair amount of work on L^p-Sobolev estimates, for $p \neq 2$. This includes both $L^p - L^p$ estimates and $L^p - L^q$ estimates with $q \neq p$, with particular emphasis on the case $q = p'$. Such estimates lead to Strichartz estimates, and have become important in the study of nonlinear wave equations. For a sampling of papers dealing with such estimates, we mention [St, L, B, Be, Pe, SSS, CC].

Rather delicate techniques have been developed, involving multiple dyadic partitions of unity. These techniques have yielded strong results, but they involve intricate reasoning. In this note we point out that several important special cases have short, simple proofs, essentially involving just FIO calculus and a look at the explicit form of some model cases, together with familiar L^p and Hardy space estimates for pseudodifferential operators. To be sure, these arguments do not work in the generality that more complicated arguments do, but the cases where they succeed do include some of the most important cases that arise in the study of wave equations.

In §1 we treat FIOs with locally invertible canonical relations. In §2 we discuss FIOs with one-sided folds, which have arisen in several contexts in scattering theory and the study of inverse problems. In §3 we treat a class of convolution operators that are also FIOs with folds. The material in §3 illustrates a situation where, in changing the geometrical hypotheses of §2, one gets better $L^p - L^p$ estimates but worse $L^p - L^{p'}$ estimates.

1. Hardy space estimates on Fourier integral operators. Assume $\mathcal{C}$ is a homogeneous canonical transformation (locally invertible) from $T^*Y \setminus 0$ to $T^*X \setminus 0$, $\dim X = \dim Y = n$. Let $T \in I^{-(n-1)/2}(X, Y, \mathcal{C})$, and assume its Schwartz kernel has compact support. We give a short demonstration that T acts on local Hardy spaces:

$$T : \mathfrak{h}^1(Y) \to \mathfrak{h}^1(X), \tag{1.1}$$

under a simplifying hypothesis. Namely we assume the Schwartz kernel of T is a conormal distribution, with singular support in $\Sigma \subset X \times Y$, a smooth

*Work supported by NSF grant DMS-9877077.

†Department of Mathematics University of North Carolina, Chapel Hill, NC 27599.

$(2n-1)$-dimensional manifold. Furthermore we assume that $\Sigma \to Y$ is a submersion. For a proof of (1.1) without these extra hypotheses, see [S].

Let S be the operator whose Schwartz kernel is surface measure on Σ. Our hypotheses imply that $S\delta_p$ is a finite measure for each $p \in Y$, in fact a smooth multiple of surface measure on

$$\Sigma_p = \{x \in X : (x,p) \in \Sigma\},$$

so we have

$$S : L^1(Y) \longrightarrow L^1(X). \tag{1.2}$$

Note that $S = S_0 + S_1$ with

$$S_j \in I^{-(n-1)/2}(X, Y, \mathcal{C}_j), \quad \mathcal{C}_0 = \mathcal{C}. \tag{1.3}$$

Furthermore, by the FIO calculus, there exists $P \in OPS^0(Y)$ such that $T = S_0 P$ (modulo a smoothing operator).

Take $f \in \mathfrak{h}^1(Y)$. Using a microlocal partition of unity, write $f = \sum f_\nu$, $f_\nu \in \mathfrak{h}^1$ having wave front set in Γ_ν (a small conic subset of $T^*Y \setminus 0$). Then $Pf_\nu \in \mathfrak{h}^1 \subset L^1$ has wave front set in Γ_ν and

$$SPf_\nu \in L^1(X), \quad SPf_\nu = Tf_\nu + S_1 P f_\nu. \tag{1.4}$$

The two terms $Tf_\nu = S_0 P f_\nu$ and $S_1 P f_\nu$ have *disjoint* singular support (if Γ_ν is small enough). Since $SPf_\nu \in L^1$, we have $Tf_\nu \in L^1$ (and $S_1 P f_\nu \in L^1$). Thus $T : \mathfrak{h}^1 \to L^1$. Since $QT \in I^{-(n-1)/2}(Y, X, \mathcal{C})$ for each $Q \in OPS^0(X)$, this yields $Q(Tf) \in L^1$ for each $Q \in OPS^0$ (if $f \in \mathfrak{h}^1$), and we have (1.1).

As a complement, note that $\Lambda^{-1} S\delta_p \in \text{bmo}(X)$ for each $p \in Y$, where $\Lambda = \sqrt{-\Delta + 1}$, Δ denoting the Laplace operator on X. Hence

$$\Lambda^{-1} S : L^1(Y) \longrightarrow \text{bmo}(X). \tag{1.5}$$

Then an argument like that above yields

$$T_2 \in I^{-(n+1)/2}(X, Y, \mathcal{C}) \Longrightarrow T_2 : \mathfrak{h}^1(Y) \longrightarrow \text{bmo}(X). \tag{1.6}$$

Interpolation of (1.1) and (1.6) with the standard result

$$T_3 \in I^0(X, Y, \mathcal{C}) \Longrightarrow T_3 : L^2(Y) \longrightarrow L^2(X), \tag{1.7}$$

using [FS], yields results on $L^p \to L^p$ and $L^p \to L^{p'}$ mapping properties that are known to be sharp, when $\mathcal{C}$ is locally bijective.

If $X = Y$ is a compact Riemannian manifold with Laplace operator Δ and $A = \sqrt{1 - \Delta}$, then the results above apply to e^{itA} for $|t|$ in a certain interval $[-T_0, T_0]$, on which no caustics are formed. In concert with dilation arguments, we get the well known estimates

$$\|A^{-1} e^{itA} f\|_{L^{p'}} \le C|t|^{-\beta} \|f\|_{L^p}, \quad |t| \le T_0, \tag{1.8}$$

with

$$p = 2\,\frac{n+1}{n+3}, \quad p' = 2\,\frac{n+1}{n-1}, \quad \beta = \frac{n-1}{n+1}. \tag{1.9}$$

Also dilation arguments yield

$$\|(1+(tA)^{\gamma})^{-1}e^{itA}f\|_{L^p} \le C_p\|f\|_{L^p}, \quad |t| \le T_0, \tag{1.10}$$

with

$$1 < p < \infty, \quad \gamma = \gamma(n,p) = (n-1)\Big|\frac{1}{p} - \frac{1}{2}\Big|, \tag{1.11}$$

and hence

$$\|A^{-\gamma}e^{itA}f\|_{L^p} \le C_p|t|^{\gamma}\|f\|_{L^p} + C_p\|A^{-\gamma}f\|_{L^p}, \quad |t| \le T_0. \tag{1.12}$$

REMARK. As mentioned, via other techniques one can show that the presence of caustics in a Fourier integral operator does not affect $L^p - L^p$ estimates. However, they do affect $L^p - L^{p'}$ estimates. Some strong positive results have been obtained in [C2].

2. Hardy space estimates when $\mathcal{C}$ has a one-sided fold. Assume $\mathcal{C}$ is a homogeneous canonical relation in $(T^*X \setminus 0) \times (T^*Y \setminus 0)$, $\dim X = \dim Y = n$. Assume $\mathcal{C} \to T^*Y \setminus 0$ is a fold. Take $T \in I^{-(n-1)/2}(X, Y, \mathcal{C})$, and assume its Schwartz kernel κ_T has compact support. We will show that T acts on local Hardy spaces:

$$T : \mathfrak{h}^1(Y) \longrightarrow \mathfrak{h}^1(X), \tag{2.1}$$

under one simplifying hypothesis. Namely we assume κ_T is a conormal distribution, with singular support in $\Sigma \subset X \times Y$, a smooth $(2n-1)$-dimensional manifold. Furthermore we assume that $\Sigma \to Y$ is a submersion. The result (2.1) is a special case of Theorem 1 of Comech-Cuccagna [CC].

We bring in operators A and B, whose Schwartz kernels are smooth multiples α and β of surface measure on Σ. We have $A = A_0 + A_1$, $B = B_0 + B_1$, with

$$A_j, B_j \in I^{-(n-1)/2}(X, Y, \mathcal{C}), \quad \mathcal{C}_0 = \mathcal{C}. \tag{2.2}$$

In fact $\mathcal{C}_0$ is one component of the conormal bundle $\nu^*\Sigma \setminus 0$, and $\mathcal{C}_1$ is the other component. We choose α to be nonvanishing on the fold set $F \subset \Sigma$ and β to vanish simply on F. Then the fold geometry guarantees that there exist $P, Q \in OPS^0$ such that

$$T = A_0P + B_0Q, \tag{2.3}$$

modulo a smoothing operator.

Take $f \in \mathfrak{h}^1(Y)$. Without loss of generality, assume $WF(f) \subset \Gamma$, a small conic neighborhood of (y_0, η_0), with $((x_0, \xi_0), (y_0, \eta_0))$ a fold point of $\mathcal{C}$. There is no loss in assuming that

$$WF(g) \subset \Gamma \Longrightarrow A_1 g, B_1 g \in C^\infty.$$

Hence

$$Tf = APf + BQf, \quad \text{mod } C^\infty. \tag{2.4}$$

Now $Pf, Qf \in \mathfrak{h}^1 \subset L^1$. On the other hand, $A\delta_p$ and $B\delta_p$ are finite measures for any p, and hence

$$A, B : L^1(Y) \longrightarrow L^1(X). \tag{2.5}$$

It follows that $f \in \mathfrak{h}^1(Y) \Rightarrow Tf \in L^1(X)$.

Finally, $RT \in I^{-(n-1)/2}(X, Y, \mathcal{C})$ for any $R \in OPS^0(X)$, so we have $f \in \mathfrak{h}^1(Y) \Rightarrow RTf \in L^1(X)$ for all such R, giving (2.1).

We also have $\mathfrak{h}^1 \to$ bmo results, arising from the following complement to (2.5). Namely, with Λ as in §1, then $\Lambda^{-1}A\delta_p$ and $\Lambda^{-1}B\delta_p$ belong to $\text{bmo}(X)$ for each $p \in Y$, so

$$\Lambda^{-1}A,\ \Lambda^{-1}B : L^1(Y) \longrightarrow \text{bmo}(X). \tag{2.6}$$

Hence, by the arguments used above,

$$T_2 \in I^{-(n+1)/2}(X, Y, \mathcal{C}) \Longrightarrow T_2 : \mathfrak{h}^1(Y) \longrightarrow \text{bmo}(X). \tag{2.7}$$

There are known L^2-results. For example

$$T_3 \in I^{-1/4}(X, Y, \mathcal{C}) \Longrightarrow T_3 : L^2(Y) \longrightarrow L^2(X). \tag{2.8}$$

If $\mathcal{C}$ has a 2-sided fold, one can replace $I^{-1/4}$ by $I^{-1/6}$. As shown in [CC], interpolating this with (2.1) produces a result that can be sharpened.

REMARK. In [C1], A. Comech has obtained $\mathfrak{h}^1 \to$ bmo estimates and $L^p \to L^{p'}$ estimates for more general FIOs with one-sided folds, of the sort treated in [CC]. He also has examples of the sharpness of the resulting $L^p - L^{p'}$ estimates, in the presence of two-sided folds.

3. Some convolution operators. Let γ be a smooth curve in $\mathbb{R}^3$, of unit speed, with nowhere vanishing curvature (κ) and torsion (τ). Let μ be a smooth, compactly supported multiple of the arclength measure on γ, and consider

$$Kf = \mu * f \tag{3.1}$$

It is not hard to show that

$$K \in I^{-1/2}(\mathbb{R}^3, \mathbb{R}^3, \mathcal{C}_\gamma), \tag{3.2}$$

where $\mathcal{C}_\gamma$ has a 2-sided fold. Let's look at the Fourier transform $\hat{\mu}$, given by

$$\hat{\mu}(\xi) = \int a(t)e^{i\varphi(t,\xi)}\,dt, \quad \varphi(t,\xi) = \xi \cdot \gamma(t). \tag{3.3}$$

We have

$$\varphi_t(t,\xi) = \xi \cdot T = 0 \ \text{ for } \ \xi \perp T, \tag{3.4}$$

where $T = \gamma'(t)$ is the unit tangent to γ. Next

$$\varphi_{tt}(t,\xi) = \xi \cdot N, \tag{3.5}$$

so a critical point for φ is degenerate if $\xi \perp T$ and also $\xi \perp N$, i.e., if ξ is parallel to the binormal $B(t)$. Furthermore,

$$\partial_t^3 \varphi(t,\xi) = \xi \cdot N' = \xi \cdot (-\kappa T + \tau B). \tag{3.6}$$

At a degenerate critical point this is equal to $\tau\xi \cdot B \neq 0$, so there is not a higher-order degeneracy, and we could provide a uniform asymptotic expansion of $\hat{\mu}(\xi)$ in terms of Airy functions. We have

$$|\hat{\mu}(\xi)| \le C(1 + |\xi|)^{-1/3}, \tag{3.7}$$

which is of course consistent with (3.2) (and the fold behavior of $\mathcal{C}_\gamma$). In particular

$$K : L^2 \longrightarrow H^{1/3,2}. \tag{3.8}$$

Also $K\delta_p$ is a measure for each $p \in \mathbb{R}^3$, so

$$K : L^1 \to L^1. \tag{3.9}$$

Furthermore, K commutes with all constant coefficient $\varphi(D) \in OPS^0$. Since $f \in \mathfrak{h}^1$ if and only if $f \in L^1$ and $\varphi(D)f \in L^1$ for all such $\varphi(D)$, we have

$$K : \mathfrak{h}^1 \longrightarrow \mathfrak{h}^1. \tag{3.10}$$

Note that the general result Theorem 1 of [CC] implies only

$$\Lambda^{-1/2}K : \mathfrak{h}^1 \longrightarrow \mathfrak{h}^1, \tag{3.11}$$

where $\Lambda = \sqrt{-\Delta + 1}$. So the special features of K yield better behavior in (3.10). Note that one only has $\Lambda^{-s}K\delta_p \in \text{bmo}$ for $s \ge 2$, so

$$\Lambda^{-2}K : L^1 \longrightarrow \text{bmo}, \quad \mathfrak{h}^1 \longrightarrow L^\infty, \tag{3.12}$$

the latter by duality. Since $\Lambda^{-2}K \in I^{-5/2}(\mathbb{R}^3, \mathbb{R}^3, \mathcal{C}_\gamma)$, this is a weaker result than the analogue (2.7), which works with different geometry.

One can apply interpolation to (3.8) and (3.10), or to (3.8) and (3.12). In [C1], A. Comech has shown that the resulting $L^p \to L^p$ and $L^p \to L^{p'}$ estimates can be improved. For other results relevant to this setting, we refer to [PS, GSW], and results cited there.

REFERENCES

[Be] M. BEALS, L^p boundedness of Fourier integral operators, Mem. AMS **264** (1982).

[B] P. BRENNER, $L_p - L_{p'}$-estimates for Fourier integral operators related to hyperbolic equations, Math. Z. **152** (1977), 273–286.

[C1] A. COMECH, Type conditions and $L_p - L_p$, $L_p - L_{p'}$ regularity of Fourier integral operators, Preprint, 2001.

[C2] A. COMECH, $L_p - L_q$ regularity of Fourier integral operators with caustics, Preprint, 2002.

[CC] A. COMECH AND S. CUCCAGNA, On L^p continuity of singular Fourier integral operators, Preprint, 2000.

[FS] C. FEFFERMAN AND E. STEIN, H^p spaces of several variables, Acta Math. **129** (1972), 137–193.

[GSW] A. GREENLEAF, A. SEEGER, AND S. WAINGER, Estimates for generalized Radon transforms in three and four dimensions, Analysis, geometry, number theory: the mathematics of Leon Ehrenpreis, AMS, Providence, R.I., 2000, 243–254.

[L] W. LITTMAN, $L^p - L^q$-estimates for singular integral operators arising from hyperbolic equations, Proc. Symp. Pure Math. **23** (1971), 479–481.

[Pe] J. PERAL, L^p estimates for the wave equation, J. Funct. Anal. **36** (1980), 114–145.

[PS] D. PHONG AND E. STEIN, Radon transforms and torsion, Internat. Math. Res. Notices **4** (1991), 49–60.

[SSS] A. SEEGER, C. SOGGE, AND E. STEIN, Regularity properties of Fourier integral operators, Ann. of Math. **134** (1991), 231–251.

[S] C. SOGGE, Fourier Integrals in Classical Analysis, Cambridge Univ. Press, Cambridge, 1993.

[St] R. STRICHARTZ, A priori estimates for the wave equation and some applications, J. Funct. Anal. **5** (1970), 218–235.

THE CAUCHY DATA AND THE SCATTERING RELATION*

GUNTHER UHLMANN†

1. Introduction. As mentioned in the preface to this volume a combination of unique continuation results with the boundary control method has led to the solution of the inverse problem of determining a metric of a Riemannian manifold (with boundary) from the dynamic Dirichlet-to-Neumann map associated with the wave equation. Although these results are very satisfactory it requires too much information. By just looking at the singularities of the dynamic Dirichlet-to-Neumann (DN) map one can determine the boundary distance function (the minimal travel time along geodesics connecting points on the boundary of a Riemannian manifold) in the case that there are no conjugate points of the metric, i.e. no caustics. This is shown in Section 3 of this article using geometrical optics expansions. A natural question to ask if one can determine the metric from this data alone; this question is at the center of the boundary rigidity problem studied in Riemannian geometry which is one of the main topics of this volume.

Microlocal analysis is also used in Section 4 of this paper to show that from the DN map for the wave equation one can determine the scattering relation (or lens map). This result does not assume that there are no caustics. Roughly speaking this encodes the information of all travel times not just the ones given by minimizing geodesics. An approach to solve the inverse problem of finding the Riemannian metric from the scattering relation proposed by Stefanov and the author is described in detail.

In Sections 5 and 6 of this paper we consider elliptic inverse boundary value problems. We study the problem of determining an anisotropic or isotropic conductivity from the corresponding DN map or set of Cauchy data which encodes all the electrical measurements that can be made at the boundary. We also survey a result of Bukhgeim and the author showing that if one measures the set of Cauchy data for isotropic conductivites on particular sets of the boundary then one can determine uniquely the conductivity. The method of proof uses new Carleman estimates.

The author believes that there is a connection between the scattering relation and the set of Cauchy data that remains to be unraveled. This explains the title of this article.

2. The hyperbolic Dirichlet-to-Neumann map. Let Ω denote a bounded domain on $\mathbf{R}^n, n \geq 2$ and let g be a Riemannian metric on Ω. In local coordinates g is represented by $g(x) = (g_{ij}(x))$ a positive definite,

---

*The research for this paper was partly supported by NSF and a John Simon Guggenheim fellowship.

†Department of Mathematics, University of Washington, Seattle, WA 98195–4350.

symmetric matrix on $\overline{\Omega}$. We assume for simplicity that the Riemannian metric g is smooth (several of the results are valid assuming finite smoothness). The Euclidean metric is denoted by $e = (\delta_{ij})$.

We denote by Δ_g the Laplace-Beltrami operator associated to the metric g, i.e. in local coordinates

$$\Delta_g = (\det g)^{-\frac{1}{2}} \sum_{i,j=1}^{n} \frac{\partial}{\partial x_i} (\det g)^{\frac{1}{2}} g^{ij} \frac{\partial}{\partial x_j} \tag{1}$$

where $(g^{ij}) = (g_{ij})^{-1}$, $\det g = \det(g_{ij})$.

We consider the solution of the initial-boundary value problem for the wave equation associated to the Laplace-Beltrami operator:

$$\begin{cases} (\partial_t^2 - \Delta_g)u &= 0 \quad \text{in } (0,T)\times\Omega, \\ u|_{t=0} = \partial_t u|_{t=0} &= 0 \quad \text{in } \Omega, \\ u|_{(0,T)\times\partial\Omega} &= f, \end{cases} \tag{2}$$

where $f \in H^2_{\text{loc}}$, $f = 0$ for $t < 0$. Denote by $\nu = \nu(x)$ the outer normal to $\partial\Omega$ at $x \in \partial\Omega$. We define the hyperbolic (dynamic) Dirichlet-to-Neumann (HDN) map Λ_g^h by

$$\Lambda_g^h f := (\det g)^{\frac{1}{2}} \sum_{i,j=1}^{3} \nu_i g^{ij} \frac{\partial u}{\partial x_j}\bigg|_{(0,T)\times\partial\Omega}.$$

The inverse problem is whether knowledge of the HDN map determines the metric g uniquely. However, any metric isometric to g gives rise to the same boundary measurements. More precisely let

$$\psi : \overline{\Omega} \longrightarrow \overline{\Omega}, \quad \psi|_{\partial\Omega} = \text{Identity}$$

be a diffeomorphism. Then

$$\Lambda_g^h = \Lambda_{\psi^* g}^h, \tag{3}$$

where $\psi^* g = (D\psi \circ g \circ {}^t D\psi) \circ \psi^{-1}$ denotes the pull back of the metric under the diffeomorphism ψ.

This is in fact the only obstruction to unique identifiability of the metric from the HDN map. This result is a consequence of the paper [6], which uses the boundary control method (BC) pioneered by Belishev (see [5] for a survey and [18] for a recent book). The article [19] in this volume has further developments on the boundary control method and applications to inverse problems. In turn the BC method depends on a Holmgren type uniqueness theorem for hyperbolic equations. This breakthrough result was proven by Tataru [43] (see also [33]), after the BC has been formulated, using very novel Carleman type estimates. For more details see the article by Tataru [44] in these proceedings.

The precise statement of the uniqueness result is:

THEOREM 2.1. *Let $g_k \in C^\infty(\bar{\Omega})$, be two Riemannian metrics on $\bar{\Omega}$ and denote by $\Lambda^h_{g_k}$, $k = 1, 2$ the corresponding HDN maps. Assume*

$$\Lambda^h_{g_1} = \Lambda^h_{g_2},$$

then there exists, for T sufficiently large, a diffeomorphism $\psi : \bar{\Omega} \to \bar{\Omega}$ with $\psi|_{\partial\Omega} = Id$, such that $\psi^* g_1 = g_2$.

The result of [6] and [43] is valid for general compact Riemannian manifolds with boundary. Some stability estimates are known in this problem but these are in weak topology [20]. Hölder type stability estimates for the case that the metrics are close to the Euclidean one were proven in [35]. It is an interesting open problem to find sharp stability estimates for the inverse problem of determining a Riemannian metric from the hyperbolic DN map.

3. The boundary distance function. The problem considered in the previous section is highly overdetermined in dimensions $n \geq 2$. The Schwartz kernel of the HDN map depends on $2n - 1$ variables and the metric g depends on n variables.

We will show in this section that if (Ω, g) is simple that by looking at the singularities of the HDN we can determine the boundary distance function $d_g(x, y), x, y \in \partial\Omega$, that is, the travel times of geodesics going through the domain. The boundary distance function is a function of $2n - 2$ variables. Thus the inverse problem of determining the Riemannian metric from the boundary distance function is formally determined in two dimensions and formally overdetermined in dimensions $n \geq 3$. Some recent results on this problem are reviewed in this volume [10]. The linearized problem consists of determining a tensor field from its integral along geodesics. Sharafutdinov surveys recent developments on this problem in the article [34] in this volume.

Under the assumptions of Theorem 2.1 we can assume that, after a change of variables which is the identity at the boundary, the two metrics g_1 and g_2 have the same Taylor series at the boundary [39]. Therefore we can extend both metrics smoothly to be equal outside outside Ω and Euclidean outside a ball of radius R. We denote the extensions to $\mathbf{R}^n$ $g_j, j = 1, 2$, as before. Let $u_j(t, x, \omega)$, be the solution of the continuation problem

$$\left\{ \begin{array}{rcl} \dfrac{\partial^2 u}{\partial t^2} - \Delta_{g_j} u_j &=& 0, \text{ in } \mathbb{R}^n \times \mathbb{R} \\ u_j(t, x) &=& \delta(t - x \cdot \omega), t < -R, \end{array} \right. \tag{4}$$

where $\omega \in \mathbb{S}^{n-1} = \{x \in \mathbf{R}^n; |x| = 1\}$.

It was shown in [39] that under the hypotheses of Theorem 2.1 the two solutions coincide outside Ω, namely

$$u_1(t, x, \omega) = u_2(t, x, \omega), x \in \mathbb{R}^n \setminus \Omega. \tag{5}$$

In the case that the manifold $(\Omega, g_j), j = 1,2$ is simple, we will use methods of geometrical optics to construct solutions of (4) to show if the HDN map of g_1 and g_2 are the same then the boundary distance functions of the metrics g_1 and g_2 coincide.

3.1. Real-geometrical optics. Let g denotes a smooth Riemannian metric which is Euclidean outside a ball of radius R.

We will construct solutions to the continuation problem for the metric g (which is either g_1 or g_2). We fix ω. Let us assume that there is a solution to Equation (4) of the form

$$u(t,x,\omega) = a(x,\omega)\delta(t-\phi(x,\omega)) + v(t,x,\omega), \quad u = 0, t < -R, \tag{6}$$

where a, ϕ are functions to be determined and $v \in L^2_{\text{loc}}$ Notice that in order to satisfy the initial conditions in (4), we require that

$$a = 1, \qquad \phi(x,\omega) = x\cdot\omega \quad \text{for } x\cdot\omega < -R. \tag{7}$$

By replacing Equation (6) in Equation (4), it follows that

$$\begin{aligned} \frac{\partial^2 u}{\partial t^2} - \Delta_g u &= A\delta''(t-\phi(x,\omega)) + B\delta'(t-\phi(x,\omega)) \\ &\quad -(\Delta_g a)\delta(t-\phi(x,\omega)) + \frac{\partial^2 v}{\partial t^2} - \Delta_g v, \end{aligned} \tag{8}$$

where

$$A = a(x,\omega)\left(1 - \sum_{i,j=1}^{n} g^{ij}\frac{\partial\phi}{\partial x^i}\frac{\partial\phi}{\partial x^j}\right) \tag{9}$$

$$B = 2\sum_{j,k=1}^{n} g^{jk}\frac{\partial a}{\partial x^k}\frac{\partial\phi}{\partial x^j} + a\Delta_g\phi\ . \tag{10}$$

We choose the functions ϕ, a in the expansion (6) to eliminate the singularities δ'' and δ' and then construct v so that

$$\frac{\partial^2 v}{\partial t^2} - \Delta_g v = (\Delta_g a)\delta(t-\phi(x,\omega)), \qquad v = 0, t < -R. \tag{11}$$

3.1.1. The Eikonal equation. In order to solve the equation $A = 0$, it is sufficient to solve the equation

$$\sum_{i,j=1}^{n} g^{ij}\frac{\partial\phi}{\partial x^i}\frac{\partial\phi}{\partial x^j} = 1, \qquad \phi(x,\omega) = x\cdot\omega, \quad x\cdot\omega < -R. \tag{12}$$

Equation (12) is known as the *Eikonal Equation*. Here we will describe a method, using symplectic geometry, to solve this equation.

Let $H_g(x,\xi) = \frac{1}{2}(\sum_{i,j=1}^n g^{ij}(x)\xi_i\xi_j - 1)$ the Hamiltonian associated to the metric g. Note that the metric induced by g in the cotangent space $T^*\mathbb{R}^n$ is given by the principal symbol of the Laplace-Beltrami operator $g^{-1}(x,\xi) = \sum_{i,j=1}^n g^{ij}(x)\xi_i\xi_j$. The equation (12) together with the initial condition can be rewritten as

$$H_g(x, d\phi) = 0, \quad \phi(x,\omega) = x\cdot\omega, x\cdot\omega < -R$$

where $d\phi = \sum_{i=1}^n \frac{\partial\phi}{\partial x^i}\, dx^i$ is the differential of ϕ.

Let $S = \{(x,\xi) : H_g(x,\xi) = 0\}$, and let $M_\phi = \{(x, \nabla\phi(x)) : x \in \mathbb{R}^n\}$, then solving Equation (12), is equivalent to finding ϕ such that

$$M_\phi \subset S, \text{ with } M_\phi = \{(x,\omega); x\cdot\omega < -R\}. \tag{13}$$

In order to find ϕ so that (13) is valid we need to find a Lagrangian submanifold L so that $L \subset S$, $L = \{(x,\omega); x\cdot\omega < -R\}$ and the projection on L of $T^*\mathbb{R}^n$ to $\mathbb{R}^n$ is a diffeomorphism [11]. We will construct such a Lagrangian manifold by flowing out from $N = \{(x,\omega) : x\cdot\omega = s \text{ and } s < -R\}$ by the geodesic flow associated to the metric g. We recall the definition of geodesic flow.

We define the Hamiltonian vector field associated to H_g

$$V_g = (\frac{\partial H_g}{\partial \xi}, -\frac{\partial H_g}{\partial x}). \tag{14}$$

The bicharacteristics are the integral curves of H_g

$$\frac{d}{ds}x_m = \sum_{j=1}^n g^{mj}\xi_j, \qquad \frac{d}{ds}\xi_m = -\frac{1}{2}\sum_{i,j=1}^n \frac{\partial g^{ij}}{\partial x_m}\xi_i\xi_j, \quad m = 1, ..., n. \tag{15}$$

The projections of the bicharacteristics contained in $H_g(x,\xi) = 0$ in the x variable are the geodesics of the metric g and the parameter s denotes arc length. We denote the associated geodesic flow by

$$X_g(s) = (x_g(s), \xi_g(s)).$$

If we impose the condition that the bicharacteristics are in S initially, then they belong to S for all time, since the Hamiltonian vector field V_g is tangent to S. The Hamiltonian vector field is transverse to N then the resulting manifold obtained by flowing N along the integral curves of V_g will be a Lagrangian manifold L contained in S. We shall write $L = X_g(N)$.

Now the projection of N into the base space is a diffeomorphism so that $L = \{(x, d_x\phi)\}$ locally near a point of N. We can construct a global solution of (13) near Ω if the manifold is simple. We recall:

DEFINITION 3.1. *Let Ω be a bounded domain of Euclidean space with smooth boundary and g a Riemannian metric on $\overline{\Omega}$. We say that (Ω, g)*

is simple if given two points on the boundary there is a unique minimizing geodesic joining the two points on the boundary and, moreover, $\partial\Omega$ is geodesically convex.

If (Ω, g) is simple then we extend the metric smoothly in a small neighborhood so that the metric g is Euclidean in a small neighborhood of Ω so that the metric is still simple. In this case we can solve the Eikonal equation globally in a neighborhood of Ω.

3.1.2. The transport equation. The equation $B = 0$ is equivalent to solving the following equation:

$$\sum_{i,j=1}^{n} g^{ij} \frac{\partial \phi}{\partial x^j} \frac{\partial a}{\partial x^i} + \frac{a}{2} \Delta_g \phi = 0. \tag{16}$$

Equation 16 is called the *transport equation.* It is a vector field equation for $a(x, \omega)$, which is solved by integrating along the integral curves of the vector field $v = \sum_{i,j=1}^{n} g^{ij} \frac{\partial \phi}{\partial x^j} \frac{\partial}{\partial x^i}$. It is an easy computation to prove that v has length 1 and that the integral curves of v are the geodesics of the metric g.

The solution of the transport equation (16) is then given by:

$$a(x, \omega) = \exp\left(-\frac{1}{2} \int_{\gamma} \Delta_g \phi \right), \tag{17}$$

where γ is the unique geodesic such that $\gamma(0) = y$, $\dot{\gamma}(0) = \omega$, $y \cdot \omega = -R$ and γ passes through x. If (Ω, g) is a simple manifold then $a \in C^{\infty}(\mathbb{R}^n)$.

To end the construction of the real-geometrical optics solutions we observe that the function $v(t, x, \omega) \in L^2_{\text{loc}}$ by using standard regularity results for hyperbolic equations.

Now we state the main result of this section:

THEOREM 3.1. *Let $(\Omega, g_i), i = 1, 2$ be simple metrics. Assume that $\Lambda^h_{g_1} = \Lambda^h_{g_2}$. Then $d_{g_1} = d_{g_2}$.*

Sketch of proof. Assume that we have two metrics g_1, g_2 with the same HDN map. Then by (5) the solutions of (4) are the same outside Ω. Therefore the main singularity of the solutions in the geometrical optics expansion must be the same outside Ω. Thus we conclude that

$$\phi_1(x, \omega) = \phi_2(x, \omega), \quad x \in \mathbf{R}^n \setminus \Omega. \tag{18}$$

Now $\phi_j(x, \omega)$ measures the geodesic distance to the hyperplane $x \cdot \omega = -R$ in the metric g. From this we can easily conclude that the geodesic distance between two points in the boundary for the two metrics is the same, that is $d_{g_1}(x, y) = d_{g_2}(x, y), x, y \in \partial\Omega$.

This type of argument was used in [31] to study a similar inverse problem for the more complicated system of elastodynamics. In particular it is proven in [31] that from the HDN map associated to the equations of isotropic elastodynamics one can determine, under the assumption of simplicity of the metrics, the lengths of geodesics of the metrics defined by

$$ds^2 = c_p^2(x)e, \quad ds^2 = c_s^2(x)e, \tag{19}$$

where $c_p(x) = \sqrt{\frac{(\lambda+2\mu)}{\rho}}$, $c_s(x) = \sqrt{\frac{\mu}{\rho}}$ denote the speed of compressional waves and shear waves respectively. Here λ, μ are the Lamé parameters and ρ the density. The article by [10] in this book examines in detail the inverse problem of determining the metric from the associated boundary distance function, the so-called boundary rigidity problem in Riemannian geometry.

Using Mukhometov's result [27], [28] we can then recover both speeds from the HDN map. This shows in particular that if we know the density one can determine the Lamé parameters from the HDN map. By using the transport equation of geometrical optics, similar to (16), and the results on the ray transform (see for instance [34]), Rachele shows that under certain a-priori conditions one can also determine the density ρ [32].

4. The scattering relation. In the presence of caustics (i.e. the exponential map is not a diffeomorphism) the expansion (6) is not valid since we cannot solve globally the Eikonal equation globally in Ω. The solution of (4) is globally a Lagrangian distribution (see for instance [16]). These distributions can locally be written in the form

$$u(t,x,\omega) = \int e^{i\phi(t,x,\omega,\theta)} a(t,x,\omega,\theta)\, d\theta \tag{20}$$

where ϕ is a phase function and $a(t,x,\omega,\theta)$ is a classical symbol.

Every Lagrangian distribution is determined (up to smoother terms) by a Lagrangian manifold and its symbol. The Lagrangian manifold associated to $u(t,x,\omega)$ is the flow out from $t = x\cdot\omega, t < -R$ by the Hamilton vector field of $p_g = \tau^2 - g^{-1}(x,\xi)$. Here (τ,ξ) are the dual variables to (t,x) respectively. The projection in the (x,ξ) variables of the flow is given by the flow out from N by geodesic flow, that is the Lagrangian submanifold L described above.

The **scattering relation** (also called lens map), $C_g \subset \big(T^*(\mathbf{R}\times\partial\Omega)\setminus 0\big) \times \big(T^*(\mathbf{R}\times\partial\Omega)\setminus 0\big)$ of a metric $g = (g^{ij})$ on $\overline{\Omega}$ with dual metric $g^{-1} = (g_{ij})$ is defined as follows. Consider bicharacteristic curves, $\gamma : [a,b] \to T^*(\overline{\Omega}\times\mathbf{R})$, of the Hamilton function $p_g(t,x,\tau,\xi) = \tau^2 - g^{-1}(x,\xi)$ which satisfy the following: $\gamma(]a,b[)$ lies in the interior, γ intersects the boundary non-tangentially at $\gamma(a)$ and $\gamma(b)$, and time increases along γ. Then the canonical projection from $(T^*_{\mathbf{R}\times\partial\Omega}(\mathbf{R}\times\Omega)\setminus 0) \times (T^*_{\mathbf{R}\times\partial\Omega}(\mathbf{R}\times\Omega)\setminus 0)$ onto $(T^*(\mathbf{R}\times\partial\Omega)\setminus 0) \times T^*(\mathbf{R}\times\partial\Omega)\setminus 0)$ maps the endpoint pair $(\gamma(b),\gamma(a))$ to a point in C_g. In other words C_g gives the geodesic distance between points in the boundary and also the points and direction of exit of the geodesic if we know the point and direction of entrance.

It is well-known that C_g is a homogeneous canonical relation on $(T^*(\mathbf{R} \times \partial\Omega) \setminus 0) \times (T^*(\mathbf{R} \times \partial\Omega) \setminus 0)$. (See [14] for the concept of a scattering relation.) C_g is, in fact, a diffeomorphism between open subsets of $T^*(\mathbf{R} \times \partial\Omega) \setminus 0$.

In analogy with Theorem 3.1 we have

THEOREM 4.1. *Let $g_i, i = 1, 2$ be Riemannian metrics on $\overline{\Omega}$ satisfying the assumptions of Theorem 2.1. Then*

$$C_{g_1} = C_{g_2}.$$

Sketch of Proof. Since by (5) we know the solutions of (4) outside Ω. Therefore the associated Lagrangian manifolds to the Lagrangian distributions u_j must be the same outside Ω. By taking the projection of these Lagrangians onto the boundary we get the desired claim.

In the case that (Ω, g) is simple then the scattering relation doesn't give any new information. In fact $((t_1, x_1, \tau, \xi_1), (t_0, x_0, \tau, \xi_0)) \in C_g$ if $t_1 - t_0 = d_g(x_1, x_0)$ and $\xi_j = -\tau \frac{\partial d_g(x_1, x_0)}{\partial x_j}, j = 0, 1$. In other words d_g is the generating function of the scattering relation.

This result was generalized in [15] to the case of the equations of elastodynamics with residual stress. It is shown that knowing the HDN map we can recover the scattering relations associated to P and S waves. For this one uses Lagrangian distributions with appropriate polarization.

The scattering relation contains all travel time data; not just information about minimizing geodesics as is the case of the boundary distance function. The natural conjecture is that this is enough to determine the metric up to isometry.

CONJECTURE 4.1. *From the scattering relation we can determine the metric up to isometry.*

We have developed in [36] an approach to this conjecture and derive the identity below. We will consider the case of subdomains of $\mathbf{R}^3$.

Let $x^{(0)} \in \Gamma, \xi^{(0)} \in S^2$ such that $\nu(x^{(0)}) \cdot g^{-1}\xi^{(0)} < 0$. Here g is either g_1 or g_2, while the initial conditions are the same for both metrics. We remark that if $\xi^{(0)} \cdot g^{-1}\xi^{(0)} = 1$, then s is the arc-length in (15).

Consider the Hamiltonian system (15) with the following initial conditions

$$\text{(21)} \quad \begin{cases} \dfrac{d}{ds}x_m = \displaystyle\sum_{j=1}^{n} g^{mj}\xi_j, & \dfrac{d}{ds}\xi_m = -\dfrac{1}{2}\displaystyle\sum_{i,j=1}^{n} \dfrac{\partial g^{ij}}{\partial x_m}\xi_i\xi_j, \; m = 1, ..., n, \\ x|_{s=-\rho} = (-\rho, z), & \xi|_{s=-\rho} = (1, 0, 0). \end{cases}$$

Here $z \in \mathbf{R}^2$, $\rho > 0$ is such that $g = e$ for $|x| > \rho$ and the solution $x = x(s, z)$, $\xi = \xi(s, z)$ depends on the parameter z. If $g = e$, then $x = (s, z) = (s, z_1, z_2)$.

Denote the solution of (15) by $x = x(s, x^{(0)}, \xi^{(0)})$, $\xi = \xi(s, x^{(0)}, \xi^{(0)})$. Let us introduce new notation

$$X := (x, \xi).$$

The solution to (15) related to g_1 and g_2, respectively, can therefore be written down as $X_{g_j} = X_{g_j}(s, X^{(0)}) = X_{g_j}(s, x^{(0)}, \xi^{(0)})$.

Set $F(s) := X_{g_2}(t - s, X_{g_1}(s, X^{(0)}))$. Here $t = t(X^{(0)})$ is the length of the geodesics issued from $X^{(0)}$ with endpoint on Γ and t is independent of $g = g_1$ or $g = g_2$. We have $F(0) = X_{g_2}(t, X^{(0)}) = X_{g_1}(t, X^{(0)}) = F(t)$. Thus

$$\int_0^t F'(s)\, ds = 0. \tag{22}$$

A calculation gives

$$\begin{aligned} F'(s) &= -V_{g_2}(X_{g_2}(t-s, X_{g_1}(s, X^{(0)}))) \\ &\quad + \frac{\partial X_{g_2}}{\partial X^{(0)}}(t-s, X_{g_1}(s, X^{(0)}))V_{g_1}(X_{g_1}(s, X^{(0)})). \end{aligned} \tag{23}$$

We claim that

$$\begin{aligned} &V_{g_2}(X_{g_2}(t-s, X_{g_1}(s, X^{(0)}))) \\ &\quad = \frac{\partial X_{g_2}}{\partial X^{(0)}}(t-s, X_{g_1}(s, X^{(0)}))V_{g_2}(X_{g_1}(s, X^{(0)})). \end{aligned} \tag{24}$$

Indeed, (24) follows from

$$\begin{aligned} 0 &= \left.\frac{d}{ds}\right|_{s=0} X(T-s, X(s, X^{(0)})) \\ &= -V(X(T, X^{(0)})) + \frac{\partial X}{\partial X^{(0)}}(T, X^{(0)})V(X^{(0)}), \quad \forall T \end{aligned}$$

after setting $T = t - s$. Therefore, (22), (23) and (24) combined together imply

$$\int_0^t \frac{\partial X_{g_2}}{\partial X^{(0)}}(t-s, X_{g_1}(s, X^{(0)}))\,(V_{g_1} - V_{g_2})\,(X_{g_1}(s, X^{(0)}))\, ds = 0. \tag{25}$$

This identity is valid whenever the scattering relation of the two metrics is the same. Roughly speaking (25) is a non-linear integral equation on the difference of the metrics $(g_1)^{-1}$ and $(g_2)^{-1}$.

Formula (25) is the main result used in [36] to prove that if two metrics have the same lengths of geodesics and they are a-priori close to the Euclidean metric in an appropriate topology then they must be isometric. The idea is to linearize this identity at the Euclidean metric and use perturbation arguments to derive the result.

As mentioned in [34] and [25] this has been extended to a semiglobal result. Namely one of the metrics satisfies a curvature assumption and the other one is sufficiently close to the Euclidean metric. The result is that if the metrics have the same boundary distance function under these conditions then they must be isometric.

5. The elliptic Dirichlet-to-Neumann map. In this section we consider the elliptic analog of the HDN map considered in Section 2. One of the motivations for this problem comes from Electrical Impedance Tomography (EIT). In this non-invasive inverse method one attempts to determine the electrical conductivity of a body by making voltage and current measurements at the boundary of the body. For anisotropic conductors the electrical properties of the medium depend on direction. Muscle tissue is a prime example of an anisotropic conductor. For instance for cardiac muscle tissue the transverse conductivity is 2.3 mho while the longitudinal one is 6.3 mho. An anisotropic conductivity is modeled as a positive definite symmetric metric $\gamma = (\gamma^{ij})$. Under the assumption of no sources or sinks of currents the equation for the potential u, given a voltage potential f on the boundary, is given by the solution of the Dirichlet problem

$$\sum_{i,j=1}^{n} \frac{\partial}{\partial x_i}\left(\gamma^{ij}\frac{\partial u}{\partial x_j}\right) = 0 \quad \text{on } \Omega; \qquad u|_{\partial\Omega} = f. \tag{26}$$

The DN map is defined by

$$\Lambda_\gamma(f) = \sum_{i,j=1}^{n} \gamma^{ij}\frac{\partial u}{\partial x_j}\nu^i \tag{27}$$

where ν^i denotes the ith component of ν.

It was observed by Luc Tartar (see [22] for an account) that if $\psi : \overline{\Omega} \to \overline{\Omega}$ is a diffeomorphism with $\psi|_{\partial\Omega} =$ Identity then $\Lambda_{\widetilde{\gamma}} = \Lambda_\gamma$ where

$$\widetilde{\gamma} = \left(\frac{(D\psi)^T \circ \gamma \circ (D\psi)}{|\det D\psi|}\right) \circ \psi^{-1} =: \psi_*\gamma. \tag{28}$$

Here $D\psi$ denotes the (matrix) differential of ψ, $(D\psi)^T$ its transpose and the composition in (28) is to be interpreted as composition of matrices.

CONJECTURE 5.1. *Let $\gamma, \widetilde{\gamma}$ be conductivities satisfying $\Lambda_\gamma = \Lambda_{\widetilde{\gamma}}$. Then there exists a diffeomorphism $\psi : \overline{\Omega} \longrightarrow \overline{\Omega}$, with $\psi|_{\partial\Omega} =$ Identity so that $\widetilde{\gamma} = \psi_*\gamma$.*

In dimensions $n \geq 3$ the problem can be reformulated as a geometric one. Given an anisotropic conductivity $\gamma = (\gamma^{ij})$ we define $g^{ij} = (\det \gamma)^{\frac{2}{n-2}}\gamma^{ij}$. The Riemannian metric is given by $(g_{ij}) = (g^{ij})^{-1}$. The conductivity equation is replaced by the Laplace–Beltrami operator. We consider the solution of the elliptic boundary value problem

$$\Delta_g u = 0 \text{ on } \Omega, \quad u\Big|_{\partial\Omega} = f. \tag{29}$$

We define the EDN map by

$$\Lambda_g(f) = \sum_{i,j=1}^{n} \nu^i g^{ij}\frac{\partial u}{\partial x_j}\sqrt{\det g}\Big|_{\partial\Omega} \tag{30}$$

where $(\nu^i) = \nu$ is the outer unit normal to $\partial\Omega$. The inverse problem is to recover g from Λ_g.

We note that under the correspondence between conductivities and Riemannian metrics indicated above the corresponding EDN maps satisfy $\Lambda_g = \Lambda_\gamma$. Similarly to (28) we have that for any diffeomorphism $\psi : \overline{\Omega} \longrightarrow \overline{\Omega}, \quad \psi|_{\partial\Omega} = \text{Identity}$

$$\Lambda_{\psi^* g} = \Lambda_g \tag{31}$$

where $\psi^* g$ denotes the pull-back of the metric g.

The problem can be formulated for arbitrary compact Riemannian manifolds with boundary. More specifically

CONJECTURE 5.2. *Let (M, g) be a compact Riemannian manifold with boundary of dimension $n \geq 3$. The pair $(\partial M, \Lambda_g)$ determines (M, g) uniquely. Of course uniquely means up to an isometric copy.*

The more general result proven about this conjecture is the following [24]

THEOREM 5.1 ($n \geq 3$). *Let (M, g) be a real-analytic compact, connected Riemannian manifold with boundary. Let $\Gamma \subseteq \partial M$ be real-analytic and assume that g is real-analytic up to Γ. Then $(\Lambda_g, \partial M)$ determines uniquely (M, g).*

This result is also valid if we measure the EDN map in an open subset of the boundary Γ (measured for functions supported on that subset) under the a-priori assumption that g is real-analytic in the interior and up to Γ.

Notice that Theorem 5.1 doesn't assume any condition on the topology of the manifold except for connectedness. An earlier result of [26] assumed that (M, g) was strongly convex and simply connected and $\Gamma = \partial M$.

In the two dimensional case there is an additional obstruction. Namely the Laplace-Beltrami operator is conformally invariant. More precisely

$$\Delta_{\alpha g} = \frac{1}{\alpha} \Delta_g$$

for any function α, $\alpha \neq 0$. Therefore we have that for $n = 2$

$$\Lambda_{\alpha(\psi^* g)} = \Lambda_g \tag{32}$$

for any smooth function $\alpha \neq 0$ such that $\alpha|_{\partial M} = 1$.

Therefore the best we can recover by knowing Λ_g is the conformal class of the metric g. That this is in fact the case was shown by Lassas and the author in [24]. We have:

THEOREM 5.2. *Let (M, g) be a compact Riemannian surface. Then the pair $(\partial M, \Lambda_g)$ determines uniquely the conformal class of (M, g). Uniquely means up to an isometric copy.*

Also this result only requires local measurements of the EDN map on an open subset of the boundary.

A basic result which is used in all the results stated in this section is the following Lemma proved in [26]:

LEMMA 5.1. *Let (M, g) be a compact Riemannian manifold with boundary. Then Λ_g determines the C^∞-jet of the metric at the boundary in the following sense. If g' is another Riemannian metric on M such that $\Lambda_g = \Lambda_{g'}$, then there exists a diffeomorphism $\varphi : M \to M$, $\varphi\big|_{\partial M} = \text{Identity}$ such that $g' = \varphi^* g$ to infinite order at ∂M.*

In other words Lemma 5.1 shows that the conjectures above are valid at thc boundary. The proof of this result is done by showing that Λ_g is a pseudodifferential operator of order 1. Its full symbol, calculated in appropriate coordinates, determines the C^∞-jet of the metric g at the boundary.

In the two dimensional case it was proven in [38], under some regularity assumptions on the conductivity, that we can reduce the anisotropic case to the isotropic one. This uses fundamentally isothermal coordinates, which are not available in dimension $n \geq 3$. An anisotropic conductivity is a conductivity $\gamma^{ij} = \alpha(x)\delta^{ij}$ where the function α is a strictly positive function. The corresponding Riemannian metric in dimension $n \geq 3$ is then conformal to the Euclidean metric.

The following are the two basic results in the isotropic case:

THEOREM 5.3. *Let $\Omega \subseteq \mathbf{R}^2$ be a bounded domain with smooth boundary. Let $\gamma^{(i)}$, $i = 1, 2$ be two isotropic conductivities in $C^2(\overline{\Omega})$. Then $\Lambda_{\gamma_1} = \Lambda_{\gamma_2}$ implies $\gamma_1 = \gamma_2$.*

THEOREM 5.4. *Let $\Omega \subseteq \mathbf{R}^n$ $n \geq 3$ be a bounded domain with smooth boundary. Let $\gamma^{(i)}$, $i = 1, 2$ be two C^2 isotropic Riemannian metrics. Then $\Lambda_{\gamma_1} = \Lambda_{\gamma_2}$ implies $\gamma_1 = \gamma_2$.*

The two dimensional case is due to Nachman [29] and the case of dimension three or higher to Sylvester and Uhlmann [40].

In the next two sections we outline the proof of these last two results.

5.1. Complex-geometrical optics solutions. In this section we outline the proof of Theorem 5.4. We actually prove a more recent result of Bukhgeim and the author [8] that states that is enough to measure the EDN map on particular subsets of the boundary. A crucial ingredient in the proof is the construction of complex geometrical optics (CGO) solutions of the Laplace-Beltrami operator when the Riemannian metric is isotropic. Another important element is the use of Carleman estimates.

The main result of CGO solutions is given by

LEMMA 5.2. *Let g be an isotropic Riemannian metric $g = \alpha e$ with α a scalar function satisfying $\alpha = 1$ outside a large ball. Let $\rho \in \mathbf{C}^n$, $\rho \cdot \rho = 0$. Then for $|\rho|$ sufficiently large, there exist solutions of $\Delta_g u = 0$ of the form*

$$u = e^{\langle x, \rho \rangle} \alpha^{-\frac{1}{2}} (1 + \psi_g(x, \rho)) \tag{33}$$

with $\psi_g \underset{|\rho| \to \infty}{\longrightarrow} 0$ uniformly in compact sets.

For more precise statements and a recent survey of other results using CGO see [45]. We remark that this method started with the pioneering work of Calderón [9].

One of the main difficulties in extending Theorem 5.4 to the general anisotropic case even in the case when M is an open subset of Euclidean space is to construct an analog of (33) in this case.

The proof of Theorem 5.4 proceeds by proving a more general result. Namely one reduces the problem to consider the set of Cauchy data for solutions of the Schrödinger equation (see [45] for more details).

Let $n \geq 3$. Let $q \in L^\infty(\Omega)$. We define the set of **Cauchy data** for the associated Schrödinger equation by

$$\mathcal{C}_q = \left\{ \left(u|_{\partial\Omega}, \frac{\partial u}{\partial \nu}\Big|_{\partial\Omega} \right) \mid (\Delta - q)u = 0 \quad \text{on } \Omega,\ u \in H^1(\Omega) \right\}. \tag{34}$$

THEOREM 5.5. *Let $q_i \in L^\infty(\Omega), i = 1, 2$. Assume*

$$\mathcal{C}_{q_1} = \mathcal{C}_{q_2}.$$

Then $q_1 = q_2$.

The proof of this result uses complex geometrical optics solutions of the Schrödinger equation. Let $q \in L^\infty(\mathbb{R}^n), n \geq 2$ have compact support. Then for $\rho \in \mathbb{C}^n$, $\rho \cdot \rho = 0$, $|\rho|$ sufficiently large, one can construct solutions to

$$(\Delta - q)u_\rho = 0$$

of the form

$$u_\rho = e^{\langle x,\rho\rangle}(1 + \psi_q(x, \rho)) \tag{35}$$

with

$$\|\psi_q(\,\cdot\,, \rho)\|_{H^s(\Omega)} \leq \frac{C}{|\rho|^{1-s}}, \qquad 0 \leq s \leq 1, \tag{36}$$

for some $C > 0$ independent of ρ.

The function $\psi_q(x, \rho)$ solves

$$\Delta_\rho \psi_q = q(1 + \psi_q), \tag{37}$$

where

$$\Delta_\rho u = e^{-\langle x,\rho\rangle} \Delta(e^{\langle x,\rho\rangle} u).$$

The Schwartz kernel G_ρ of Δ_ρ^{-1} is the so-called Faddeev's Green kernel [12]. The following estimate was proved in [41] ($n = 2$), [40] ($n \geq 3$) for $-1 < \delta < 0$ and $\rho \in \mathbb{C}^n - 0, \rho \cdot \rho = 0$:

$$\|G_\rho f\|_{H^s_\delta} \leq C \frac{\|f\|_{L^2_{\delta+1}}}{|\rho|^{1-s}}. \tag{38}$$

Here H^s_α denotes the Sobolev space associated to the weighted L^2 space with norm given by

$$\|f\|^2_{L^2_\alpha} = \int |f(x)|^2(1+|x|^2)^\alpha dx.$$

A natural question is whether one can determine the potential by measuring the Cauchy data on strict subsets of the boundary. The only result known beyond the case of real-analytic potentials was proven in [8]. We describe the result below.

We first modify the set of Cauchy data to allow for more singular distributions on the boundary. We define the function space

$$H_\Delta(\Omega) = \{u \in \mathcal{D}'(\Omega) \mid u \in L^2(\Omega),\ \Delta u \in L^2(\Omega)\};$$

$H_\Delta(\Omega)$ is a Hilbert space with the norm

$$\|u\|^2_{H_\Delta(\Omega)} = \|u\|^2_{L^2(\Omega)} + \|\Delta u\|^2_{L^2(\Omega)}.$$

For $u \in H_\Delta(\Omega)$, we have $u|_{\partial\Omega} \in H^{-\frac{1}{2}}(\partial\Omega)$ and $\frac{\partial u}{\partial\nu}\big|_{\partial\Omega} \in H^{-\frac{3}{2}}(\partial\Omega)$. We define the set of modified Cauchy data for $q \in L^\infty(\Omega)$ by

$$C_q = \left\{\left(u|_{\partial\Omega}, \frac{\partial u}{\partial\nu}\Big|_{\partial\Omega}\right) \in H^{-\frac{1}{2}}(\partial\Omega) \times H^{-\frac{3}{2}}(\partial\Omega) \mid (\Delta - q)u = 0 \text{ in } \Omega,\ u \in H_\Delta(\Omega)\right\}.$$

If 0 is not a Dirichlet eigenvalue of $\Delta - q$ in Ω then C_q contains the graph of the Dirichlet-to-Neumann map Λ_q conventionally defined on $H^{1/2}(\partial\Omega)$ by the relation $\Lambda_q(f) = \frac{\partial u}{\partial\nu}\big|_{\partial\Omega}$, where $u \in H^1(\Omega)$ is a solution to the problem

$$(\Delta - q)u = 0 \quad \text{in } \Omega, \quad u|_{\partial\Omega} = f;$$

i.e., $\{(f, \Lambda_q(f)) \mid f \in H^{1/2}(\partial\Omega)\} \subset C_q$.

Fix $\xi \in S^{n-1} = \{\xi \in \mathbf{R}^n, |\xi| = 1\}$. We define

$$\partial\Omega_+(\xi) = \{x \in \partial\Omega \mid \langle\nu,\xi\rangle > 0\}, \quad \partial\Omega_-(\xi) = \{x \in \partial\Omega \mid \langle\nu,\xi\rangle < 0\} \tag{39}$$

and for $\varepsilon > 0$

$$\partial\Omega_{+,\varepsilon}(\xi) = \{x \in \partial\Omega \mid \langle\nu,\xi\rangle > \varepsilon\}, \quad \partial\Omega_{-,\varepsilon}(\xi) = \{x \in \partial\Omega \mid \langle\nu,\xi\rangle < \varepsilon\}. \tag{40}$$

We also define the set of restricted Cauchy data

$$C_{q,\varepsilon} = \left\{\left(u|_{\partial\Omega}, \frac{\partial u}{\partial\nu}\Big|_{\partial\Omega_{-,\varepsilon}(\xi)}\right) \Bigm| (\Delta - q)u = 0 \text{ in } \Omega,\ u \in H_\Delta(\Omega)\right\}.$$

The main result of [8] is

THEOREM 5.6. *Let $n \geq 3$ and $q_i \in L^\infty(\Omega)$, $i = 1, 2$. Given $\xi \in S^{n-1}$ and $\varepsilon > 0$, assume that $C_{q_1,\varepsilon} = C_{q_2,\varepsilon}$. Then $q_1 = q_2$.*

Theorem 5.6 has an immediate consequence in Electrical Impedance Tomography. We assume here now γ is an isotropic conductivity, i.e, $\gamma_{ij} = \gamma(x)\delta_{ij}$ with $\gamma \in C^2(\overline{\Omega})$ a strictly positive function on $\overline{\Omega}$.

As a direct consequence of Theorem 5.6 we prove

COROLLARY 5.1. *Let $\gamma_i \in C^2(\overline{\Omega})$, $i = 1, 2$, be strictly positive. Given $\xi \in S^{n-1}$ and $\varepsilon > 0$, assume that $\gamma_1|_{\partial\Omega} = \gamma_2|_{\partial\Omega}$ and*

$$\Lambda_{\gamma_1}(f)|_{\partial\Omega_{-,\varepsilon}(\xi)} = \Lambda_{\gamma_2}(f)|_{\partial\Omega_{-,\varepsilon}(\xi)} \quad \forall f \in H^{-\frac{1}{2}}(\partial\Omega).$$

Then $\gamma_1 = \gamma_2$.

As far as we know, Theorem 5.6 (Corollary 5.1) is the first global uniqueness result for the Schrödinger equation (conductivity equation) in which the Cauchy data are given only on part of the boundary, beyond the case of a real-analytic potential.

A natural way to attack the problem of finding a potential from partial information of the Cauchy data is to construct solutions of the form (35) with $\psi_q = 0$ on part of the boundary. As it is shown in [17] it is impossible in general to solve the Dirichlet problem for (37) with ψ_ρ decaying (or even polynomially bounded in ρ.) In [8] it is shown that we can prescribe Dirichlet conditions for ψ_ρ on particular subsets of the boundary. More precisely we have

LEMMA 5.3. *Let $n \geq 2$. Let $\rho \in \mathbb{C}^n$ with $\langle \rho, \rho \rangle = 0$ and $\rho = \tau(\xi + i\eta)$ with $\xi, \eta \in S^{n-1}$. Suppose that $f(\cdot, \rho/|\rho|) \in W^{2,\infty}(\Omega)$ satisfies $\partial_\xi f = \partial_\eta f = 0$, where ∂_ξ denotes the directional derivative in the direction ξ. Then we can find solutions to $(\Delta - q)u = 0$ in Ω of the form*

$$u(x, \rho) = e^{\langle x, \rho \rangle}\Big(f\Big(x, \frac{\rho}{|\rho|}\Big) + \psi(x, \rho)\Big), \quad \psi|_{\partial\Omega_-(\xi)} = 0,$$

with

$$\|\psi(\,\cdot\,, \rho)\|_{L^2(\Omega)} \leq \frac{C}{\tau}, \quad \tau \geq \tau_0,$$

for some $C > 0$ and $\tau_0 > 0$.

5.2. Carleman estimates. The proof of Theorem 5.6 and Lemma 5.3 uses Carleman estimates for the operator Δ_ρ, which is not an elliptic operator if we consider the dependence on the parameter ρ, to construct the solutions and prove the main result. The use of a linear phase function in these Carleman estimates gives rise to the restriction on measuring the Cauchy data on particular subsets of the boundary.

THEOREM 5.7. *For $q \in L^\infty(\Omega)$ there exist $\tau_0 > 0$ and $C > 0$ such that for all $u \in C^2(\bar{\Omega})$, $u|_{\partial\Omega} = 0$, and $\tau \geq \tau_0$ we have the estimate*

$$\tau^2 \int_\Omega |e^{-\tau\langle x,\xi\rangle} u|^2 \, dx + \tau \int_{\partial\Omega_+} \langle \xi, \nu\rangle |e^{-\tau\langle x,\xi\rangle} \partial_\nu u|^2 \, dS$$
$$\leq C\left(\int_\Omega |e^{-\tau\langle x,\xi\rangle}(\Delta - q)u|^2 \, dx - \tau \int_{\partial\Omega_-} \langle \xi, \nu\rangle |e^{-\tau\langle x,\xi\rangle} \partial_\nu u|^2 \, dS\right).$$

Sketch of the Proof of Theorem 5.6. As before we let $\xi \in S^{n-1}$. Fix $k \in \mathbf{R}^n$ such that $\langle \xi, k\rangle = 0$. Using Lemma 5.2, we choose a solution $u_2 \in H_\Delta(\Omega)$ to $(\Delta - q_2)u_2 = 0$ in Ω of the form

$$u_2 = e^{\langle x,\rho_2\rangle}(1 + \psi_{q_2}(x, \rho_2))$$

with

$$\rho_2 = \tau\xi - i\frac{k+l}{2},$$

where $\langle l, k\rangle = \langle l, \xi\rangle = 0$ and $|k+l|^2 = 4\tau^2$ (with these conditions $\langle \rho_2, \rho_2\rangle = 0$). In dimension $n \geq 3$ we can always choose such a vector l. Since $C_{q_1,\varepsilon} = C_{q_2,\varepsilon}$, there is a solution $u_1 \in H_\Delta(\Omega)$ to $(\Delta - q_1)u_1 = 0$ in Ω such that

$$u_1|_{\partial\Omega} = u_2|_{\partial\Omega}, \qquad \frac{\partial u_1}{\partial\nu}\bigg|_{\partial\Omega_{-,\varepsilon}(\xi)} = \frac{\partial u_2}{\partial\nu}\bigg|_{\partial\Omega_{-,\varepsilon}(\xi)}.$$

Let us denote $u := u_1 - u_2$ and $q := q_1 - q_2$. We have

$$(\Delta - q_1)u = qu_2 \quad \text{in } \Omega, \qquad u|_{\partial\Omega} = 0.$$

Now it is easy to see that $u|_{\partial\Omega} = 0$ and $u \in H_\Delta(\Omega)$ implies that $u \in H^2(\Omega)$. Also Green's formula is valid for $v \in H_\Delta(\Omega)$. Thus we obtain

$$\int_\Omega (\Delta - q_1)u\bar{v} \, dx = \int_\Omega qu_2\bar{v} \, dx = \int_\Omega u(\Delta - q_1)\bar{v} \, dx + \int_{\partial\Omega} \frac{\partial u}{\partial\nu}\bar{v} \, dS; \tag{41}$$

Now, we choose

$$\bar{v} = e^{\langle x,\rho_1\rangle}(1 + \psi_{q_1}(x, \rho_1))$$

as in (35) to be a solution to $(\Delta - q_1)\bar{v} = 0$, where

$$\rho_1 = -\tau\xi - i\frac{k-l}{2}$$

with ξ, k, and l as before so that $\langle \rho_1, \rho_1\rangle = 0$. Notice that with this choice of ρ_j, $j = 1, 2$, we have

$$\rho_1 + \rho_2 = -ik.$$

With these choices of u_2 and v, the identity (41) now reads

$$\int_\Omega q u_2 \bar{v} = \int_{\partial\Omega} \frac{\partial u}{\partial \nu} \bar{v}\, dS. \tag{42}$$

The final step in the proof is to show that the right hand side of (42) goes to 0 as $\tau \to \infty$.

By hypothesis,

$$\partial_\nu u|_{\partial\Omega_{-,\varepsilon}(\xi)} = 0.$$

Then we have

$$\int_{\partial\Omega} \frac{\partial u}{\partial \nu} \bar{v}\, dS = \int_{\partial\Omega \backslash \partial\Omega_{-,\varepsilon}} \frac{\partial u}{\partial \nu} \bar{v}\, dS = \int_{\partial\Omega_{+,\varepsilon}} \frac{\partial u}{\partial \nu} \bar{v}\, dS.$$

The Cauchy-Schwarz inequality and the estimate (38) yields

$$\left| \int_{\partial\Omega} \frac{\partial u}{\partial \nu} \bar{v}\, dS \right| = \left| \int_{\partial\Omega_{+,\varepsilon}} \frac{\partial u}{\partial \nu} e^{\langle x, \rho_1 \rangle} (1 + \psi_{q_1}(x, \rho_1))\, dS \right| \tag{43}$$

$$\leq \int_{\langle \xi, \nu \rangle \geq \varepsilon} \left| \frac{\partial u}{\partial \nu} e^{-\tau \langle \xi, x \rangle} (1 + \psi_{q_1}(\,\cdot\,, \rho_1)) \right| dS \tag{44}$$

$$\leq \left(\int_{\langle \xi, \nu \rangle \geq \varepsilon} |e^{-\tau \langle \xi, x \rangle} \partial_\nu u|^2\, dS \right)^{\frac{1}{2}} \tag{45}$$

for some $C > 0$. Now we use the Carleman estimate of Theorem 5.7 to obtain

$$\tau\varepsilon \int_{\partial\Omega_{+,\varepsilon}} |e^{-\tau\langle \xi, x \rangle} \partial_\nu u|^2\, dS \leq \tau \int_{\partial\Omega_+} \langle \xi, x \rangle |e^{-\tau\langle \xi, x \rangle} \partial_\nu u|^2\, dS \tag{46}$$

$$\leq \int_\Omega |e^{-\tau\langle \xi, x \rangle} (\Delta - q_1) u|^2 dx = \int_\Omega |e^{-\tau\langle \xi, x \rangle} q u_2|^2\, dx \tag{47}$$

$$\leq 2\big(\|q_1\|_{L^\infty(\Omega)} + \|q_2\|_{L^\infty(\Omega)}\big)^2 \big(1 + \|\psi_2\|^2_{L^2(\Omega)}\big). \tag{48}$$

Hence, we have proved that

$$\left| \int_{\partial\Omega} \frac{\partial u}{\partial \nu} \bar{v}\, dS \right| \leq C\tau^{-1} \to 0, \quad \tau \to \infty.$$

Now letting $\tau \to \infty$ gives

$$\int_\Omega e^{-i\langle x, k \rangle} q(x)\, dx = 0$$

for all $k \perp \xi$. Changing $\xi \in S^{n-1}$ in a small conic neighborhood and using the fact that $\widehat{q}(k)$ is analytic we get that $q = 0$ finishing the proof of Theorem 5.6.

Sketch of Proof of Corollary 5.1. It is well known that we can reduce the problem to the case of the Schrödinger equation using the transformation $w = \gamma^{\frac{1}{2}}u$. If u solves the conductivity equation $\operatorname{div}(\gamma\nabla u) = 0$, then w solves

$$(\Delta - q)w = 0 \quad \text{in } \Omega$$

with $q = \frac{\Delta\sqrt{\gamma}}{\sqrt{\gamma}}$. It is easy to see that

$$\Lambda_q(f) = \gamma^{-\frac{1}{2}}|_{\partial\Omega}\Lambda_\gamma(\gamma^{-\frac{1}{2}}|_{\partial\Omega}f) + \frac{1}{2}\left(\gamma^{-1}\frac{\partial\gamma}{\partial\nu}\right)\Big|_{\partial\Omega} f.$$

Now Kohn and Vogelius showed in [23] that given any open subset Γ of $\partial\Omega$, if we know $\Lambda_\gamma(f)|_\Gamma$ for all f then we can determine $\gamma|_\Gamma$ and $\frac{\partial\gamma}{\partial\nu}|_\Gamma$, reducing therefore the proof of Corollary 5.1 to Theorem 5.6.

CONJECTURE 5.3. *It is natural to expect that one needs to only measure the following subset of the Cauchy data to recover the potential. Let Γ be an arbitrary open subset of the boundary. We define*

$$C_{q,\Gamma} = \left\{\left(u|_{\partial\Omega}, \frac{\partial u}{\partial\nu}\Big|_\Gamma\right) \;\Big|\; (\Delta - q)u = 0 \quad \text{in } \Omega,\; u \in H_\Delta(\Omega)\right\}. \tag{49}$$

The conjecture is that if we know $C_{q,\Gamma}$ then we can recover the potential q. It would also be interesting to prove stability estimates and give a reconstruction of the potential under the conditions of Theorem 5.6.

6. The $\overline{\partial}\partial$ system. In this section we describe an extension of Nachman's result to $C^1(\overline{\Omega})$ conductivities due to Brown and the author [7]. Instead of reducing the conductivity equation to the Schrödinger equation we reduce it to a first order system (the $\overline{\partial}\partial$ system). We construct CGO solutions for this system. We combine this with the inverse scattering method developed for this system by Beals and Coifman [4] and L. Sung [37].

The main result of [7] is:

THEOREM 6.1. *Let $n = 2$. Let $\gamma \in C^1(\overline{\Omega})$, γ strictly positive on $\overline{\Omega}$. Assume $\Lambda_{\gamma_1} = \Lambda_{\gamma_2}$. Then $\gamma_1 = \gamma_2$ in Ω.*

Theorem 6.1 was extended to complex conductivities with small imaginary part in [13], using the $\overline{\partial}\partial$ method. Complex conductivities with small imaginary part arise naturally when considering Maxwell's equations for time harmonic waves with small frequency.

Logarithmic type stability estimates were proven in [3] for $C^{1+\epsilon}(\overline{\Omega})$ conductivities. A reconstruction method based on the uniqueness proof of [7] was developed by Knudsen and Tamasan [21] for $C^{1+\epsilon}(\overline{\Omega})$ conductivities.

As mentioned earlier, the proof of theorem 6.1 first reduces the conductivity equation to a first order system. We define

$$q = -\frac{1}{2}\partial \log\gamma \tag{50}$$

and a matrix potential Q by

$$Q = \begin{pmatrix} 0 & q \\ \overline{q} & 0 \end{pmatrix}. \tag{51}$$

Let also D be the operator

$$D = \begin{pmatrix} \overline{\partial} & 0 \\ 0 & \partial \end{pmatrix}, \tag{52}$$

where $\overline{\partial} = \frac{1}{2}(\partial_{x_1} - i\partial_{x_2})$.

An easy calculation shows that, if u satisfies the conductivity equation $\operatorname{div}(\gamma \nabla u) = 0$, then

$$\begin{pmatrix} v \\ w \end{pmatrix} = \gamma^{\frac{1}{2}} \begin{pmatrix} \partial u \\ \overline{\partial} u \end{pmatrix} \tag{53}$$

solves the system

$$D\begin{pmatrix} v \\ w \end{pmatrix} - Q\begin{pmatrix} v \\ w \end{pmatrix} = 0. \tag{54}$$

In [7] Brown and Uhlmann construct matrix solutions of (54) of the form

$$\psi(z,k) = m(z,k)\begin{pmatrix} e^{izk} & 0 \\ 0 & e^{-i\overline{z}k} \end{pmatrix}. \tag{55}$$

where $z = x_1 + ix_2$, $k \in \mathbb{C}$ with $m \to 1$ as $|z| \to \infty$ in a sense to be described below. A simple calculation shows that m from (55) satisfies in Ω the following equation

$$D_k m - Qm = 0, \tag{56}$$

where D_k is the operator

$$D_k = \begin{pmatrix} (\overline{\partial} - i\overline{k}) & 0 \\ 0 & (\partial + ik) \end{pmatrix}.$$

In order to explain the construction of m we need a few more definitions. Let

$$\Lambda_k(z) = \begin{pmatrix} e(z,k) & 0 \\ 0 & e(z,-\overline{k}) \end{pmatrix}, \quad e(z,k) = e^{i(zk+\overline{z}\overline{k})}$$

and for any matrix A, define the following operator

$$E_k A = E_k \begin{pmatrix} a_{11} & a_{12} \\ a_{21} & a_{22} \end{pmatrix} = \begin{pmatrix} a_{11} & e(z,-\overline{k})a_{12} \\ a_{21}e(z,k) & a_{22} \end{pmatrix}.$$

Notice that

$$D_k = E_k^{-1} D E_k. \tag{57}$$

Let D^{-1} be the operator

$$D^{-1} = \begin{pmatrix} \overline{\partial}^{-1} & 0 \\ 0 & \partial^{-1} \end{pmatrix},$$

where

$$\overline{\partial}^{-1} f(z) = \frac{1}{2\pi i} \int \frac{f(w)}{z-w} dw \wedge d\overline{w}$$

and

$$\partial^{-1} f(z) = \frac{1}{2\pi i} \int \frac{f(w)}{\overline{z}-\overline{w}} dw \wedge d\overline{w}.$$

We have from (57) that $D_k^{-1} = E_k^{-1} D^{-1} E_k$. We look for solutions of (56) among the solutions of the integral equation

$$(I - D_k^{-1} Q) m(z) = I, \tag{58}$$

where I is the 2×2 identity matrix. For a 2×2 matrix A, let A^d and A^{off} denote its diagonal respectively off-diagonal part. If

$$J = \frac{1}{2} \begin{pmatrix} -i & 0 \\ 0 & i \end{pmatrix}$$

we define the operator $\mathcal{J}$ by

$$\mathcal{J} A = [J, A] = 2 J A^{off} = -2 A^{off} J. \tag{59}$$

To end with the preliminary notation, we recall the definition of the weighted L^p space

$$L^p_\alpha(\mathbb{R}^2) = \{f; \int (1+|x|^2)^\alpha |f(x)|^p dx < \infty\}.$$

The next result gives the solvability of (56) in an appropriate space.

THEOREM 6.2. *Let $Q \in L^\infty(\mathbb{R}^2)$ and compactly supported. Assume that Q is a hermitian matrix. Choose r so that $r < 2$ and then β so that $\beta r > 2$. Then the operator $(I - D_k^{-1} Q)$ is invertible in $L^r_{-\beta}$. Moreover the inverse is differentiable in k in the strong operator topology.*

Theorem 6.2 implies the existence of solutions of the form (55) with $m - 1 \in L^r_{-\beta}(\mathbb{R}^2)$.

We remark that the proof of Theorem 6.2 consists in showing that the integral equation (58) is of Fredholm type in $L^r_{-\beta}$. The fact that it has been a trivial kernel follows by showing that if $(I - D_k^{-1}Q)n(z,k) = 0$, then $n \in L^p$, for all $p > 2$, satisfies a pseudoanalytic equation in the z-variable. By the standard Liouville's theorem for pseudoanalytic equations with coefficients in $L^p, p > 2$ it follows that $n = 0$.

Next we compute $\frac{\partial}{\partial \bar{k}} m(z,k)$.

THEOREM 6.3. *Let m be the solution of (56) with $m - 1 \in L^r_{-\beta}(\mathbb{R}^2)$. Then*

$$\frac{\partial}{\partial \bar{k}} m(z,k) - m(z,\bar{k})\Lambda_k(z) S_Q(k) = 0 \tag{60}$$

where the scattering data S_Q is given by [4]

$$S_Q(k) = -\frac{1}{\pi}\mathcal{J}\int_{\mathbb{R}^2} E_k Q m \, d\mu, \tag{61}$$

where $d\mu$ denotes Lebesgue measure in $\mathbb{R}^2$.

A further calculation shows that

$$S_Q(k) = \frac{i}{\pi}\int_{\mathbb{R}^2} \begin{pmatrix} 0 & e(z,-\bar{k})q(z)m_{22}(z,k) \\ -e(z,k)\overline{q}(z)m_{11}(z,k) & 0 \end{pmatrix} d\mu. \tag{62}$$

The behavior of m in the k variable is given by the following result:

THEOREM 6.4. *Let $Q \in L^\infty(\mathbb{R}^2)$ and compactly supported. Then there exists $R = R(Q)$ so that for all $q > 2$*

$$sup_z \|m(z,\cdot) - 1\|_{L^q\{k;|k|>R\}} \le C\|Q\|^2_{L^\infty}$$

where the constants depend on q and the diameter of the support of Q.

Outline of proof of Theorem 6.1 We know [2], [23], [30], [42] that if $\gamma_i \in C^1(\overline{\Omega})$ and $\Lambda_{\gamma_1} = \Lambda_{\gamma_2}$, then $\partial^\alpha \gamma_1\big|_{\partial\Omega} = \partial^\alpha \gamma_2\big|_{\partial\Omega}$ $\forall\, |\alpha| \le 1$. Therefore we can extend $\gamma_i \in C^1(\mathbb{R}^2)$, $\gamma_1 = \gamma_2$ in $\mathbb{R}^2 \setminus \Omega$ and $\gamma_i = 1$ outside a large ball. Thus $Q_i \in L^\infty(\mathbb{R}^2), i = 1,2$. The proof follows the following steps.

Step 1. $\Lambda_{\gamma_1} = \Lambda_{\gamma_2} \Rightarrow S_{Q_1} = S_{Q_2} =: S$. With these extensions, we observe that for each j the scattering data $S_{Q_j}(k), j = 1,2$, has the representation

$$\begin{aligned} S_{Q_j}(k) &= -2J\int_{\mathbf{R}^2}\begin{pmatrix} 0 & \bar{\partial}\psi_j^{12}e^{-iz\bar{k}} \\ \partial\psi_j^{21}e^{i\bar{z}k} & 0 \end{pmatrix} d\mu(z) \\ &= -2J\left[\int_{\mathbf{R}^2\setminus\Omega}\begin{pmatrix} 0 & \bar{\partial}\psi_j^{12}e^{-iz\bar{k}} \\ \partial\psi_j^{21}e^{i\bar{z}\bar{k}} & 0 \end{pmatrix} d\mu(z) \right. \\ &\quad \left. + \int_{\partial\Omega}\begin{pmatrix} 0 & \bar{\nu}\psi_j^{12}e^{-iz\bar{k}} \\ \nu\psi_j^{21}e^{i\bar{z}\bar{k}} & 0 \end{pmatrix} d\mu(z)\right]. \end{aligned}$$

The formula for S_{Q_j} uses the complexified normal to the boundary

$$\nu(z) = \nu_1(z) + i\nu_2(z), \ \overline{\nu}(z) = \nu_1(z) - i\nu_2(z) \tag{63}$$

with $(\nu_1(z), \nu_2(z))$ the unit outer normal at $z \in \partial\Omega$.

From this expression for $S_{Q_j}, j = 1, 2$, we see that if we can show

$$\psi_1(z,k) = \psi_2(z,k) \quad \text{in } \mathbf{R}^2 \backslash \bar{\Omega}, \tag{64}$$

then $S_{Q_1} = S_{Q_2}$.

The last formula follows by using a very similar argument to Lemma 2.6 in [41].

Step 2. Let $\widetilde{m} = m_1 - m_2$. Using the $\overline{\partial}$-equation (60) and Step 1 we conclude that

$$\frac{\partial}{\partial \overline{k}} \widetilde{m}(z,k) - \widetilde{m}(z,\overline{k}) \Lambda_k(z) S(k) = 0 \tag{65}$$

With the elements of $\widetilde{m}$ we form the following four functions

$$\begin{aligned} u_\pm(k,z)) &= \widetilde{m}_{11}(z,k) \pm \overline{\widetilde{m}_{12}(z,\overline{k})} \\ v_\pm(k,z)) &= \widetilde{m}_{21}(z,k) \pm \overline{\widetilde{m}_{22}(z,\overline{k})} \end{aligned}$$

each of which lies in $L^q(\mathbb{R}^2)$ in the k-variable and satisfies, for a fixed z, a pseudoanalytic equation in the k-variable,

$$\frac{\partial}{\partial \overline{k}} w(z,k) = r(z,k) \overline{w(z,k)}. \tag{66}$$

where $r(z, k)$ is some component of S multiplied by a complex coefficient of norm 1.

Step 3. In [7] it was shown that, for $Q \in L_c^\infty$ with $Q^* = Q$, we have that

$$\int tr S_Q S_Q^* \le \int tr Q Q^*.$$

This shows that $S_Q \in L^2$. Consequently, for each fixed z we have the map $k \to r(z,k)$ is in $L^2(\mathbb{R}^2)$.

Step 4. Prove that $u_\pm = v_\pm = 0$, hence $\widetilde{m} = 0$ or $m_1 = m_2$. Then it is easy to show $Q_1 = Q_2$ and therefore $\gamma_1 = \gamma_2$.

To do this we need the following generalization of Liouville Theorem for pseudoanalytic functions proven in [7].

LEMMA 6.1. *Let $f \in L^2(\mathbb{R}^2)$ and $w \in L^p(\mathbb{R}^2)$ for some finite p. Assume that $we^{\overline{\partial}^{-1} f}$ is analytic. Then $w = 0$.* Let us define

$$\begin{aligned} \widetilde{u}_\pm &= u_\pm e^{\overline{\partial}^{-1} r} \\ \widetilde{v}_\pm &= v_\pm e^{\overline{\partial}^{-1} r}. \end{aligned} \tag{67}$$

It is easy to check that $\widetilde{u}_\pm$ and $\widetilde{v}_\pm$ are analytic. By the lemma above we conclude that $u_\pm = v_\pm = 0$ which in turn gives $m_1 = m_2$. It is easy to show $Q_1 = Q_2$ and therefore $\gamma_1 = \gamma_2$, concluding the proof of the Theorem 6.1.

The idea of the proof of Lemma 6.1 is the observation that since $r \in L^2(\mathbb{R}^2)$, $f = \overline{\partial}^{-1} r$ is in $VMO(\mathbb{R}^2)$ (the space of functions with vanishing mean oscillation) and thus is $O(\log|z|)$ as $|z| \to \infty$. Hence $e^f w \in L^{\widetilde{p}}$ for $\widetilde{p} > p$. By Liouville's theorem it follows that $e^f w = 0$. The details can be found in [7].

7. Final remarks. It was shown in Section 2 that from the hyperbolic Dirichlet- to-Neumann map we can recover the boundary distance function, assuming that the domain is geodesically convex. Is there any connection between the elliptic Dirichlet to Neumann map and the boundary distance function d_g? To know the EDN map is the same as knowing the set of Cauchy data

$$C_g = \left\{ \left(u|_{\partial\Omega}, \Lambda_g(u|_{\partial\Omega}) \right) \right\}. \tag{68}$$

This set is resemblant of the scattering relation defined in Section 4. The scattering relation is a Lagrangian manifold in finite dimensions and the set of Cauchy data a Lagrangian manifold in infinite dimensions. It would be very interesting to find a more direct relationship between these two sets without, of course, constructing the metric first.

REFERENCES

[1] G. Alessandrini, *Stable determination of conductivity by boundary measurements*, App. Anal. **27** (1988), 153–172.

[2] G. Alessandrini, *Singular solutions of elliptic equations and the determination of conductivity by boundary measurements*, J. Diff. Equations **84** (1990), 252–272.

[3] J. A. Barceló, T. Barceló, and A. Ruiz, *Stability of the inverse conductivity problem in the plane for less regular conductivities*, J. Diff. Equations **173** (2001), no. 2, 231–270.

[4] R. Beals and R. Coifman, *The spectral problem for the Davey-Stewarson and Ishimori hierarchies*, in *Nonlinear evolution equations: Integrability and spectral methods*, Manchester University Press (1988), 15–23.

[5] M. Belishev, *Boundary control in reconstruction of manifolds and metrics (the BC method)*, Inverse Problems **13** (1997), no. 5, R1–R45.

[6] M. Belishev and Y. Kurylev, *To the reconstruction of a Riemannian manifold via its spectral data (BC-method)*, Comm. PDE **17** (1992), 767–804.

[7] R. Brown and G. Uhlmann, *Uniqueness in the inverse conductivity problem for nonsmooth conductivities in two dimensions*, Comm. Partial Diff. Equations **22** (1997), 1009–1027.

[8] A. Bukhgeim and G. Uhlmann, *Determining a potential from partial Cauchy data*, Comm. PDE **27** (2002), 653-668.

[9] A.P. Calderón, *On an inverse boundary value problem*, Seinar on Numerical Analysis and its Applications to Continuum Physics, Soc. Brasileira de Matemática, Río de Janeiro (1980), 65–73.

[10] C. Croke, This IMA Volume in Mathematics and its Applications, *Geometric methods in inverse problems and PDE control*, Editors: Christopher B. Croke, Irena Lasiecka, Gunther Uhlmann, and Michael Vogelius. Publisher: Springer-Verlag, New York.

[11] J.J. Duistermaat, *Fourier Integral Operators*, Birkhäuser.

[12] L. FADDEEV, *Growing solutions of the Schrödinger equation*, Dokl. Akad. Nauk SSSR **165** (1965), 514–517 (translation in Sov. Phys. Dokl. 10, 1033).

[13] E. FRANCINI, *Recovering a complex coefficient in a planar domain from the Dirichlet -to-Neumann map*, Inverse Problems **16** (2000), 107–119.

[14] V. GUILLEMIN, *Sojourn times and asymptotic properties of the scattering matrix*, Proceedings of the Oji Seminar on Algebraic Analysis and the RIMS Symposium on Algebraic Analysis (Kyoto Univ., Kyoto, 1976), Publ. Res. Inst. Math. Sci. **12** (1976/77), supplement, 69–88.

[15] S. HANSEN AND G. UHLMANN, *Propagation of polarization for the equations in elastodynamics with residual stress and travel times*, to appear Math. Annalen.

[16] L. HÖRMANDER, *The Analysis of Linear Partial Differential Operators, Vol III*, Springer-Verlag Berlin Heidelberg New York Tokyo (1985).

[17] M. IKEHATA, *The enclosure method and its applications*, International Society for Analysis, Applications and Computation, Vol. 9, Kluwer Academic Pub., DORDRECHT (2001), 87-203 (edited by S. Saito, N. Hayashi, and M. Yamamoto.)

[18] A. KATCHALOV, Y. KURYLEV, AND M. LASSAS, *Inverse boundary spectral problems*, Chapman&Hall/CRC, 2001.

[19] A. KATCHALOV, Y. KURYLEV, AND M. LASSAS, This IMA Volume in Mathematics and its Applications, *Geometric methods in inverse problems and PDE control*, Editors: Christopher B. Croke, Irena Lasiecka, Gunther Uhlmann, and Michael Vogelius. Publisher: Springer-Verlag, New York.

[20] A. KATSUDA, Y. KURYLEV, AND M. LASSAS, *Stability on inverse boundary spectral problem*, in: New Geom. and Anal. Meth. in Inv. Probl. (Eds. Y. Kurylev and E. Somersalo), *Springer Lect. Notes*, to appear.

[21] K.KNUDSEN AND A. TAMASAN, *Reconstruction of less regular conductivities in the plane*, MSRI preprint series, Berkeley, 2001.

[22] R. KOHN AND M. VOGELIUS, *Identification of an unknown conductivity by means of measurements at the boundary*, in Inverse Problems, edited by D. McLaughlin, SIAM-AMS Proc. no. 14, Amer. Math. Soc, Providence (1984), 113–123.

[23] R. KOHN AND M. VOGELIUS, *Determining conductivity by boundary measurements*, Comm. Pure Appl. Math. **37** (1984), 289–298.

[24] M. LASSAS AND G. UHLMANN, *On determining a Riemannian manifold from the Dirichlet-to-Neumann map*, Annales Scientifiques de l' Ecole Normale Superiéure **34** (2001), 771–787.

[25] M. LASSAS, V. SHARAFUTDINOV, AND G. UHLMANN, *Semiglobal boundary rigidity for Riemannian metrics*, to appear Math. Annalen.

[26] J. LEE AND G. UHLMANN, *Determining anisotropic real-analytic conductivities by boundary measurements*, Comm. Pure Appl. Math. **42** (1989), 1097–1112.

[27] R.G. MUKHOMETOV, *A problem of reconstructing a Riemannian metric*, Siberian Math. J. **22** (1982), 420-433.

[28] R.G. MUKHOMETOV, *The reconstruction problem of a two-dimensional Riemannian metric, and integral geometry (Russian)*, Dokl. Akad. Nauk SSSR **232** (1977), no. 1, 32–35

[29] A. NACHMAN, *Global uniqueness for a two-dimensional inverse boundary value problem, Ann. of Math. (2)* **143** (1996), 71–96.

[30] A. NACHMAN, *Reconstructions from boundary measurements*, Annals of Math. 128, (1988), 531–587.

[31] L. RACHELE, *An inverse problem in elastodynamics: Determination of the wave speeds in the interior*, J. Diff Eqs. **162** (2000), 300–325.

[32] L. RACHELE, *Uniqueness of the density in an inverse problem for isotropic elastodynamics*, to appear Trans. AMS.

[33] L. ROBIANNO AND ZUILLY, *Uniqueness in the Cauchy problem for operators with partially holomorphic coefficients*, Invent. Math. **131** (1998), 493–539.

[34] V.A. SHARAFUTDINOV, This IMA Volume in Mathematics and its Applications, *Geometric methods in inverse problems and PDE control*, Editors: Christopher B. Croke, Irena Lasiecka, Gunther Uhlmann, and Michael Vogelius. Publisher: Springer-Verlag, New York.

[35] P. STEFANOV AND G. UHLMANN, *Stability estimates for the hyperbolic Dirichlet to Neumann map in anisotropic media*, *J. Funct. Anal.* **154** (1998), 330–358.

[36] P. STEFANOV AND G. UHLMANN, Rigidity for metrics with the same lengths of geodesics, Math. Res. Lett. **5** (1998), 83–96.

[37] L. SUNG, *An inverse scattering transform for the Davey-Stewartson II equations. I, II, III*, J. Math. Anal. Appl. **183** (1994), 121–154, J. Math. Anal. Appl. **183** (1994), 289–325. J. Math. Anal. Appl. **183** (1994), 477–494.

[38] J. SYLVESTER, *An anisotropic inverse boundary value problem*, Comm. Pure Appl. Math. (1990), 201–232.

[39] J. SYLVESTER AND G. UHLMANN, *Inverse problems in anisotropic media*, Contemp. Math. **122** (1991), 105–117.

[40] J. SYLVESTER AND G. UHLMANN, *A global uniqueness theorem for an inverse boundary value problem*, Ann. of Math. **125** (1987), no. 2, 153–169.

[41] J. SYLVESTER AND G. UHLMANN, *A uniqueness theorem for an inverse boundary value problem in electrical prospection*, Comm. Pure Appl. Math. **39** (1986), 91–112

[42] J. SYLVESTER AND G. UHLMANN, *Inverse boundary value problems at the boundary – continuous dependence*, Comm. Pure Appl. Math. **41** (1988), 197–221.

[43] D. TATARU, *Unique continuation for solutions to PDE's; between Hörmander's theorem and Holmgren's theorem*, Comm. Partial Differential Equations **20** (1995), 855–884.

[44] D. TATARU. This IMA Volume in Mathematics and its Applications, *Geometric methods in inverse problems and PDE control*, Editors: Christopher B. Croke, Irena Lasiecka, Gunther Uhlmann, and Michael Vogelius. Publisher: Springer-Verlag, New York.

[45] G. UHLMANN, *Developments in inverse problems since Calderón's foundational paper*, chapter 19 of *Essays in Harmonic Analysis and Partial Differential Equations*, University of Chicago Press, edited by M. Christ, C. Kenig, C. Sadosky (1999).

INVERSE RESONANCE PROBLEM FOR $\mathbb{Z}_2$-SYMMETRIC ANALYTIC OBSTACLES IN THE PLANE*

STEVE ZELDITCH†

Abstract. We given an exposition of a proof that a mirror symmetric configuration of two convex analytic obstacles in $\mathbb{R}^2$ is determined by its Dirichlet resonance poles. It is the analogue for exterior domains of the proof that a mirror symmetric bounded simply connected analytic plane domain is determined by its Dirichlet eigenvalues. The proof uses 'interior/exterior duality' to simplify the argument.

1. Introduction. This article is part of a developing series [Z1, Z2] concerned with the inverse spectral problems for analytic plane domains. It is essentially the lecture we presented at the IMA workshop on Inverse Spectral Problems in July, 2001. We prove an analogue for exterior domains of the result proved in [Z2] for interior ones, namely that a mirror symmetric configuration of two convex analytic obstacles in $\mathbb{R}^2$ is determined by its Dirichlet or Neumann resonance poles. The proof combines the known result that wave invariants of an exterior domain are resonance invariants (see §3) with the method of [Z1, Z2] for calculating the wave invariants explicitly in terms of the boundary defining function. In keeping with the expository nature of the lecture, we give a detailed exposition of the background results on the Poisson relation for exterior domains and of the main steps in [Z1, Z2] in the calculation of wave invariants (§6) and on determining the domain from its wave invariants (§7).

The motivating problem is whether analytic domains are determined by their spectra for Dirichlet or Neumann boundary conditions. This inverse spectral problem can be posed for both interior and exterior domains. For the interior problem, we assume the domain is a bounded simply connected plane domain and ask whether its Dirichlet or Neumann spectrum determines the boundary up to rigid motion. For the exterior problem, we assume the domain is the complement of two bounded, simply connected obstacles and ask whether we can determine the pair of obstacles from the resonance poles (or scattering phase) of the exterior Dirichlet or Neumann Laplacian. Since the method is similar for both boundary conditions, we assume the boundary conditions are Dirichlet in the exterior domain.

In the previous paper [Z2], the interior inverse spectral problem is studied for analytic domains with a certain mirror symmetry σ. The symmetry is assumed to fix a bouncing ball orbit (as a set) and to reverse its orientation; i.e. if the domain is translated and rotated so that the bouncing ball orbit lies along the y-axis with its midpoint at the origin,

*Research partially supported by NSF grant #DMS-0071358.

†Department of Mathematics, Johns Hopkins University, Baltimore, MD 21218, USA (zelditch@math.jhu.edu).

then $\sigma(x, y) = (x, -y)$. In addition, we assume the length of the bouncing ball orbit is a fixed number L and that the orbit satisfies a non-degeneracy condition. The main result (see Theorem (1.4)) is that the interior domain satisfying some generic conditions is determined by its Dirichlet spectrum among other such domains. We note that this result is stronger than the one in [Z3] in having eliminated one of the two symmetries. We should emphasize that the proof does not immediately imply the result for the other mirror symmetry $\sigma(x, y) = (-x, y)$, which fixes γ pointwise and preserves the orientation.

Our main purpose in this paper is to extend the result to connected exterior domains of the form

$$\Omega^c := \mathbb{R}^2 - \Omega, \quad \text{with } \Omega = \{\mathcal{O} \cup \tau_{x,L}\mathcal{O}\} \tag{1}$$

where $\mathcal{O}$ is a bounded, simply connected analytic domain and where

$$\tau_{x,L} = \text{ reflection through the line } \langle \nu_x, x + 1/2L\nu_x - y \rangle = 0.$$

Here, $x \in \mathcal{O}$, ν_x is the outward unit normal to $\mathcal{O}$ at x and $L > 0$ is a given positive number. Thus, the obstacle consists of two (non-intersecting) isometric components which are mirror images of each other across a common orthogonal segment of length L. Such an obstacle is called a (symmetric) 2-component scatterer. The segment is the projection to $\mathbb{R}^2$ of a bouncing ball orbit γ of the exterior billiard problem. Throughout, the notation Ω refers to a bounded domain and Ω^c is the unbounded complement.

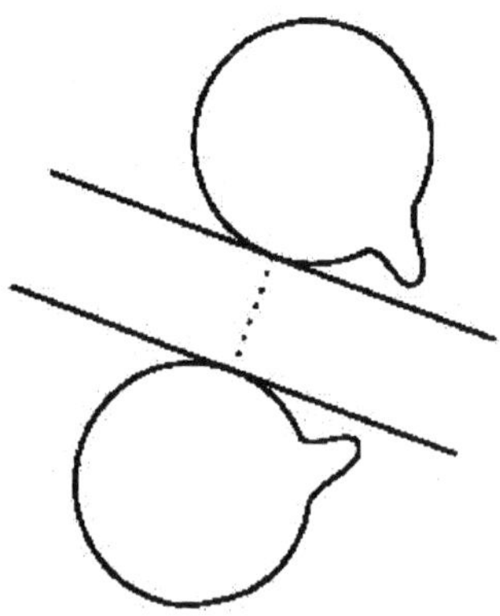

FIG. 1. *Mirror symmetric 2-obstacle scatterer.*

The problem is thus to recover the obstacle $\mathcal{O} \cup \tau_{x,L}\mathcal{O}$ from the set of resonance poles $\{\lambda_j\}$ of the exterior Dirichlet Laplacian $\Delta_D^{\Omega^c}$, i.e. the poles of the analytic continuation of its resolvent

$$R_D^{\Omega^c}(k + i\tau) = (\Delta_D^{\Omega^c} + (k + i\tau)^2)^{-1}$$

to the logarithmic plane. To be precise, we consider the class $\mathcal{OBSTA}$–$\mathcal{CLE}_{\mathbb{Z}_2,L}$ of two-component obstacles satisfying:

- (i) $\mathcal{O}$ is simply connected and real analytic;
- (ii) γ is a non-degenerate bouncing ball orbit, whose length L is isolated in $Lsp(\Omega) \cup Lsp(\Omega^c)$.

We denote by $\mathcal{RES}(\Omega^c)$ the set of resonance poles of the Laplacian $\Delta_D^{\Omega^c}$ of the domain Ω^c with Dirichlet boundary conditions.

We now state the main results.

THEOREM 1.1. *$\mathcal{RES} : \mathcal{OBSTACLE}_{\mathbb{Z}_2,L} \mapsto \mathbb{C}^{\mathbf{N}}$ is 1-1.*

In fact, in combination with a result of M. Zworski [Zw], the proof shows more: the obstacle is determined by the resonances close to the real axis which are associated to the bouncing ball orbit γ of the exterior billiard problem. We note that assumption (ii) is stronger than the one in the interior case (cf. Theorem (1.4) in demanding multiplicity one for L in the combined interior and exterior length spectra. By using the original proof in [Z1, Z2], one could remove $Lsp(\Omega)$ from the assumption and just demand multiplicity one in $Lsp(\Omega^c)$. We do not do so, because we wish to present a simpler proof than the one in [Z1, Z2], as we will explain below in the introduction.

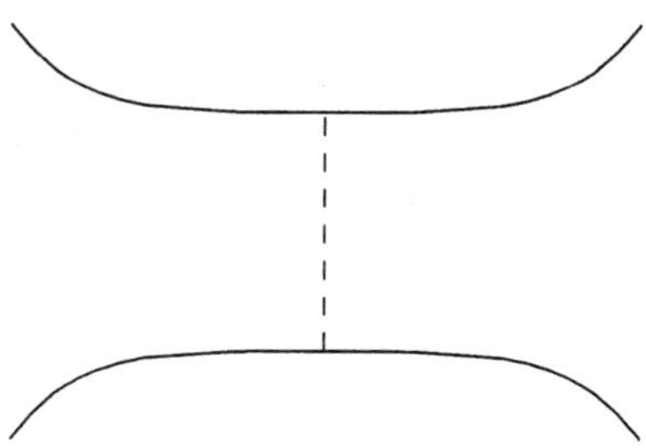

FIG. 2. *Bouncing ball orbit.*

As a corollary, we prove that the obstacle is determined by its scattering phase. We denote by $S_D(\lambda)$ the Dirichlet scattering operator for Ω^c and by $s_D(\lambda) = \det S_D(\lambda)$ the scattering phase.

COROLLARY 1.2. *If two exterior domains Ω in $\mathcal{OBSTACLE}_{\mathbb{Z}_2,L}$ have the same scattering phase $s_D(\lambda)$, then they are isometric.*

We note that the length L of the segment between the components has been marked, since apriori it is not a resonance invariant. Under some additional assumptions, it is a resonance invariant and one does not have to mark it. For instance, if one assumes that $\mathcal{O}$ is convex, then the bouncing ball orbit γ is the unique periodic reflecting ray of the exterior domain up to iterates, so its length is a resonance invariant. Thus, let $\mathcal{C}OBSTACLE_{\mathbb{Z}_2}$ denote the convex component obstacles in $\mathcal{O}BSTACLE_{\mathbb{Z}_2}$. We then have:

COROLLARY 1.3. *$\mathcal{R}ES : \mathcal{C}OBSTACLE_{\mathbb{Z}_2} \mapsto \mathbb{C}^{\mathbf{N}}$ is 1-1.*

For the sake of completeness, we also recall the precise statement of the interior result of [Z2]. Let Spec(Ω) denote the spectrum of the interior Laplacian Δ_B^Ω of the domain Ω with either Dirichlet or Neumann boundary conditions.

THEOREM 1.4. *Let $\mathcal{D}_L$ denote the class of simply connected real-analytic plane domains Ω satisfying:*

- *(i) There exists an isometric involution σ of Ω;*
- *(ii) σ 'reverses' a non-degenerate bouncing ball orbit $\gamma \to \gamma^{-1}$;*
- *(iii) The lengths $2rL$ of all iterates γ^r have multiplicity one in $Lsp(\Omega)$, and the eigenvalues of the linear Poincare map P_γ are not roots of unity;*

Then: Spec: $\mathcal{D}_{1,L} \mapsto \mathbb{R}_+^{\mathbf{N}}$ is 1-1.

To be cautious, we should point out that these results are part of a work-in-progress which has not yet reached its final form. They have been independently verified when there is an additional mirror symmetry (i.e. each obstacle is itself 'left-right' mirror symmetric), and in this case the relevant calculations of wave invariants are easy. When there is only one mirror symmetry, the calculations become somewhat messy (see [Z2]) and the proof therefore becomes rather 'unstable'; at this time of writing, some details have been independently verified (by R. Bacher [B]), but some have not.

Let us briefly describe the proof and the organization of the paper. As emphasized by Zworski, there is nothing really new in Theorem 1.1 beyond the result of [Z1, Z2] in the interior case (i.e. Theorem 1.4). We are just combining the known fact that wave trace invariants of the exterior domain are resonance invariants with the calculation of the wave trace invariants in terms of the germ of the defining function of $\partial\Omega$ at the endpoints of the bouncing ball orbit. This calculation was done for the germ of any kind of bouncing ball orbit in [Z2], developing a method originating in the work of Balian-Bloch [BB1, BB2]. However, it seems to us worthwhile to collect all the facts one needs for the proof of Theorem (1.1) in one place and to explain the main steps without all the details in [Z1, Z2].

Moreover, our exposition has one novel point: in §4, we use the so-called interior/exterior duality to simplify formula for the resolvent trace in [Z1]. Interior/exterior duality is a distributional trace formula, which we informally write for $\tau >> 0$ as

$$(2) \qquad \begin{aligned} &Tr_{\mathbb{R}^2}[R_D^{\Omega^c}(k+i\tau) \oplus R_N^{\Omega}(k+i\tau) - R_0(k+i\tau)] \\ &= \frac{d}{dk} \log \det \Big(I + N(k+i\tau) \Big). \end{aligned}$$

Here, the determinant is the Fredholm determinant, R_N^X, R_D^X denote Neumann (resp. Dirichlet) resolvents on X and the notation Tr_X indicates the space on which the trace is taken. We write $L^2(\mathbb{R}^2) = L^2(\Omega) \oplus L^2(\Omega^c)$ and view $R_D^{\Omega^c}(k+i\tau) \oplus R_N^{\Omega}(k+i\tau)$ as an operator on this space. For notational simplicity, we do not put in the explicit projections $1_\Omega, 1_{\Omega^c}$ (i.e. the characteristic functions). Also, $N(k+i\tau)$ is a boundary integral operator which will be defined in §4. The correct statement and proof of (2) will be given in Proposition (4.1). It is essentially a differentiated version of the formula

$$(3) \qquad \frac{d}{dE} \log \det S_D(E) + \pi \frac{d}{dE} N_N(E) = \Im \frac{d}{dE} \log \zeta_{D,N}(E+i0)$$

of Eckmann-Pillet (cf. [EP, EP2]), relating the scattering phase of the exterior Dirichlet problem, the eigenvalue counting function of the interior Neumann problem and a spectral zeta function for the integral operator N along the boundary.

This reduction to the boundary trace on the right side of (2) simplifies some of the technical details of [Z1] and allows one to deduce the inverse spectral result simultaneously for the interior and exterior problems. In addition, it brings our calculations into closer contact with the physics literature, where the Balian-Bloch approach is now almost always applied to the boundary trace (see e.g. [AG, THS, THS2, GP]. The price we pay is that the combination of inside and outside requires us to make the additional multiplicity free assumption on the combined inner and outer length spectra to obtain the inverse results in Theorem 1.1. As mentioned above, this assumption could be eliminated; but we feel that the simplification in the proof is worth the extra assumption. To simplify the exposition, we also assume in the last step that $\mathcal{O}$ is convex. Hence, we only complete the proof for hyperbolic bouncing ball orbits. For the general case, we refer to [Z2].

Let us give a brief outline of the method of [Z1, Z2] and this paper for determining an analytic domain from its spectrum. It is based on the use of certain spectral (i.e. resonance) invariants known as the wave trace invariants at the bouncing ball orbit. As the name implies, the wave trace invariants at a periodic reflecting ray γ are coefficients of the singularity expansion of the relative trace of the wave group $E_D^{\Omega^c}(t) = \cos t\sqrt{\Delta_D^{\Omega^c}}$ at $t = L_\gamma$ (the length). The inverse results are proved by explicitly calculating the wave trace invariants at a bouncing ball orbit in terms of the Taylor coefficients of the boundary defining function at the endpoints, and then determining the boundary from these Taylor coefficients.

Rather than studying the wave trace per se in [Z1, Z2] we follow [BB1, BB2] in studying an essentially equivalent expansion involving the resolvent. Potential theory gives an exact formula for the Dirichlet (or Neumann) resolvent of a domain in terms of the free resolvent of $\mathbb{R}^2$ and of the layer potentials associated to the domain. Formally, one may derive an infinite series expansion known as the 'multiple reflection expansion' of [BB1, BB2], whose Mth term corresponds intuitively to M reflections on the boundary. The trace of each term is, again formally, an oscillatory integral corresponding to M-fold periodic reflecting rays.

The advantage of this approach is that the formula is exact and does not require a microlocal paramatrix construction, which is messy and complicated for bounded domains and therefore very hard to use in inverse spectral theory. Moreover, the terms of the multiple reflection expansion are 'canonical' in that the integrands are the same for all domains; the only difference lies in the domain one integrates over. Once it is 'legalized', it gives an explicit and even routine algorithm for calculating wave invariants. At this time of writing, no other approach seems to provide a workable algorithm for doing the calculations. The disadvantages are that the legalization has two complicated parts: one needs to estimate the remainder in the infinite series and to regularize the oscillatory integral defined by the Mth term.

Our formulae for the wave invariants come from applying stationary phase to the regularized traces. One of the principal results is that, for each iterate γ^r, only the term $M = rm$ of the multiple reflection expansion contributes important inverse spectral data, because only this term contains the maximum number of derivatives of the defining function of Ω at a given order of k^{-j}. The heart of the matter is then the calculation and analysis of the wave invariants. Since one needs wave invariants of all orders, one seemingly runs into an infinite jungle of complicated expressions. To tame the jungle, we enumerate and evaluate the terms of the wave trace invariants using Feyman diagram techniques. It turns out that only five diagrams play an important role (this has been confirmed by R. Bacher [B]). The remaining issue is the calculation of the amplitudes and their dependence on r. It is this latter dependence that is crucial in allowing one to remove a symmetry of the domain.

The calculation of wave trace invariants is the same for periodic reflecting rays of interior and exterior domains since it depends only on the germ of the boundary at the reflection points. The material in §5 - §7 summarizes the calculation in [Z1, Z2]. We hope that this guide clarifies the calculation and makes it easier to check the details.

We close the introduction with some remarks on open problems in the inverse spectral problem for analytic plane domains. The most obvious one is whether one can eliminate the remaining symmetry assumption. It may be that one can recover the domain from wave invariants at one closed billiard orbit, or that one has to combine information from several orbits. A key problem is, how much of the Taylor expansion of the boundary defining function one can recover from wave invariants at one orbit?

The method of this paper and [Z2] gives detailed but somewhat undigested information about wave invariants in terms of the Taylor coefficients of the boundary defining function at endpoints of a bouncing ball orbit. Another approach, suggested originally by Colin de Verdiere [CdV], is to focus on the Birkhoff normal form of the billiard map at the closed orbit. As was proved by V. Guillemin in the boundaryless case [G], the Birkhoff normal form at a closed orbit is a spectral invariant. In [Z3] the author generalized this to the boundary case. Colin de Verdiere's observation that the Birkhoff normal form at a bouncing ball orbit determines an analytic plane domain with two symmetries then gave the solution of the inverse spectral problem for analytic plane domains with these symmetries [Z3]. We found it difficult however to calculate the normal form coefficients by this method without assuming two symmetries. That motivated us to try the Balian-Bloch approach.

In the recent paper [ISZ], Iantchenko-Sjostrand-Zworski give another proof of the inverse result with two symmetries using Birkhoff normal forms of the billiard map and quantum monodromy operator rather than the Laplacian; the method is quite elegant and flexible. However, we believe that their normal form of the monodromy operator is the same as the normal form in [Z3] of what is called there the semiclassical wave group. We changed to the present approach because we found it too difficult to calculate the latter without the two symmetry assumption. However, the possibility remains that the Birkhoff normal form could be a better way to organize the inverse spectral data than the wave invariants themselves. The question is, how much of the boundary defining function is determined by the Birkhoff normal form of the wave group or the monodromy operator at a periodic orbit? It is known that the classical Birkhoff normal form by itself will not determine the Taylor coefficients of the domain unless there are two symmetries.

Acknowledgements. We are heavily indebted to M. Zworski for many helpful conversations on the contents of this article, particularly during a visit at the Erwin Schrodinger Institute during June 2001. In addition, we thank the ESI for making the visit possible. We also thank J. Ralston for confirming that various results of [BGR] which we need are also valid in the even-dimensional case. Finally, we thank the IMA for the opportunity to present the results in their final form.

2. Billiards and the length functional. Throughout this paper, we will assume that our obstacle is up-down symmetric in the sense that there is an isometric involution σ of Ω which interchanges the endpoints of a bouncing ball orbit (extremal diameter). However, we only use the assumption in the last step of calculating wave invariants and determining the domain. Elsewhere, we make it for notational convenience. We now extend the notation and terminology of [Z1] so that it applies to exterior domains which are complements of obstacles with two components.

We align the obstacles so that the bouncing ball orbit γ projects to the line segment $\overline{a_- a_+}$ along the y-axis and so that its midpoint lies at the orgin $0 \in \mathbb{R}^2$ and its endpoints are $a_\pm = (0, \pm L/2)$. We refer to the top component of Ω as the $+$ component $\mathcal{O}_+$ and the bottom one as the $-$ component $\mathcal{O}_-$.

There are two convenient parametrizations of $\partial\Omega$. First, we denote by $q_\pm(\theta)$ the arc-length parameterization of $\mathcal{O}_\pm$ with $q_\pm(0) = a_\pm$. At the end we will also use graph parametrizations: In a small strip $T_\epsilon(\gamma)$ around $\overline{a_1 a_2}$, the boundary consists of two components which are symmetric graphs over the x-axis. We write the graphs in the form $y = \pm f(x)$ near $a_\pm$.

In the multiple reflection expansion of the exterior resolvent $R_D^{\Omega^c}(k + i\tau)$, we will encounter multiple integrals $\int_{(\partial\Omega)^M}$. Since $\partial\Omega$ consists of two components, each such term breaks up into 2^M multiple integrals over the circle $\mathbf{T}$. The terms can be enumerated by maps $\sigma : \{1, \ldots, M\} \to \{\pm\}$. We needed the same enumeration in [Z2] to denote the two local components of Ω at the ends of the bouncing ball orbit.

2.1. Length functional. For each map $\sigma : \{1, \ldots, M\} \to \{\pm\}$ as above, we define a length functional on $\mathbf{T}^M$ by:

$$\begin{aligned} L_\sigma(\varphi_1, \ldots, \varphi_M) \\ = |q_{\sigma_1}(\varphi_1) - q_{\sigma_2}(\varphi_2)| + \cdots + |q_{\sigma_{M-1}}(\varphi_{M-1}) - q_{\sigma_M}(\varphi_M)|. \end{aligned} \tag{4}$$

It is clear that L_σ is a smooth function if $\sigma(j) \neq \sigma(j+1)$ for any j. When there do exist such j, then L_σ is singular on the 'large diagonal' $\Delta_{j,j+1} := \{\varphi_j = \varphi_{j+1}\}$, where it has $|x|$ singularities . A standard (and easy) calculation shows that

$$\begin{aligned} \frac{\partial}{\partial \varphi_j} L_\sigma &= \sin \angle(q_{\sigma_{j+1}}(\varphi_{j+1}) - q_{\sigma_j}(\varphi_j), \nu_{q(\varphi_{j+1})}) \\ &\quad - \sin \angle(q_{\sigma_j}(\varphi_j) - q_{\sigma_{j-1}}(\varphi_{j-1}), \nu_{q_{\sigma_j}(\varphi_j)}). \end{aligned} \tag{5}$$

Here we denote the acute angle between the link $q_{\sigma_{j+1}}(\varphi_{j+1}) - q_{\sigma_j}(\varphi_j)$ and the inward unit normal $\nu_{q_{\sigma_{j+1}}(\varphi_{j+1})}$ by $\angle(q_{\sigma_{j+1}}(\varphi_{j+1}) - q_{\sigma_j}(\varphi_j), \nu_{q_{\sigma_j}(\varphi_j)})$. The condition that $\frac{\partial}{\partial \varphi_j} L = 0$ is thus that the 2-link defined by the triplet $(q_{\sigma_{j-1}}(\varphi_{j-1}, q_{\sigma_j}(\varphi_j), q_{\sigma_{j+1}}(\varphi_{j+1})$ is *Snell* at φ_j, i.e. satisfies the law of equal angles at this point. A smooth critical point of L on $\mathbf{T}^M$ is thus the same as an M-link Snell polygon. Note that Snell polygons include polygons which have links inside the obstacles. They are known as 'ghost orbits' in the physics literature, since they cross the boundary between the interior and exterior domains. One may also include singular critical points of L_σ which correspond to polygons in which an edge is collapsed.

2.2. Length spectrum. We will denote by $Lsp(\Omega^c)$, resp. $Lsp(\Omega)$, the lengths of periodic billiard trajectories in Ω^c, resp. Ω. We do not include ghost orbits. By periodic trajectory, we mean closed bicharacteristic curve; for the definition see [GM, PS]. Special periodic trajectories are m-link periodic reflecting ray, i.e. m-link Snell polygons. Their lengths are isolated in the length spectrum.

When Ω^c is the complement of a pair of disjoint, strictly convex obstacles, the only exterior transversal reflecting rays are iterates γ^r of the bouncing ball orbit γ between the components, i.e. the line segment between the points $a_\pm \in \mathcal{O}_\pm$. When $\mathcal{O}$ is not assumed convex, there may be additional periodic reflecting rays and the shortest periodic orbit of Ω^c need not be the invariant bouncing ball orbit. Recall that we assume throughout that L_γ is isolated in $Lsp(\Omega) \cup Lsp(\Omega^c)$.

It is part of the folklore of the Balian-Bloch approach that contributions to the trace formula due to ghost orbits are exactly cancelled by other ghost orbits. We will not prove that this cancellation occurs since the rigorous Poisson relation already implies that only γ^r contribute to the wave trace. Additionally, there are so-called diffraction orbits, which have points of tangential intersection with the boundary and may glide along the boundary. They are not critical points of the length functionals, and correspond not to smooth singularities in the wave trace, but to analytic singularities.

3. Poisson relation for the exterior problem. We now recall the Poisson relation for exterior domains in dimension 2.

The exterior *Green's kernel* $G_B^{\Omega^c}(k+i\tau, x, y) \in \mathcal{D}'(\Omega^c \times \Omega^c)$ with boundary condition $B = D$, or N (Dirichlet or Neumann) is the kernel of the exterior resolvent $R_B^{\Omega^c}(k+i\tau) = -(\Delta_B^{\Omega^c} + (k+i\tau)^2)^{-1} : H^s(\Omega^c) \to H^{s+2}(\Omega^c)$. Here, $k \in \mathbb{R}$ and $\tau > 0$. Its kernel (the exterior Green's function) may be characterized as the unique solution of the boundary problem:

$$(6) \quad \begin{cases} -(\Delta_B^{\Omega^c} + (k+i\tau)^2) G_B^{\Omega^c}(k+i\tau, x, y) = \delta_y(x), \quad (x, y \in \Omega) \\[2ex] BG_B^{\Omega^c}(k+i\tau, x, y) = 0, \quad x \in \partial\Omega \\[2ex] \dfrac{\partial G_B^{\Omega^c}(k+i\tau, x, y)}{\partial r} - i(k+i\tau) G_B^{\Omega^c}(k+i\tau, x, y) = o(\dfrac{1}{r}), \text{ as } r \to \infty. \end{cases}$$

Here, the boundary operator could be either $Bu = u|_{\partial\Omega^c}$ (D) or $Bu = \partial_\nu u|_{\partial\Omega^c}$ (N). We use a similar notation for the interior resolvent and boundary conditions. The boundary conditions D and N are in a certain sense complementary, and we will write B' for the complementary boundary condition to B.

We are interested in the regularized distribution trace of the combined operator $R_B^{\Omega^c}(k+i\tau) \oplus R_{B'}^{\Omega}(k+i\tau)$, more usually regarded as a distribution trace of the combined wave group $E_B^{\Omega^c}(t) \oplus E_{B'}^{\Omega}(t)$, where $E_B^{\Omega^c}(t) = \cos t\sqrt{-\Delta_B^{\Omega^c}}$). Let $\hat{\rho} \in C_0^\infty(\mathbb{R}^+)$ and define

$$
\begin{aligned}
&R_{\rho B}^{\Omega^c}(k+i\tau) \oplus R_{\rho B'}^{\Omega}(k+i\tau) \\
&\qquad = \int_{\mathbb{R}} \rho(k-\mu)(\mu+i\tau) R_B^{\Omega^c}(\mu+i\tau) \oplus R_{\rho B'}^{\Omega}(\mu+i\tau)d\mu.
\end{aligned} \tag{7}
$$

From the resolvent identity (e.g.)

$$
R_B^{\Omega^c}(\mu+i\tau) = \frac{1}{\mu+i\tau}\int_0^\infty e^{i(\mu+i\tau)t} E_B^{\Omega^c}(t)dt,
$$

it follows that

$$
R_{\rho B}^{\Omega^c}(k+i\tau) \oplus R_{\rho B'}^{\Omega}(k+i\tau) = \int_0^\infty \hat{\rho}(t) e^{i(k+i\tau)t} E_B^{\Omega^c}(t) \oplus E_{B'}^{\Omega}(t)dt. \tag{8}
$$

3.1. Poisson relation and scattering phase. The following is the Birman-Krein formula for the relative trace of the wave group of an exterior problem, as generalized by Lax-Phillips and Bardos-Guillot-Ralston (see [BGR], Theoreme 3) : Let $\hat{\varphi} \in C_0^\infty(\mathbb{R})$. Then:

$$
Tr \int_{\mathbb{R}} \hat{\varphi}(t)[E_B^{\Omega^c}(t) \oplus E_{B'}^{\Omega}(t) - E_0(t)]dt = \frac{1}{2\pi}\int_{\mathbb{R}} \varphi(\lambda)[ds_B(\lambda) + dN_{B'}(\lambda)].
$$

Putting $\hat{\varphi}(t) = \hat{\rho}(t)e^{i(k+i\tau)t}$ and rewriting in terms of the resolvent, we get:

$$
\begin{aligned}
&Tr[R_{B\rho}^{\Omega^c}(k+i\tau) \oplus R_{B'\rho}^{\Omega}(k+i\tau) - R_{0\rho}(k+i\tau)] \\
&\qquad = \frac{1}{2\pi}\int_{\mathbb{R}} \rho(k+i\tau-\lambda)[ds_B(\lambda) + dN_{B'}(\lambda)].
\end{aligned} \tag{9}
$$

Here, $s_B(\lambda) = \log\det S_B(\lambda)$ where $S_B(E)$ is the scattering operator and $N_{B'}$ is the interior Weyl counting function with the indicated boundary conditions.

The singular support of the regularized wave trace $Tr[E_B^{\Omega^c}(t) - E_0(t)]$ is contained in the set of lengths of exterior periodic generalized billiard trajectories. That is, by [BGR], Theoreme 5, we have

$$
singsupp\, Tr[E_B^{\Omega^c}(t) - E_0(t)] \subset Lsp(\Omega^c).
$$

Similarly for the interior wave group with $Lsp(\Omega)$ replacing $Lsp(\Omega^c)$. When L_γ is the length of a non-degenerate periodic reflecting ray γ, and when L_γ is not the length of any other generalized periodic orbit, then $Tr[E_B^{\Omega^c}(t) - E_0(t)]$ is a Lagrangian distribution in the interval $(L_\gamma - \epsilon, L_\gamma + \epsilon)$ for sufficiently small ϵ, hence $Tr[E_{\rho B}^{\Omega^c}(k+i\tau) - E_{\rho 0}(t)]$ has a complete asymptotic expansion in powers of k^{-1}. Let us recall the statement in the exterior case (see [GM], Theorem 1, and [PS] Theorem 6.3.1 for the interior case, and [BGR], §6 for the exterior case). Let γ be a non-degenerate billiard trajectory whose length L_γ is isolated and of multiplicity one in $Lsp(\Omega)$. Let Γ_L be a sufficiently small conic neighborhood of $\mathbb{R}^+\gamma$ and let χ be a microlocal cutoff to Γ_L. Then for t near L_γ, the trace of the wave group has the singularity expansion

$$(10)\quad \begin{aligned} Tr\chi[E_B^{\Omega^c}(t)-E_0(t)] \;\sim\; & a_\gamma(t-L_\gamma+i0)^{-1}+a_{\gamma 0}\log(t-L_\gamma+i0) \\ & +\sum_{k=1}^{\infty} a_{\gamma k}(t-L_\gamma+i0)^k\log(t-L_\gamma+i0) \end{aligned}$$

where the coefficients $a_{\gamma k}$ are calculated by the stationary phase method from a Lagrangian parametrix. In the interior case, one of course omits the term $E_0(t)$.

The result may be re-stated as follows: Let $\hat{\rho} \in C_0^\infty(L_\gamma - \epsilon, L_\gamma + \epsilon)$, equal to one on $(L_\gamma - \epsilon/2, L_\gamma + \epsilon/2)$ and with no other lengths in its support. Then the interior trace $TrR^{\Omega}_{B\rho}(k + i\tau)$ and the exterior trace $Tr[R^{\Omega^c}_{B\rho}(k + i\tau) - R_{0\rho}(k + i\tau)]$ admit complete asymptotic expansions of the form

$$(11)\quad \begin{cases} Tr[R^{\Omega^c}_{B\rho}(k+i\tau) - R_{0\rho}(k+i\tau)] \sim \sum_{j=0}^{\infty} B_{\gamma;j}k^{-j} \\ TrR^{\Omega}_{B\rho}(k+i\tau) \sim e^{i(k+i\tau)\tau L_\gamma}\sum_{j=0}^{\infty} B_{\gamma;j}k^{-j}, \end{cases}$$

whose coefficients $B_{\gamma;j}$ are canonically related to the wave invariants $a_{\gamma;j}$ of periodic (internal, resp. external) billiard orbits. We have removed the cutoff operator χ since there are no singularities of the trace at $t = L_\gamma$ in the microsupport of $I - \chi$. The coefficients depend on the choice of boundary condition but we do not indicate this in the notation.

3.2. Poisson relation and resonances. We now recall the Poisson relation in dimension 2. Let $\{\lambda_j\}$ denote the resonances of $\Delta_B^{\Omega^c}$, i.e. the poles of the analytic continuation of $R_B^{\Omega^c}(k + i\tau)$ from $\{\tau > 0\}$ to the logarithmic plane Λ. Let θ denote a (small) angle, and let Λ_θ denote the conic neighborhood of the real axis in Λ consisting of points with angular coordinates in $(-\theta, \theta)$. Further, let $\hat{\varphi} \in C_0^\infty(\mathbb{R}^+)$ and let $E_0(t)$ denote the free even wave kernel. The global Poisson formula (asserts the following:

PROPOSITION 3.1. *(see Zworski [Zw], Theorem 1)*

$$\begin{aligned} Tr(\int_{\mathbb{R}} \hat{\varphi}(t)(E_B^{\Omega^c}(t) - E_0(t))dt = & \sum_{\lambda_j\in\Lambda_\theta} m(\lambda_j)\varphi(\lambda_j) + m(0)\varphi(0) \\ & + 2\int_0^\infty \psi(\lambda)\varphi(\lambda)\frac{ds_B}{d\lambda}d\lambda + \int_0^\infty \hat{\varphi}(t)\nu_{\theta,\psi}(t)dt, \end{aligned}$$

$$\textit{with}\quad \nu_{\theta,\psi} \in C^\infty(\mathbb{R}\backslash\{0\}),\;\; \partial_t^k\nu_{\theta,\psi} = O(t^{-N}),\;\; |t| \to \infty,$$

where s_B *denotes the scattering phase of* $\Delta_B^{\Omega^c}$, *and where* $m(\lambda)$ *denotes the multiplicity of the resonance* λ. *Also,* $\psi \in C_0^\infty(\mathbb{R})$ *is a cutoff which equals* 1 *for* t *near* 0.

(We note that the sum over eigenvalues term in the formula stated in [Zw] (Theorem 1) is absent in the case of an exterior domain.)

Substituting $\hat{\varphi}(t) = \hat{\rho}(t)e^{i(k+i\tau)t}$, with $\hat{\rho} \in C_0^\infty(\mathbb{R}^+)$ with supp$\hat{\rho}$ sufficiently close to rL for some r, into (3.1) and using (8), we obtain:

$$(12)\quad \begin{aligned} &Tr[R_{\rho B}^{\Omega^c}(k+i\tau) - R_{0\rho}(k+i\tau)] \\ &\qquad = \sum_{\lambda_j \in \Lambda_0} m(\lambda_j)\rho(\lambda + k + i\tau) + O(|k|^{-M}), \quad (k \to \infty). \end{aligned}$$

In the remainder estimate we use that

$$m(0)\rho(k+i\tau), \quad \int_0^\infty \psi(\lambda)\rho(\lambda+k+i\tau)\frac{ds_B}{d\lambda}d\lambda, \quad \langle \nu_{\theta,\psi}, \hat{\rho}(t)e^{i(k+i\tau)t}\rangle = O(k^{-\infty}).$$

In the case of the first two terms, this follows immediately from the fact that $\varphi(\lambda) = \rho(k+i\tau-\lambda)$ and that both terms are integrals over a compact set of values of λ. For the third term we additionally use that $\hat{\rho}$ vanishes in a neighborhood of $t=0$; integration by parts and the estimate on on $\nu_{\theta,\psi}$ then gives the rapid decay in k of the term. It follows that the asymptotics of the regularized resolvent trace $Tr[R_{\rho B}^{\Omega^c}(k+i\tau) - R_{0\rho}(k+i\tau)]$ are an invariant of the resonance poles.

The following well-known proposition, explained to the author by M. Zworski, implies more: only poles in a logarithmic neighborhood $\{|\Im\lambda_j| > N\log|\lambda_j|\}$ of the real axis cause singularities in the wave trace:

PROPOSITION 3.2. *For any* $N \geq 0$, *there exists* $k(N)$ *with* $k(N) \to \infty$ *as* $N \to \infty$, *such that*

$$\sum_{\lambda_j : |\Im\lambda_j| < \rho|\lambda_j|, |\Im\lambda_j| > N\log|\lambda_j|} e^{i\lambda_j t} \in C^{k(N)}(\mathbb{R} - \{0\}).$$

Proof. We have

$$(13)\quad \begin{aligned} &|\partial_t^k \sum_{\lambda_j : |\Im\lambda_j| < \rho|\lambda_j|, |\Im\lambda_j| > N\log|\lambda_j|} e^{i\lambda_j t}| \\ &\qquad \leq \sum_{\lambda_j : |\Im\lambda_j| < \rho|\lambda_j|, |\Im\lambda_j| > N|\lambda_j|} |\lambda_j|^k |\lambda_j|^{-Nt} \\ &\qquad \leq \int_{\mathbb{R}^+} r^{k-tN} dN(r) \leq C \int_{\mathbb{R}^+} r^{n+k-1-tN} dr. \end{aligned}$$

In the last line we used a polynomial bound on the number of resonance poles (cf. [Zw] (2.1) for the estimate). Thus, the portion of the wave trace coming from poles outside of sufficiently large logarithmic neighborhoods of the real axis is as smooth as desired. □

4. Neumann expansion and Poisson relation. In the Balian-Bloch approach to the Poisson relation, one uses the classical theory of layer potentials ([TII] §7. 11) to express $G_D^{\Omega^c}(k+i\tau,x,y)$ in terms of the 'free' Green's function $G_0(k+i\tau,x,y)$ of $\mathbb{R}^2$, i.e. the kernel of the free resolvent $(\Delta_0 - k + i\tau^2)^{-1}$ of the Laplacian Δ_0 on $\mathbb{R}^2$. A classical reference for this approach is the paper of Pleijel [P]. This approach is also used in [M] to make a reduction to the boundary in studying resonance poles and in the duality result of (3). We review the necessary background and notation.

4.1. Potential theory. We recall that the free Green's function in dimension two is given by: $G_0(k+i\tau,x,y) = H_0^{(1)}(k+i\tau|x-y|)$, where $H_0^{(1)}(z)$ is the Hankel function of index 0. We recall its small and large distance asymptotics ([TI], Chapter 3, §6):

$$H_\nu^{(1)}(r) = \begin{cases} -\dfrac{1}{2\pi}\ln(r) \text{ as r} \to 0, \text{ if } \nu = 0 \\[2ex] -\dfrac{i\Gamma(\nu)}{\pi}(\dfrac{2}{r})^\nu \text{ as r} \to 0, \text{ if } \nu > 0 \\[2ex] \dfrac{e^{i(k+i\tau r-\nu\pi/2-\pi/4)}}{r^{1/2}}\left(1+O\Big(\dfrac{1}{|r|^3}\Big)\right) \text{ as } |\text{k}+\text{i}\tau|\text{r} \to \infty. \end{cases} \tag{14}$$

Here it is assumed that $\tau > 0$. Thus, $G_0(k+i\tau,x,y)$ has two kinds of asymptotics: a semiclassical asymptotics for large $(k+i\tau)|x-y|$ and a singularity asymptotics along the diagonal. It is the first kind of asymptotics which accounts for the connection to billiard trajectories. The singularity along the diagonal will have to be regularized.

4.1.1. Layer potentials and boundary integral operators. The double layer potential is the operator

$$\mathcal{D}\ell(k+i\tau)f(x) = \int_{\partial\Omega}\frac{\partial}{\partial\nu_y}G_0(k+i\tau,x,q)f(q)ds(q), \tag{15}$$

from $H^s(\partial\Omega) \to H^{s+1/2}_{loc}(\Omega)$, where $ds(q)$ is the arc-length measure on $\partial\Omega$, where ν is the interior unit normal to Ω, and where $\partial_\nu = \nu\cdot\nabla$. It induces the boundary operator

$$N(k+i\tau)f(q) = 2\int_{\partial\Omega}\frac{\partial}{\partial\nu_y}G_0(k+i\tau,q,q')f(q')ds(q') \tag{16}$$

which map $H^s(\partial\Omega) \to H^{s+1}(\partial\Omega)$.

By the explicit formula we have:

$$\begin{aligned} \frac{1}{2}N(k+i\tau,q(\varphi),q(\varphi')) &= \partial_{\nu_y}G_0(\mu,q(\varphi),q(\varphi')) \\ &= -(k+i\tau)H_1^{(1)}(k+i\tau|q(\varphi)-q(\varphi')|) \\ &\quad \times \cos\angle(q(\varphi)-q(\varphi'),\nu_{q(\varphi)}). \end{aligned}$$

We will need the 'jump' formula of potential theory ([TI] Chapter 7.11, (11.7)),

$$(\mathcal{D}\ell(k+i\tau)u)_{\pm}(q) = \frac{1}{2}(N(k+i\tau) \mp I)u(q), \tag{17}$$

where

$$\begin{cases} f_+(x) := \lim_{x\to q, x\in\mathcal{O}} f(x), \\ \\ f_-(x) = \lim_{x\to q, x\in\Omega} f(x). \end{cases} \tag{18}$$

Note that the sign of each term in (17) depends on the choice of the interior/exterior: in $N(k+i\tau)$ it depends on the choice of interior/exterior unit normal and in the identity operator term it depends on whether the limit is taken from the interior or exterior. In the interior case we have $(\mathcal{D}\ell(k+i\tau)u)_{\pm}(q) = \frac{1}{2}(N(k+i\tau) \pm I)u$ where $N(k+i\tau)$ is defined using the interior normal and where $\pm$ have the same meanings as in the interior case.

4.2. Interior/Exterior duality. In [Z1, Z2] we used the classical reduction of the Dirichlet problem to the boundary to study the Dirichlet resolvent (as in [BB1, BB2]). We now use a similar method to obtain a convenient formula for the Fredholm determinant, $\det((I+N(k)))$. Similar ideas can be found in [THS, THS2, AG, GP].

The following formula is sometimes referred to as interior/exterior duality. Combined with the Birman-Krein formula (9) it expresses the growth rate of the inside plus outside spectra in terms of the determinant of a boundary integral operator:

PROPOSITION 4.1. *For any $\tau \geq 0$, the operator $(I+N(k+i\tau))$ is of trace class and has a well-defined Frehdolm determinant, and we have:*

$$\begin{aligned} &Tr_{\mathbb{R}^2}[R^{\Omega^c}_{\rho D}(k+i\tau) \oplus R^{\Omega}_{N\rho} - R_{0\rho}(k+i\tau)] \\ &\qquad = \int_{\mathbb{R}} \rho(k-\lambda)\frac{d}{d\lambda}\log\det(I+N(\lambda+i\tau)d\lambda,. \end{aligned}$$

where $\det(I+N(\lambda+i\tau)$ *is the Fredholm determinant.*

Proof. We first argue formally. The interior/exterior resolvent kernels can be constructed in the classical way, as follows (see e.g. [P]). For simplicity, we consider the interior Dirichlet resolvent, but the construction for the exterior resolvent or with a change to Neumann boundary conditions is almost the same. We have:

$$\begin{aligned} R^{\Omega^c}_D(k+i\tau) &= 1_{\Omega^c}[R_0(k+i\tau) \\ &\qquad - \mathcal{D}\ell(k+i\tau)(I+N(k+i\tau))^{-1}\gamma R_0(k+i\tau)]1_{\Omega^c}, \\ (R^{\Omega}_N(k+i\tau))^{tr} &= 1_{\Omega}[R_0(k+i\tau) \\ &\qquad - \mathcal{D}\ell(k+i\tau)(I+N(k+i\tau))^{-1}\gamma R_0(k+i\tau)]1_{\Omega}, \end{aligned} \tag{19}$$

where γ denotes the restriction to the boundary taken from within the relevant domain. When we take the regularized trace, we first subtract from $R_N^{\Omega}(k+i\tau)\oplus R_D^{\Omega^c}(k+i\tau)$ the free operator $R_0(k+i\tau)$, which removes the first terms on the right. We then cycle the factor $\gamma R_0(k+i\tau)$ from the right to the left side, obtaining an operator on $\partial\Omega$. The inside and outside terms add up to the kernel

$$\int_{\mathbb{R}^2} G_0(k+i\tau,q',y)\partial_{\nu_q}G_0(k+i\tau,y,q)dy. \tag{20}$$

This indeed is why the interior Dirichlet and exterior Neumann problems were combined and explains the sense in which they are complementary. We now claim that the resulting kernel equals $\frac{1}{2k}\frac{d}{dk}N(k+i\tau,q',q)$. To prove this we note that

$$\frac{d}{d\lambda}R_0(\lambda) = 2\lambda(\Delta+\lambda^2)^{-2} = 2\lambda R_0(\lambda)^2.$$

Hence,

$$\frac{d}{d\lambda}N(\lambda) = \frac{d}{d\lambda}\gamma\partial_{\nu_q}R_0(\lambda)\gamma = 2\lambda\gamma\partial_{\nu_q}R_0(\lambda)\circ R_0(\lambda)\gamma$$

which is precisely the kernel (20). Hence the right side of (19) equals $\frac{1}{k+i\tau}\frac{d}{dk}\log(I+N(k+i\tau))$. The final formula follows by putting $k=\lambda$, integrating against $\rho(k-\lambda)(\lambda+i\tau)$ and noting the cancellation of the second factor. □

To justify the formal manipulations, we need to show that the relevant traces and determinant are well defined. This will be done in a sequence of Lemmas.

LEMMA 4.2. *For any τ, $N(k+i\tau)\in\mathcal{I}_1(\partial\Omega)$, the ideal of trace class operators on $L^2(\partial\Omega)$. Hence, $\det(I+N(k+i\tau))$ is well-defined as a Fredholm determinant.*

Proof. In [Z1] it is proved that $N(k+i\tau)\in\Psi^{-2}(\partial\Omega)$ and this immmediately implies that it is of trace class. In fact, $N(k+i\tau,q,q')$ has just a $|q-q'|\log|q-q'|$ singularity on the diagonal. That it is trace class also follows from the Hille-Tamarkin theorem (see [Z1] for references).

It is then a classical remark that the Fredholm determinant $\det(I+N(k+i\tau))$ is well-defined, in fact $|\det(I+N(k+i\tau))|\leq\exp(||N(k+i\tau)||_1)$, where $||\cdot||_1$ is the trace norm. See [S], Lemma 3.3. □

REMARK 4.3.

1. *Note that the usual statement (cf. [TII]) is that $N(k+i\tau)$ is of order -1. However, the principal symbol vanishes in dimension 2.*

2. *We recall (see [S], Theorem 3.10) that the Fredholm determinant is given by:*

$$\det(I+N(k+i\tau)) = \sum_{n=0}^{\infty}\frac{1}{n!}\int_{\partial\Omega^n}\det[N(k+i\tau,q_i,q_j)]_{1\leq i,j\leq n}ds(q_1)\cdots ds(q_n).$$

Again combining classically known facts, we have:

LEMMA 4.4. *For any $\tau > 0, \log\det N(k+i\tau)$ is well-defined and differentiable in k, $(I+N(k+i\tau))^{-1}N'(k+i\tau)$ is of trace class and we have:*

$$\frac{d}{dk}\log\det N(k+i\tau) = Tr_{\partial\Omega}(I+N(k+i\tau))^{-1}N'(k+i\tau).$$

Proof. First we note that $(I+N(k+i\tau))$ is invertible on $L^2(\partial\Omega)$ if $\tau > 0$, so $\det(I+N(k+i\tau)) \neq 0$. Hence its logarithm is well-defined. Differentiability of $\log\det(I+N(k+i\tau))$ and the formula for the derivative follows from the general fact that $\det(I+A)$ is Frechet differentiable on $\mathcal{I}_1$ with derivative equal to $(I+A)^{-1}$ if $-1 \notin \sigma(A)$ (here, $\sigma(A)$ is its spectrum; see [S], Corollary 5.2).

The singularity on the diagonal of $N(k+i\tau, q, q')$ is independent of $k+i\tau$, so the derivative is at least as regular. Hence, $N'(k+i\tau) \in \mathcal{I}_1$. The statement that $(I+N(k+i\tau))^{-1}N'(k+i\tau) \in \mathcal{I}_1$ follows from the fact that $(I+N(k+i\tau)^{-1} \in \Psi^0(\partial\Omega)$. $\square$

LEMMA 4.5. *The operators*

$$\begin{cases} (i) \quad 1_\Omega \int_{\mathbb{R}} \rho(k-\lambda)\mathcal{D}\ell(\lambda+i\tau)(I+N(\lambda+i\tau))^{-1}\gamma R_0(\lambda+i\tau)]d\lambda 1_\Omega, \\ (ii) \quad 1_{\Omega^c} \int_{\mathbb{R}} \rho(k-\lambda)\mathcal{D}\ell(\lambda+i\tau)(I+N(\lambda+i\tau))^{-1}\gamma R_0(\lambda+i\tau)]d\lambda 1_{\Omega^c}, \end{cases}$$

are of trace class and the sum of their traces equals $Tr_\Omega(I+N(k+i\tau))^{-1} \times N'(k+i\tau)$.

Proof. We expand $(I+N(\lambda+i\tau))^{-1} = I - N(\lambda)(I+N(\lambda+i\tau))^{-1}$. The I term gives us $\mathcal{D}\ell(\lambda+i\tau)\gamma R_0(\lambda+i\tau)]$. We break up each of $\mathcal{D}\ell(\lambda+i\tau)$ and $\gamma R_0(\lambda+i\tau)]$ into regular (continuous) and singular parts using the small z expansion (14) of the Hankel functions $H_0^1(z), H_z^1(z)$. The singular parts of the kernels are independent of $(k+i\tau)$. Hence the product of the singular parts gets multiplied by $\hat{\rho}(0) = 0$ upon integration against $\rho(k-\lambda)$. Therefore only the regular parts survive the convolution, and these are of trace class. The remaining term is a bounded operator composed with trace class operator for each λ, so it is of trace class for each λ.

We may then cycle the factor of $\mathcal{D}\ell(\lambda+i\tau)$ (or each of its regular and singular parts) to the right side of the traces. After doing so, we reassemble the regular and singular parts into $\gamma R_0(\lambda+i\tau)]\mathcal{D}\ell(\lambda+i\tau)$and we reassemble $I - N(\lambda)(I+N(\lambda+i\tau))^{-1}$ into $(I+N(\lambda+i\tau))^{-1}$. The calculation at the beginning of the proof then gives the stated formula. $\square$

This concludes the proof of Proposition 4.1.

COROLLARY 4.6. *Suppose that L_γ is the only length in the support of $\hat{\rho}$. Then,*

$$\int_{\mathbb{R}} \rho(k-\lambda)\frac{d}{d\lambda}\log\det(I+N(\lambda+i\tau))d\lambda \sim \sum_{j=0}^{\infty} B_{\gamma;j}k^{-j},$$

where $B_{\gamma;j}$ are the wave invariants of γ in (11).

To prove Theorems (1.1) and (1.4), it thus suffices to determine Ω from the integrals in Corollary (4.6). Since we have combined the interior/exterior, we emphasize that our inverse result only assumes knowledge of the resonances: Indeed, suppose the resonance poles of Ω^c are known. Then by Proposition (3.1), the asymptotics of the exterior resolvent trace in terms of exterior periodic orbits are known. But by Proposition (4.1), these are the same as the asymptotics of the integrals in Corollary (4.6).

5. Trace asymptotics. We now explain how to use Corollary (4.6) to calculate the coefficients $B_{\gamma,j}$. We write

$$\frac{d}{d\lambda}\log\det(I+N(\lambda+i\tau)) = Tr_{\partial\Omega}(I+N(\lambda+i\tau))^{-1}N'(k+i\tau),$$

and then expand $(I+N(k+i\tau))^{-1}$ in a finite geometric series plus remainder:

$$\begin{aligned}(21)\qquad &(I+N(\lambda+i\tau))^{-1}\\ &\quad=\sum_{M=0}^{M_0}(-1)^M\,N(\lambda)^M+(-1)^{M_0+1}\,N(\lambda)^{M_0+1}(I+N(\lambda+i\tau))^{-1}.\end{aligned}$$

We now argue that for each order k^{-J} in the trace expansion of Corollary (4.6) there exists $M_0(J)$ such that

$$\begin{aligned}(22)\qquad &(i)\quad \sum_{M=0}^{M_0}(-1)^M Tr\int_{\mathbb{R}}\rho(k-\lambda)\,N(\lambda)^M N'(k+i\tau)d\lambda\\ &\qquad\qquad=\sum_{j=0}^{J}B_{\gamma;j}k^{-j}+O(k^{-J-1}),\\ &(ii)\quad Tr\int_{\mathbb{R}}\rho(k-\lambda)N(\lambda)^{M_0+1}(I+N(\lambda+i\tau))^{-1}N'(k+i\tau)d\lambda\\ &\qquad\qquad=O(k^{-J-1}).\end{aligned}$$

We will sketch the proof of (i) in some detail since it explains how to calculate the coefficients $B_{\gamma^r,j}$. For a discussion of the remainder (in an analogous but not identical calculation) we refer to [Z2].

To simplify the notation, we integrate by parts in (i) to throw the derivative onto ρ, and then the issue is to analyse the traces:

$$\begin{aligned}(23)\qquad &Tr\int_{\mathbb{R}}\rho(k-\lambda)N(\lambda+i\tau)^M d\lambda\\ &\quad=\int_{\mathbb{R}}\int_{(\partial\Omega)^M}\rho(k-\lambda)[\Pi_{j=1}^{M}N(\lambda+i\tau,q_j,q_{j+1})ds(q_j)]d\lambda\\ &\qquad\qquad\text{(where } q_{M+1}=q_1);\end{aligned}$$

Since

$$N(k+i\tau):L^2(\mathcal{O}_+)\oplus L^2(\mathcal{O}_-)\to L^2(\mathcal{O}_+)\oplus L^2(\mathcal{O}_-)$$

we write it as

$$N(k) = \begin{pmatrix} N_{++}(k) & N_{+-}(k) \\ N_{-+}(k) & N_{--}(k) \end{pmatrix}.$$

Then (23) is a sum of terms $\sum_{\sigma:\{1,\dots,M\}\to\{\pm1\}} I^{\sigma}_{M,\rho}$ with

$$(24)\quad I^{\sigma}_{M,\rho} = \int_{\mathbb{R}} \int_{\mathbf{T}^M} \rho(k-\lambda)[\Pi_{j=1}^{M} N_{\sigma_j,\sigma_{j+1}}(\lambda+i\tau, q_{\sigma_j}(\varphi_j), q_{\sigma_{j+1}}(\varphi_{j+1})d\varphi_j]d\lambda.$$

Here and hereafter, indices are understand modulo M. Using the asymptotics of the free Green's function in (14)-(27), the σth term is formally an oscillatory integral with phases given by the length functional (4). We would like to calculate the trace asymptotically by the stationary phase method.

As discussed extensively in [Z1], we cannot immediately do so because of the singularities along the diagonals of the integrand. We therefore first need to de-singularize the integrals. One of the main advantages of the present inside/outside reduction to the boundary is that it simplifies the regularization procedure by eliminating the integral over Ω in [Z1]. We briefly outline the method and refer the reader to [Z1] for further details.

5.1. Boundary integral operators as quantized billiard maps. The operators $N_{-+}(k+i\tau), N_{+-}(k+i\tau)$ are semiclassical Fourier integral operators with phases $|q_+(\varphi) - q_-(\varphi')|$. The phase (and amplitude) is non-singular since the boundary components are disjoint. The phase in the $+-$ case parametrizes the graph of the following multi-valued billiard map from $B^*\mathcal{O}_+$ to $B^*\mathcal{O}_-$: given $(q_+, v_+) \in T_q\mathcal{O}_+$ with $|v| < 1$, add a multiple of the unit outward normal to turn v into an outward unit vector and proceed along the straight line in that direction and let q_- be a point of intersection of this line with $\mathcal{O}_-$. Let v_- be the tangential projection of the terminal velocity vector at q_- , and put $\beta_{+-}(q_+, v_+) = (q_-, v_-)$. We have (deliberately) described β_{+-} in an ambiguous way: when q_- is the first intersection point, we have the usual exterior billiard map; but the phase actually parametrizes the canonical relation which includes intersection points which occur after the line enters the interior of $\mathcal{O}_-$. As mentioned above, these are 'ghost orbits' of the billiard flow and they cancel out of the trace formula and do not contribute to the wave invariants we are calculating.

The operators $N_{++}(k+i\tau)$ and $N_{--}(k+i\tau)$ are more complicated, and also, as it turns out, less important. Each is a combination of a homogeneous pseudodifferential operator of order -1 with singularity on the diagonal and a Fourier integral operator of order 0 which 'quantizes' the billiard map of the interiors of Ω and Ω^c. They are exactly the kind of operators discussed extensively in [Z1, HZ]. There is a further transition region between these two regimes in which it behaves like an Airy operator, but this region will not be important for our problem.

To separate the two basic regions (tangential and transverse) of N_{++} and N_{--}, we introduce a cutoff of the form $\chi(k^{1-\delta}(\varphi-\varphi'))$ in terms of the arc-length coordinate φ used in the parameterization $q(\varphi)$. Here, χ a smooth bump function which is supported on $[-1,1]$ and equals 1 on $[-1/2,1/2]$. We then write (dropping the subscripts)

$$(25)\quad \begin{cases} N(k+i\tau)=N_0(k+i\tau)+N_1(k+i\tau), \\ N_0(k+i\tau,q(\varphi),q(\varphi'))=\chi(k^{1-\delta}(\varphi-\varphi'))N(k+i\tau,q(\varphi),q(\varphi')), \\ N_1(k+i\tau,q(\varphi),q(\varphi'))=(1-\chi(k^{1-\delta}(\varphi-\varphi'))N(k+i\tau,q(\varphi),q(\varphi')). \end{cases}$$

Each term is a (non-standard) Fourier integral operator with amplitudes belonging to the following symbol class: We denote by $S^p_\delta(\mathbf{T}^m)$ the class of symbols $a(k,\varphi_1,\ldots,\varphi_m)$ satisfying:

$$(26)\qquad |(k^{-1}D_\varphi)^\alpha a(k,\varphi)|\leq C_\alpha|k|^{p-\delta|\alpha|},\quad (|k|\geq 1).$$

This follows from the asymptotics of Hankel functions (14), which may be described as follows: there exist amplitudes a_0,a_1 satisfying

$$(1-\chi(k^{1-\delta}z))a_1((k+i\tau)z)\in S^0_\delta(\mathbb{R}),$$
$$(1-\chi(k^{1-\delta}z))a_0((k+i\tau)z)\in S^{-1/2}_\delta(\mathbb{R})$$

and such that

$$(27)\quad \begin{cases} (i)\quad H_0^{(1)}((k+i\tau)z)=e^{i((k+i\tau)z}a_0((k+i\tau)z), \\ (ii)\quad (k+i\tau)H_1^{(1)}((k+i\tau)z)=(k+i\tau)^{\frac{1}{2}}e^{-i((k+i\tau)z}a_1((k+i\tau)z). \end{cases}$$

The N_0 term has a singularity on the diagonal of a pseudodifferntial operator of order -2. The N_1 term is manifestly an oscillatory integral operator of order 0 with phase $|q(\varphi)-q(\varphi')|$. As discussed in the $-+$ case, it generates the interior billiard map (interpreted in the multivalued sense of the $-+$ case).

5.2. Microlocalization to γ. So far, our considerations have been global. But our goal is to calculate wave invariants at a bouncing ball orbit, and for this purpose we can microlocalize the trace to a neighborhood of γ. This obvious sounding statement is somewhat problematic in the present approach since the relevant operators are not standard pseudodifferential or Fourier integral operators.

The billiard map β_{+-} is a cross section of the billiard flow, and in this cross section γ corresponds to the periodic points $(0,0)\in B^*\mathcal{O}_+$ and $(0,0)\in B^*(\mathcal{O}_-)$ of period 2 for β_{+-}. Here, we choose the parametrizations of $\mathcal{O}_\pm$ such that $\varphi = 0$ at each endpoint of the bouncing ball orbit; it is normal to $\mathcal{O}_\pm$, so its tangential part is the vector $(0,0)$. To microlocalize to this orbit we introduce a (block diagonal) semiclassical pseudodifferential cutoff operator

$$\chi(\varphi, k^{-1}D_\varphi) = \begin{pmatrix} \chi_{++} & 0 \\ 0 & \chi_{--} \end{pmatrix}$$

with complete symbol $\chi(\varphi,\eta)$ supported in $V_\epsilon := \{(\varphi,\eta) : |\varphi|, |\eta| \le \epsilon\}$. We also write

$$\rho * \frac{d}{d\lambda} \log\det(I + N(k+i\tau)) = Tr\rho * (I + N(k+i\tau))^{-1} \circ \frac{d}{dk} N(k+i\tau).$$

Recall that to test whether $(L,\tau) \in WF(u)$ (wave front set) for $u(t) \in \mathcal{S}'(\mathbb{R})$, one checks whether $\mathcal{F}(\hat{\rho}u)(k) \sim 0$ for all $\hat{\rho} \in C_0^\infty(\mathbb{R})$ satisfying $\hat{\rho} \equiv 1$ in some interval $(L-\epsilon, L+\epsilon)$. Here, $\mathcal{F}u = \hat{u}$ denotes the Fourier transform. Equivalently, one checks whether $\rho * \hat{u} \sim 0$. We now show that in calculating wave invariants we can microlocalize our non-standard operators to γ in the standard way.

As a preliminary step, we prove the simpler relation

$$\rho * Tr[\mathcal{D}(k+i\tau)(I + N(k))^{-1tr}\mathcal{S}(k+i\tau)^{tr} \circ (1 - \tilde{\chi}_\gamma(k)] \sim 0 \tag{28}$$

where $\tilde{\chi}_\gamma(k)$ is a semiclassical cutoff in Ω^c to a microlocal neighborhood of γ which has the form

$$\tilde{\chi}_\gamma(k) = \tilde{\chi}(r, y, k^{-1}D_y)$$

near the boundary. Here, (r,y) are Fermi normal coordinates in Ω^c near the endpoints of γ , with r the distance to the boundary. Thus, differentiations are only in tangential directions. Note that (28) is equivalent to saying that

$$Tr[R_{\rho D}^{\Omega^c} - 1_{\Omega^c} R_{0\rho}] \circ (I - \tilde{\chi}_\gamma(k)) \sim 0, \tag{29}$$

In view of (19), (29)-(28) follow from standard wave front set (WF) considerations for the exterior even wave group $E_D^{\Omega^c}(t,x,y)$, and from our assumption that γ^r is the only exterior periodic orbit of its length. Recall that $WF(E_D^{\Omega^c}(t,x,y))$ is the space-time graph

$$\Gamma = \{(t,\tau,x,\xi,x',\xi') \in T^*(\mathbb{R}\times\Omega^c\times\Omega^c) : \tau = -|\xi|, \ G^t(x,\xi) = (x',\xi')\}, \tag{30}$$

of the generalized billiard flow in the exterior. The only unusual feature is our particular choice of cutoff operator to microlocalize to γ. To verify that it behaves correctly like a microlocal cutoff to γ we write

$$R_{\rho D}^{\Omega^c}(k)\tilde{\chi}_\gamma(k) = \int_0^\infty \hat{\rho}(t) E_D^{\Omega^c}(t) \tilde{\chi}_\gamma(r, y, |D_t|^{-1}D_y) e^{i(k+i\tau)t} dt. \tag{31}$$

Now in Fermi coordinates $(r, \rho dr, y, \eta dy), \mathbb{R}^+\gamma$ is ray in the direction of dr. Since the space-time graph of the cotangent bundle along γ may be described in normal coordinates as a neighborhood of

$$\{(t, \tau, t, \tau, 0, 0)\}$$

a conic neighborhood may be described in these coordinates by $|y| \leq \epsilon, |\eta/\tau| \leq \epsilon$. This is precisely the set to which $\tilde{\chi}_\gamma(r, y, |D_t|^{-1}D_y)$ microlocalizes. Thus, (29) and hence (28) are correct.

We now show that this empty WF relation remains true after moving the cutoff to the left of $\mathcal{S}\ell(k)^{tr}$ as a cutoff pseudodifferential operator on the boundary to a neighborhood of the billiard map orbit. As mentioned above, this would be straightforward if $\mathcal{D}\ell(k)$ and the other operators in the composition were standard FIO's. However in addition to their semi-classical canonical relations they have singularities on the diagonal and both contribute to their WF. We will shorten the notation for the cutoff operator to χ_0.

LEMMA 5.1. *We have:*

$$Tr\rho * (I + N(k+i\tau))^{-1} \circ \tfrac{d}{dk}N(k+i\tau)$$
$$\sim Tr\rho * (I + N(k+i\tau))^{-1} \circ \chi_0(k) \circ \tfrac{d}{dk}N(k+i\tau).$$

Proof. We first express the various layer potentials in terms of the wave group:

$$\begin{cases} \mathcal{S}\ell(k+i\tau)^{tr} = \gamma G_0(k+i\tau) = \displaystyle\int_0^\infty e^{it(k+i\tau)}\gamma E_0(t)dt, \\ 1_{\Omega^c}(\mathcal{D}\ell(k+i\tau) \circ (I + N(k+i\tau))^{-1}(x,q) = P_N^{\Omega^c}(k+i\tau, x, q) \\ \qquad = \gamma_{\nu 2} G_N^{\Omega^c}(k+i\tau, x, q) = \displaystyle\int_0^\infty e^{it(k+i\tau)}\gamma_{\nu 2} E_N^{\Omega^c}(t)dt, \end{cases}$$

where $P_N^{\Omega^c}$ is the Neumann Poisson kernel of the exterior domain and where $\gamma_\nu u = \partial_\nu u|_{\partial\Omega^c}, \gamma u = u|_{\partial\Omega^c}$. We have subscripted the restriction operators to clarify which variables they operate on. The composition $\mathcal{D}\ell(k) \circ (I + N(k))^{-1} \circ (1 - \chi_\gamma(k)) \circ \mathcal{S}\ell(k)^{tr}$ may be written (with the relevant value of k) as

$$\int_0^\infty e^{it(k+i\tau)}\left\{\int_0^t \gamma_{\nu 2} E_N^{\Omega^c}(t-s) \circ (I - \chi_\gamma)(y, |D_t|^{-1}D_y)\gamma \circ E_0(s)ds\right\} dt. \tag{32}$$

We therefore have:

$$\begin{aligned} &\rho * \mathcal{S}\ell(k) \circ (I + N(k))^{-1tr} \circ (1 - \chi_\gamma(k)) \circ \mathcal{D}\ell(k)^{tr} \\ &= \int_0^\infty \hat{\rho}(t) \Bigg[\int_0^t \gamma_{\nu 2} E_N^{\Omega^c}(t-s) \circ (I - \chi_\gamma)(y, |D_t|^{-1}D_y)\gamma_{\nu 1} \\ &\qquad \circ E_0(s)ds \Bigg] e^{i(k+i\tau)t}dt. \end{aligned} \tag{33}$$

Our goal is to show that

$$WF\left[Tr\int_0^t \gamma_{\nu 2}E_N^{\Omega^c}(t-s)\circ(I-\chi_\gamma)(y,|D_t|^{-1}D_y)\gamma \circ E_0(s)ds\right]\cap[rL_\gamma-\epsilon, rL_\gamma+\epsilon]=\emptyset. \tag{34}$$

This follows as long as

$$V(s,t)=Tr\gamma_{\nu 2}E_N^{\Omega^c}(t-s)\circ(I-\chi_\gamma)(y,|D_t|^{-1}D_y)\gamma\circ E_0(s)$$

is a smooth function for $t\in(rL_\gamma-\epsilon, rL_\gamma+\epsilon)$ and for $s\in(0,rL_\gamma+\epsilon)$. To prove that it is smooth, we observe that the singular support consists of (s,t) such that there exists a closed billiard orbit of length t outside a phase space neighborhood of γ^r which consists of a straight line segment of length s from a point $x\in\Omega^c$ to a boundary point q, followed by a generalized billiard orbit of length $t-s$ from q back to x. By our assumption on the length spectrum, the only possible orbit with length $t\in(rL_\gamma-\epsilon, rL_\gamma+\epsilon)$ is γ^r of length r; but the cutoff has removed this orbit. This completes the proof. □

5.3. Regularization. In the preceding section we explained that the trace can be microlocalized to γ. However, it is not true that the operator N can be microlcalized to γ, that is, we cannot replace N by $\chi_\gamma N\chi_\gamma$. This is due to the singularities on the diagonal. We now explain how to regularize them.

When we expand

$$(N_{++0}+N_{++1}+N_{+-}+N_{--0}+N_{--1}+N_{-+})^M, \tag{35}$$

we get a sum of 'strings' of factors. The factors of N_{++0} or N_{--0} forbid the immediate application of the stationary phase method due to the singularities of its phase and amplitude on the diagonal. We now explain how to remove such factors and regularize the term except for presence of N_{++0}^M or N_{--0}^M, which has to be handled separately. We will only explain how to remove one factor of N_{++0} next to a smooth factor such as N_{++1}. In general, one just repeats the procedure one pair at a time.

5.4. The compositions $N_0\circ N_1$. We begin by characterizing the composition $N_0\circ N_1$. Essentially the same result holds for $N_1\circ N_0$ by the same argument.

The composed kernel equals

$$\begin{aligned} &N_0\circ N_1(k+i\tau,\varphi_1,\varphi_2)\\ &\quad := (k+i\tau)^2\int_{\mathbf{T}}\chi(k^{-1+\delta}(\varphi_1-\varphi_3))(1-\chi(k^{-1+\delta}(\varphi_2-\varphi_3)))\\ &\quad\quad\times H_1^{(1)}((k\mu+i\tau)|q(\varphi_3)-q(\varphi_1)|)\cos\angle(q(\varphi_3)-q(\varphi_1),\nu_{q(\varphi_3)})\\ &\quad\quad\times H_1^{(1)}((k\mu+i\tau)|q(\varphi_3)-q(\varphi_2)|)\cos\angle(q(\varphi_2)-q(\varphi_3),\nu_{q(\varphi_2)})d\varphi_3. \end{aligned} \tag{36}$$

The next proposition shows that these compositions are semiclassical Fourier integral kernels. Thus, the singularity on the diagonal is 'regularized'. The crucial point is that the composition with the singular factor N_0 decreased the order by one unit. We state in a somewhat simplified form and refer to [Z1] for a more detailed statement.

PROPOSITION 5.2. *$N_0 \circ N_1 \circ \chi_0(k + i\tau, \varphi_1, \varphi_2))$ defines a semiclassical Fourier integral operator on $\partial\Omega$ of order -1 associated to the billiard map.*

Proof. We only sketch the proof and refer to [Z1] for further details.

The proof is an explicit calculation. Following [AG], we change variables

$$\varphi_3 \to \vartheta = \varphi_1 - \varphi_3,$$

and then to change variables $\vartheta \to u$, with:

$$u := \begin{cases} |q(\varphi_3) - q(\varphi_1)|, & \varphi_1 \geq \varphi_3 \\ -|q(\varphi_3) - q(\varphi_1)|, & \varphi_1 \leq \varphi_3 \end{cases} = \begin{cases} |q(\varphi_1 - \vartheta) - q(\varphi_1)|, & \vartheta \geq 0 \\ -|q(\varphi_1 - \vartheta) - q(\varphi_1)|, & \vartheta \leq 0. \end{cases} \tag{37}$$

In other words, we change from the intrinsic distance along $\partial\Omega$ to chordal distance. The purpose of this change of variables is to simplify the difficult factor in (36):

$$H_1^{(1)}((k\mu + i\tau)|q(\varphi_3) - q(\varphi_1)|) \to H_1^{(1)}((k\mu + i\tau)|u|).$$

We then substitute the asymptotic WKB formula for the 'easy' factor $H_1^{(1)}((k\mu + i\tau)|q(\varphi_1) - q(\varphi_3|)$ which is valid on the support of the cutoff. The key points are then that

$$\begin{cases} (i) \ \cos(\angle q(\varphi_2) - q(\varphi_2 + \vartheta), \nu_{q(\varphi_2)}) \to |u| K(\varphi_2, u), \\ \quad \text{with } K \text{ smooth in } u; \\ (ii) \ H_1^{(1)}((k\mu + i\tau)|q(\varphi_1) - q(\varphi_2 - \vartheta)|) \to e^{ik|q(\varphi_1) - q(\varphi_2)|} e^{ikua} \\ \qquad\qquad \times A(k + i\tau, \varphi_1, \varphi_2, u), \\ \quad \text{where } A_k \text{ is a symbol in } k \text{ of order } -1/2 \text{ and smooth in } u. \end{cases}$$

Here,

$$a = \sin \vartheta_{1,2}, \quad \text{with } \vartheta_{1,2} = \angle(q(\varphi_2) - q(\varphi_1), \nu_{q(\varphi_2)}). \tag{38}$$

It follows that the composed kernel (36) can be expressed in the form $Ae^{i(k+i\tau)|q(\varphi_1) - \varphi_2)|}$ (further composed with χ_0) with

$$A(k+i\tau, \varphi_1, \varphi_2) = \int_{-\infty}^{\infty} \tilde{\chi}(k^{1-\delta}u)(1 - \chi(k^{1-\delta}(\varphi_2 - \varphi_1 - u)) \times |u| e^{ikau} H_1^{(1)}((k+i\tau)|u|) G((k+i\tau), u, \varphi_1, \varphi_2) du, \tag{39}$$

where $G((k+i\tau), u, \varphi_1, \varphi_2)$ is a symbol in k of order $-1/2$ and smooth in u, and where $\tilde{\chi}(k^{1-\delta}u) = \chi(k^{1-\delta}(\varphi_1 - \varphi_{3)})$.

We now change variables again, $u' = ku$ (and then drop the prime), to get

$$\begin{aligned} A(k+i\tau,\varphi_1,\varphi_2) = k^{-2}\int_{-\infty}^{\infty} \tilde{\chi}(k^{-\delta}u)(1-\chi(k^{1-\delta}(\varphi_2-\varphi_1-k^{-1}u)))|u|e^{iau} \\ \times H_1^{(1)}(b|u|)G(k+i\tau,\frac{u}{k},\varphi_1,\varphi_2)du, \end{aligned} \tag{40}$$

with $b = 1 + i(\tau/k)$. Since $|u| \leq k^{\delta}$ on the support of the cutoff, we have $|\frac{u}{k}| \leq k^{-1+\delta}$. The Taylor expansion of $G(k+i\tau, u, \varphi_1, \varphi_2)$ at $u = 0$ produces an asymptotic series in k. The main point is that the change of variables introduced a factor of k^{-2}, which cancels the original factor of k^2 in (36). We are left with an amplitude of order $-1/2$, which defines a Fourier integral operator of order -1 with phase $|q(\varphi_1) - \varphi_2)|$ as long as the integral (40) has the same order as G.

After Taylor expanding, we end up with terms of the form

$$\int_{-\infty}^{\infty} \tilde{\chi}(k^{-\delta}u)|u|u^n e^{iau} H_1^{(1)}(b|u|)du$$

times simple functions of (φ_1, φ_2). The integral may be expressed in the form

$$\frac{\partial}{\partial b}\frac{\partial^n}{\partial a^n}\int_{-\infty}^{\infty} \tilde{\chi}(k^{-\delta}u)e^{iau}H_0^{(1)}(b|u|)du|_{a=\sin\vartheta_{1,2},b=(1+i\tau/k)}. \tag{41}$$

Using the cosine transform of the Hankel function, one can explicitly evaluate (41) as $\frac{\partial^n}{\partial a^n}(1-a^2)^{-3/2}$, modulo lower order terms. After composition with the cutoff χ_0 this factor is bounded above, so one gets that A has order $-1/2$. □

We iterate this proposition to deal with the full strings in (35). The only cases where this is not possible are the ones which have no factors of $N_{+-}(k), N_{-+}(k)$. However, they are still composed with the cutoff operator χ_γ and a somewhat similar calculation shows that this composition produces a semiclassical pseudodifferential operator, whose order is decreased from 0 by one unit of k for each factor of N_0,

5.5. Final remarks and remainder estimate. After removing the N_0 factors from the terms (35) of (21, we end up with a finite sum of ordinary oscillatory integrals plus a complicated remainder. In the next section, we will describe the result of applying stationary phase to the finite sum. We now make a few comments on the remainder, referring to [Z1] for further discussion.

The estimate of the remainder is complicated because the operator norm $||N(k+i\tau)||$ does not decrease with increasing τ. To obtain a small remainder we redefine $\tau \to \tau \log k$. This changes the wave trace expansions by k^{-CrL_γ} and hence for R sufficiently large a remainder estimate of $O(k^{-R})$ is sufficient to separate a finite part of the wave trace from the remainder. We then estimate the remainder

$$Tr \int_{\mathbb{R}} \rho(k-\lambda)N(\lambda)^{M_0+1}(I+N(\lambda+i\tau))^{-1}\, \chi_\gamma\, N'(k+i\tau)d\lambda$$

by applying the Schwarz inequality for the Hilbert-Schmidt inner product. We further use standard estimates on the Poisson kernel to remove the factor of $(I+N(\lambda+i\tau))^{-1}$. Unfortunately, this estimate will also remove the cutoff ρ and replaces N^M by $N^M N^{*M}$. If one regularizes these products as above, we find that the critical points correspond to closed circuits of M-links which begin at some point x, end at some point x' and then return to x by traversing the links in reverse order. The cutoffs χ_0, χ_γ prohibit a proliferation of small links (grazing rays) and force the links in critical paths to point in the direction of γ and hence to be of length roughly ML_γ. The imaginary part $i\tau \log k$ of the semiclassical parameter then contributes a damping factor of $e^{-\tau ML_\gamma \log k}$ for each link. The links correspond to the N_1 factors. Thus, for each string, we have one k^{-1} for each N_0 factor and one $e^{-\tau ML_\gamma \log k}$ for each N_1 factor. For sufficiently large τ these combine to give a factor of k^{-R} for any prescribed R for each term of (35).

6. Calculating coefficients. We now apply the stationary phase method directly to each regularized integrals. Additionally, we have to prove that the remainder is as small as claimed. The result is the following explict formula for the wave invariants:

PROPOSITION 6.1. *We have:*

$$B_{\gamma^r,j} + B_{\gamma^{-r},j-1} = b_{r,j,2j} f^{2j}(0) + b_{r,j,2j-1} f^{2j-1}(0) + \tilde{B}_{\gamma^r,j},$$

with

$$\begin{aligned} &b_{r,j,2j} f^{2j}(0) + b_{r,j,2j-1} f^{2j-1}(0) \\ &\quad = r\Big\{ 2(h^{11})^j f^{(2j)}(0) + \Big\{ 2(h^{11})^j \frac{1}{2-2\cos\alpha/2} \\ &\qquad + (h^{11})^{j-2} \sum_{q=1}^{2r} (h^{1q})^3 \Big\} f^{(3)}(0) f^{(2j-1)}(0) \Big\} \end{aligned}$$

where $\tilde{B}_{\gamma^r,j}$ depends only on the $2j-2$ Taylor polynomial of f at $x=0$.

We review the calculation of the coefficients from [Z2] and adapt it to the exterior case. We also take advantage of the reduction to the boundary to simplify the arguments.

6.1. Setting things up. We recall that, in a small strip $T_\epsilon(\gamma)$ around $\overline{a_+a_-}$, the boundary $\partial\Omega$ consists of two components which are symmetric graphs over the x-axis. We write the graphs in the form $y = \pm f(x)$ near $a_\pm$.

As in (23), integrals over $(\Omega \cap T_\epsilon(\gamma))^M$ consist of 2^M terms, corresponding to a choice of an element σ of

$$\{\pm\}^M := \{\sigma : \mathbb{Z}_M \to \{\pm\}\}.$$

The length functional in Cartesian coordinates for a given assignment σ of signs is given by

$$\mathcal{L}_\sigma(x_1, \ldots, x_{2r}) = \sum_{j=1}^{2r-1} \sqrt{(x_{j+1} - x_j)^2 + (f_{\sigma(j+1)}(x_{j+1}) - f_{\sigma(j)}(x_j))^2}. \tag{36}$$

If we write out the integrals $I^\sigma_{M,\rho}$ of (24) in Cartesian coordinates, we obtain

$$\begin{aligned} I^\sigma_{M,\rho}(k+i\tau) = \int_{(-\epsilon,\epsilon)^M} \int_0^\infty \int_{-\infty}^\infty \hat{\rho}(t) \Big\{ \Pi_{p=1}^{M-1} H_1^{(1)}\big((k\lambda + i\tau) \\ \times |(x_{p+1} - x_p, f_{\sigma(p+1)}(x_{p+1}) - f_{\sigma(p)}(x_p))|\big) \\ \times \frac{(x_{p+1} - x_p) - f'_{\sigma(p+1)}(x_{p+1})(f_{\sigma(p+1)}(x_{p+1}) - f_{\sigma(p)}(x_p))}{\sqrt{(x_{p+1} - x_p)^2 + (f_{\sigma(p+1)}(x_{p+1}) + f_{\sigma(p)}(x_{p+1})(x_p) + L)^2}} \Big\} \\ \times \chi(1-\lambda) e^{ik(1-\lambda)t} d\lambda dt dx_1 \ldots dx_M. \end{aligned} \tag{37}$$

We now regularize the integrals (37) as above and obtain classical oscillatory integrals in $x-y$ coordinates. We also eliminate the (t, λ) variables by stationary phase. The Hessian in these variables is easily seen to be non-degenerate, and the Hessian operator equals $-\frac{\partial^2}{\partial t \partial \lambda}$. Since the amplitude depends on t only in the factor $\hat{\rho}(t)$, which is constant in a neighborhood of the critical point, only the zeroth order term in the Hessian operator survives the method of stationary phase. We may therefore eliminate the $ds d\mu$ integrals and replace the amplitude and phase by their evaluations at $\mu = 1, t = rL_\gamma$. This simplifies the above to

$$\begin{aligned} k^{-\nu s} \int_{[-\epsilon,\epsilon]^{2r}} e^{ik\mathcal{L}_\pm(x_1,\ldots,x_{2m})} \hat{\rho}(\mathcal{L}_\pm(x_1, \ldots, x_{2m})) \\ \times a(k, x_1, x_2, \ldots, x_{2r}) dx_1 \cdots dx_{2r}\ . \end{aligned} \tag{38}$$

An examination of the amplitudes and of the regularization procedure leads to the following conclusions:

- (i) The regularized integral $I^\sigma_{M,\rho}((k + i\tau))$ is negligeable as $k \to \infty$ unless $M = 2r$ (where rL_γ is the unique length in the support of $\hat{\rho}$), and where only strings of N_{-+} and N_{+-} are left upon regularization. Otherwise there are no critical points.
- (ii) The amplitudes of these oscillatory integrals have the form

$$\sum_{n=0}^\infty k^{-n} A_n(x),$$

where A_n depends only on the first $n+2$ derivatives of f_σ.

- There is a *special term*, namely the initially regular one in which $2r$ factors of N_{-+} and N_{+1} alternate, corresponding to $2r$ bounces of the bouncing ball orbit. The phase is simply $\mathcal{L}_{\pm}$ and the amplitude is

$$
\begin{aligned}
&a^0_{\pm}(k, x_1, \ldots, x_{2r}) \\
&= \Pi_{p=1}^{2r} a_1((k+i\tau)\mu\sqrt{(x_p - x_{p+1})^2 + (f_{\sigma\pm(p)}(x_p) - f_{\sigma\pm(p+1)}(x_{p+1})^2)} \\
&\times \frac{(x_p - x_{p+1})f'_{\sigma\pm(p)}(x_p) - (f_{\sigma\pm(p)}(x_p) - f_{\sigma\pm(p+1)}(x_{p+1}))}{\sqrt{(x_p - x_{p+1})^2 + (f_{\sigma\pm(p)}(x_p) - f_{\sigma\pm(p+1)}(x_{p+1}))^2}}.
\end{aligned}
\tag{39}
$$

We only need the principal terms in the symbols a_1, but to simplify notation we omit this step.

This amplitude and phase of the special term have certain key attributes which will be used extensively below

$$
\left\{
\begin{aligned}
&(i) \quad \text{In its dependence on } f, \text{ the amplitude has the form } \mathcal{A}(x, y, f, f'). \\
&(ii) \quad \text{In its dependence on the variables } x_p, \text{ the amplitude has the form:} \\
&\qquad a^0(k, x_1, \ldots, x_{2r}) = \Pi_{p=1}^{2r} A_p(x_p, x_{p+1}) \quad (2p+1 \equiv 1) \\
&(iii) \; D_{x_p}^{(2j-1)} \mathcal{L} \equiv ((x_p - x_{p+1})^2 + (f_{\epsilon_p}(x_p) \\
&\qquad - f_{\epsilon_{p+1}}(x_{p+1}))^2)^{-1/2}(f_{\epsilon_p}(x_p) \\
&\qquad - f_{\epsilon_{p+1}}(x_{p+1}))f_{\epsilon_p}^{(2j-1)}(x_p) \\
&\qquad \Longrightarrow D_{x_p}^{(2j-1)} \mathcal{L}|_{x=0} \equiv \epsilon_p f_{\epsilon_p}^{(2j-1)}(0); \\
&(iv) \; \nabla a^0(k, x_1, \ldots, x_{2r})_{x=0} = 0.
\end{aligned}
\right.
\tag{40}
$$

Here, the notation $\equiv$ means equivalence modulo terms with fewer derivatives of $f_{\pm}$. The last statement follows from the values

$$
f(0) = L, f'(0) = x|_{x=0} = 0, \angle(A - B, \nu_B) = \pi/2,
\tag{41}
$$

where A, B denote the endpoints.

We now use this information to determine where the data $f^{2j}_{\pm}(0)$, $f^{(2j-1)}_{\pm}$ first appears in the stationary phase expansion for $I_k(a_{\mathcal{S}_0}, \mathcal{L}_{\pm})$.

6.2. Stationary phase expansion. We now apply the method of stationary phase to these integrals and keep careful track of the dependence on the number of derivatives of $f_{\pm}$. To analyze the jungle of terms in the stationary phase expansion, we use the diagrammatic approach (see e.g. [AG]). We give a brief review.

Consider a general oscillatory integral $Z_k = \int_{\mathbb{R}^n} a(x)e^{ikS(x)}dx$ where $a \in C_0^\infty(\mathbb{R}^n)$ and where S has a unique critical point in supp$a$ at 0. Let us write H for the Hessian of S at 0. The stationary phase expansion takes the form:

$$Z_k = \left(\frac{2\pi}{k}\right)^{n/2} \frac{e^{i\pi sgn(H)/4}}{\sqrt{|detH|}} e^{ikS(0)} Z_k^{h\ell},$$

$$\text{where} \quad Z_k^{h\ell} = \sum_{j=0}^{\infty} k^{-j} \Big\{ \sum_{(\Gamma,\ell):\chi_{\Gamma'}=j} \frac{I_\ell(\Gamma)}{S(\Gamma)} \Big\}.$$

Here, the sum runs over the set $\mathcal{G}_{V,I}$ of labelled graphs (Γ, ℓ) with V closed vertices of valency ≥ 3 (each corresponding to the phase), with one open vertex (corresponding to the amplitude), and with I edges. Further, the graph Γ' is defined to be Γ minus the open vertex, and $\chi_{\Gamma'} = V - I$ equals its Euler characteristic. We note that there are only finitely many graphs for each χ because the valency condition forces $I \geq 3/2V$. Thus, $V \leq 2j, I \leq 3j$.

The function ℓ 'labels' each end of each edge of Γ with an index $j \in \{1, \dots, n\}$. Also, $S(\Gamma)$ denotes the order of the automorphism group of Γ, and $I_\ell(\Gamma)$ denotes the 'Feynman amplitude' associated to (Γ, ℓ). By definition, $I_\ell(\Gamma)$ is obtained by the following rule: To each edge with end labels j, k one assigns a factor of $\frac{-1}{ik} h^{jk}$ where $H^{-1} = (h^{jk})$. To each closed vertex one assigns a factor of $ik \frac{\partial^\nu S(0)}{\partial x^{i_1} \cdots \partial x^{i_\nu}}$ where ν is the valency of the vertex and $i_1 \dots, i_\nu$ at the index lables of the edge ends incident on the vertex. To the open vertex, one assigns the factor $\frac{\partial^\nu a(0)}{\partial x^{i_1} \dots \partial x^{i_\nu}}$, where ν is its valence. Then $I_\ell(\Gamma)$ is the product of all these factors. To the empty graph one assigns the amplitude 1. In summing over (Γ, ℓ) with a fixed graph Γ, one sums the product of all the factors as the indices run over $\{1, \dots, n\}$.

6.2.1. The special term: The data $f_\pm^{2j}(0)$. We first claim that $f_\pm^{(2j)}(0)$ appears first in the k^{-j+1} term. This is because any labelled graph (Γ, ℓ) for which $I_\ell(\Gamma)$ contains the factor $f_\pm^{(2j)}(0)$ must have a closed vertex of valency $\geq 2j$, or the open vertex must have valency $\geq 2j - 1$. The minimal absolute Euler characteristic $|\chi(\Gamma')|$ in the first case is $1 - j$. Since the Euler characteristic is calculated after the open vertex is removed, the minimal absolute Euler characteristic in the second case is $-j$ (there must be at least j edges.) Hence such graphs do not have minimal absolute Euler characteristic. It follows that the only labelled graph (Γ, ℓ) with $-\chi(\Gamma') = j - 1$ with $I_\ell(\Gamma)$ containing $f_\pm^{(2j)}(0)$ is given by:

- $\mathcal{G}_{1,j}^{2j,0} \subset \mathcal{G}_{1,j}$with $V = 1, I = j$; the unique graph has no open vertex, one closed vertex and j loops at the closed vertex. The only labels producing the desired data are those ℓ_p which assign all endpoints of all edges labelled the same index p.

- Assuming the up/down symmetry, the corresponding Feynman amplitude $I_{\ell_p}(\Gamma)$ has the form $2rL(h^{11})^j f^{(2j)}(0) + \cdots$, where again...refers to terms with $\leq 2j-1$ derivatives.

To prove the last statement about the Feynman amplitude, we first observe that $f^{(2j)}(0)_\pm$ appears linearly in $I_{\ell_p}(\Gamma)$. Indeed, the only labelled graphs which produce this datum give the same label to all endpoints of all edges, corresponding to applying only derivatives in a single variable $\frac{\partial}{\partial x_k}$. Using (41), an examination of (39) shows that the coefficient of $f^{(2j)}(0)$ of $I_k(a_{\mathcal{S}_0}, \mathcal{L}_+)$ equals

$$\sum_{p=1,p\equiv 1}^{2r} (h_+^{pp})^j \left(\frac{\partial}{\partial x_p}\right)^{2j} \mathcal{L}_+(\ldots, x_k, \ldots x_{2r}) = L\Bigg[\sum_{p=1,p\equiv 1}^{2r} (h_+^{pp})^j\Bigg] f_+^{(2j)}(0).$$

To complete the proof it suffices to observe that the diagonal matrix elements $h_\pm^{pp}$ are constant in p and in the sign $\pm$.

Had we assumed that the obstacles were both up-down and right-left symmetric as in [Z3], then we would already have solved the inverse spectral problem in this step, since all odd Taylor coefficients of f vanish at $x = 0$ and we will be able to recover all even ones.

6.2.2. The special term: The data $f_\pm^{(2j-1)}(0)$. We now consider the more difficult odd coefficients $f_\pm^{(2j-1)}(0)$, which will require the attributes of the amplitude (39) detailed in (40).

We again claim that the Taylor coefficients $f_\pm^{(2j-1)}(0)$ appear first in the term of order k^{-j+1}, and we enumerate the labelled graphs which have a combinatorial structure capable of producing $f_\pm^{(2j-1)}(0)$ as a factor in $I_\ell(\Gamma)$ in the k^{-j+1} term. In fact, only the main term $I^{\sigma\pm}_{M,\rho,C}$ will produce such a term. The following is proved in [Z2]:

LEMMA 6.2. *We have:*

(i) *There are no labelled graphs with* $-\chi' := -\chi(\Gamma') \leq j-1$ *for which* $I_\ell(\Gamma)$ *contains the factor* $f_\pm^{(2j-1)}(0)$.

(ii) *There are exactly two types of labelled diagrams* (Γ, ℓ) *with* $\chi(\Gamma') = -j+1$ *such that* $I_\ell(\Gamma)$ *contains the factor.* $f_\pm^{(2j-1)}(0)$. *They are given by:*

- $\mathcal{G}^{2j-1,3,0}_{2,j+1} \subset \mathcal{G}_{2,j+1}$ *with* $V = 2, I = j+1)$: *Two closed vertices,* $j-1$ *loops at one closed vertex,* 1 *loop at the second closed vertex, one edge between the closed vertices; no open vertex. Labels* $\ell_{p,q}$: *All labels at the closed vertex with valency* $2j-1$ *must be the same index* p *and all at the closed vertex must the be same index* q. *Feynman amplitude:* $(h_\pm^{pp})^{j-1} h_\pm^{qq} h_\pm^{pq} D^{2j-1}_{x_p} \mathcal{L}_\pm D^3_{x_q} \mathcal{L}_\pm \sim (h_\pm^{pp})^{j-1} h_\pm^{qq} h_\pm^{pq} f^{(2j-1)}(0) f^{(3)}(0)$.
- $\mathcal{G}^{2j-1,3,0}_{2,j+1} \subset \mathcal{G}_{2,j+1}$ *with* $V = 2, I = j+1$: *Two closed vertices, with* $j-2$ *loops at one closed vertex, and with three edges be-*

tween the two closed vertices; no open vertex. Labels $\ell_{p,q}$: All labels at the closed vertex with valency $2j-1$ must be the same index p and all at the closed vertex must the be same index q; $(h^{pp}_{\pm})^{j-2}(h^{pq}_{\pm})^3 D^{2j-1}_{x_p}\mathcal{L}_{\pm} D^3_{x_q}\mathcal{L}_{\pm} \sim (h^{pp}_{\pm})^{j-2}(h^{pq}_{\pm})^3 f^{(2j-1)}(0) f^{(3)}(0)$.

This Lemma requires a careful search through the diagrams and an explicit calculation of the amplitudes. It is the most 'unstable' step of the proof, in which any calculational error could ruin the result. For instance, it turns out that in addition to the two diagrams which contribute, there are three more which have a combinatorial structure capable of contributing the datum $f^{2j-1}(0)$; but the corresponding amplitudes turn out to vanish due to special attributes (40) of the Taylor expansion of the amplitude at the critical point. We should note other attributes in addition to (40) are needed to show that the Feynman amplitude for these diagrams vanishes.

6.2.3. Other terms. We also have:

LEMMA 6.3. *Other terms do not contribute the data $f^{(2j)}(0)$, $f^{(2j-1)}(0)$ in the term of order k^{-j+1}.*

The proof of this is easy. It merely requires checking how the regularization compares to the stationary phase expansion. As noted above, the regularized integrals are lower order yet do not involve higher derivatives of the amplitude or phase.

7. Recovering the domain. In this section, we sketch the proof of the main result, Theorem 1.1. The proof is virtually identical to the proof that one can determine a $\mathbb{Z}_2$-symmetric analytic domain from the wave invariants of a $\mathbb{Z}_2$- symmetric hyperbolic bouncing ball orbit. In fact, the proof requires fewer assumptions, since L_γ is a resonance invariant (being the least non-zero singular time in the wave trace.) We may then inspect the wave invariants (i.e. the Balian-Bloch trace invariants) at the iterates γ^r of γ. Give the expressions above for the wave invariants, it suffices to analyse the Hessian coefficients.

7.1. Poincare map and Hessian of the length functional at γ. We will need to recollect certain facts about the Hessian H_{2r} of the length function at critical points corresponding to the iterates γ^r. In most respects,the discussion is identical to the interior case, so we refer the reader to [Z1, Z2] for background on the linear Poincare map P_γ and its relation to the Hessian of the length function. We will only consider the $\mathbb{Z}_2$-symmetric case since it is the one relevant to our result.

Henceforth we assume $\mathcal{O}$ is convex, so that the bouncing γ is necessarily hyperbolic and the eigenvalues of its Poincare map P_γ are of the form $\{e^{\pm\lambda}\}$. They are related to the geometry by

$$\cosh \alpha/2 = \left(1 - \frac{L}{R}\right), \tag{43}$$

where R is the common radius of curvature at the endpoints a_j. For details, see [PS][KT]. When $\mathcal{O}$ is not convex, γ could be elliptic. We refer to [Z2] for the calculation of the wave invariants in that case.

We will need formulae involving matrix elements of the inverse Hessian matrix $H_{2r}^{-1} = (h^{pq})$ at the $\mathbb{Z}_2$-symmetric bouncing ball orbit. As in [Z2], the Hessian in $x - y$ coordinates is the *circulant* matrix

$$H_{2r} = C(2\cosh\alpha/2, 1, 0, \ldots, 0, 1), \tag{44}$$

that is, the matrix whose rows (or columns) are obtained by cycling the first row (or column) (see [Z1, Z2] for background). Therefore H_{2r} is diagonalized by the finite Fourier matrix F of rank $2r$ (see [Z1, Z2] for its definition and background). Exactly as in [Z1, Z2], we have

$$H_{2r} = F^* \text{diag}\left(2\cos\alpha/2 + 2, \ldots, 2\cos\alpha/2 + 2\cos\frac{(2r-1)\pi}{r}\right)F. \tag{45}$$

An important role in the inverse problem is played by sums of powers of elements of columns of H_{2r}^{-1}. From (45), it follows that

$$H_{2r}^{-1} = F^*\left(\text{diag}\left(\frac{1}{2\cos\alpha/2 + 2}, \ldots, \frac{1}{2\cos\alpha/2 + 2\cos\frac{(2r-1)\pi}{r}}\right)\right)F. \tag{46}$$

Hence, the first row $[H_{2r}^{-1}]_1 = (h^{11}, \ldots, h^{12r})$ (or column) of the inverse is given by:

$$h^{1q} = \sum_{k=0}^{2r-1} \frac{w^{(q-1)k}}{p_{\alpha,r}(w^k)}, \tag{47}$$

where $w = e^{\frac{2\pi i}{2r}}$ and where $p_{\alpha,r}(z) = 2\cosh\alpha/2 + z + z^{n-1}$, in [Z1, Z2] we will need that, for any p,

$$\sum_{q=1}^{2r} h^{pq} = \frac{1}{2 + \cosh\alpha/2}. \tag{48}$$

7.2. Domain recovery. We now prove by induction on j that the Taylor coefficients $f^{2j-1}(0), f^{2j}(0)$ can be determined from $B_{\gamma^r,j}$ as r varies over $r = 1, 2, 3, \ldots$. At first, we assume $f^3(0) \neq 0$.

It suffices to separately determine the two terms

$$\begin{aligned} &2(h_{2r}^{11})^2\left\{f^{(2j)}(0) + \frac{1}{2 - 2\cos\alpha/2}\, f^{(3)}(0)f^{(2j-1)}(0)\right\}, \\ &\text{and } \left\{\sum_{q=1}^{2r}(h_{2r}^{1q})^3\right\}f^{(3)}(0)f^{(2j-1)}(0). \end{aligned} \tag{49}$$

As discussed in [Z2], the terms decouple as r varies if and only if $F_r(\cos\alpha/2) := \sum_{q=1}^{2r}(h_{2r}^{1q})^3$ is non-constant in $r = 1, 2, 3, \ldots$

By the explicit calculation in [Z2], we have:

$$\sum_{q=1}^{2r}(h^{pq})^3 =$$

$$2r\sum_{k_1,k_2=0}^{2r}\frac{1}{(\cosh\alpha/2+\cos\frac{k_1\pi}{r})(\cosh\alpha/2+\cos\frac{k_2\pi}{r})(\cos\alpha/2+\cosh\frac{(k_1+k_2)\pi}{r})}.$$

It is obvious that the sum is strictly increasing as r varies over even integers.

We now begin the inductive argument. From the $j = 0$ term we determine $f''(0)$. Indeed, $(1 - Lf^{(2)}(0) = \cos(h)\alpha/2$ and α is a wave trace invariant. From the $j = 2$ term we recover $f^3(0), f^4(0)$. The induction hypothesis is then that the Taylor polynomial of f of degree $2j - 2$ has been recovered by the $j - 1$st stage. By the decoupling argument we can determine $f^{2j}(0), f^{2j-1}(0)$ as long as $f^3(0) \neq 0$.

As discussed in [Z2], the domain can still be recovered if $f^3(0) = 0$, but $f^5(0) \neq 0$. The argument is essentially the same. Similarly if $f^3(0) = f^5(0) = 0$, but $f^7(0) \neq 0$, and so on. If all odd derivatives vanish, then one still recovers the domain as noted above.

This completes the sketch of the proof of Theorem (1.1).

REFERENCES

[AG] D. Alonso and P. Gaspard, $\hbar$ expansion for the periodic orbit quantization of chaotic systems. Chaos **3** (1993), No. 4, 601–612.

[AM] K. G. Andersson and R. B. Melrose, The propagation of singularities along gliding rays. Invent. Math. 41 (1977), No. 3, 197–232.

[B] R. Bacher (unpublished note, 2002).

[BB1] R. Balian and C. Bloch, Distribution of eigenfrequencies for the wave equation in a finite domain I: three-dimensional problem with smooth boundary surface, Ann. Phys. **60** (1970), 401–447.

[BB2] R. Balian and C. Bloch, Distribution of eigenfrequencies for the wave equation in a finite domain. III. Eigenfrequency density oscillations. Ann. Physics **69** (1972), 76–160.

[BGR] C. Bardos, J. C. Guillot and J. Ralston, La relation de Poisson pour l'quation des ondes dans un ouvert non born. Application la thorie de la diffusion, Comm. Partial Differential Equations **7** (1982), No. 8, 905–958.

[CdV] Y. Colin de Verdire, Sur les longueurs des trajectoires priodiques d'un billard, *South Rhone seminar on geometry, III* (Lyon, 1983), 122–139, Travaux en Cours, Hermann, Paris, 1984.

[EP] J.P. Eckmann and C.A. Pillet, Zeta functions with Dirichlet and Neumann boundary conditions for exterior problems, Helv. Phys. Acta **70** (1997), 44–65.

[EP2] J.P. Eckmann and C.A. Pillet, Scattering phases and density of states for exterior domains. Ann. Inst. H. Poincar Phys. Thor. **62** (1995), No. 4, 383–399.

[F] G.B. Folland, *Introduction to Partial Differential Equations*, Math. Notes. 17, Princeton U. Press, Princeton (1976).

[GP] B. GEORGEOT AND R.E. PRANGE, Exact and quasiclassical Fredholm solutions of quantum billiards. Phys. Rev. Lett. **74** (1995), No. 15, 2851–2854.

[G] V. GUILLEMIN, Wave-trace invariants. Duke Math. J. **83** (1996), No. 2, 287–352.

[GM] V. GUILLEMIN AND R.B. MELROSE, The Poisson summation formula for manifolds with boundary. Adv. in Math. **32** (1979), No. 3, 204–232.

[HZ] A. Hassell and S. Zelditch, Quantum ergodicity of boundary values of eigenfunctions (arXiv preprint math.SP/0211140).

[ISZ] A. IANTCHENKO, J. SJOSTRAND AND M. ZWORSKI, Birkhoff normal forms in semi-classical inverse problems, (arXiv.org/abs/math.SP/0201191).

[KT] V.V. KOZLOV AND D.V. TRESHCHEV, *Billiards: A Genetic Introduction to the Dynamics of Systems with Impacts*, Translations of Math. Monographs 89, AMS publications, Providence, R.I. (1991).

[M] R.B. MELROSE, Polynomial bounds on the distribution of poles in scattering by an obstacle, Journees "Equations aux Derivee Partielles", Saint-Jean de Monts (1984).

[MS] R. B. Melrose and J. Sjöstrand, Singularities of boundary value problems. I. Comm. Pure Appl. Math. 31 (1978), No. 5, 593–617.

[PS] V.M. PETKOV AND L.N. STOYANOV, *Geometry of Reflecting Rays and Inverse Spectral Problems*, John Wiley and Sons, N.Y. (1992).

[THS] T. HARAYAMA, A. SHUDO, AND S. TASAKI, Semiclassical Fredholm determinant for strongly chaotic billiards. Nonlinearity **12** (1999), No. 4, 1113–1149.

[THS2] T. HARAYAMA, A. SHUDO, AND S. TASAKI, Interior Dirichlet eigenvalue problem, exterior Neumann scattering problem, and boundary element method for quantum billiards. Phys. Rev. E (3) **56** (1997), No. 1, part A, R13–R16.

[P] A. PLEIJEL, A study of certain Green's functions with applications in the theory of vibrating membranes, Arkiv for Math 2 (1952), 553–569.

[S] B. Simon, *Trace ideals and Their Applications*, London Math. Soc. Lecture Note Series 35, Cambridge Univ. Press, Cambridge (1979).

[TI] M.E. TAYLOR, *Partial differential equations. I. Basic theory.* Applied Mathematical Sciences, 115. Springer-Verlag, New York, 1996.

[TII] M.E. TAYLOR, *Partial differential equations. II. Qualitative studies of linear equations.* Applied Mathematical Sciences, 116. Springer-Verlag, New York, 1996.

[Z1] S. ZELDITCH, Inverse spectral problem for analytic plane domains I: Balian-Bloch trace formula (preprint, 2001).

[Z2] S. ZELDITCH, Inverse spectral problem for analytic plane domains II: domains with symmetry (preprint, 2001).

[Z3] S. ZELDITCH, Spectral determination of analytic bi-axisymmetric plane domains. Geom. Funct. Anal. **10** (2000), No. 3, 628–677.

[Zw] M. ZWORSKI, Poisson formula for resonances in even dimensions. Asian J. Math. **2** (1998), No. 3, 609–617.

LIST OF WORKSHOP PARTICIPANTS

- Chi Young Ahn, Department of Mathematics, Brain 21 Korea/ Yonsei University
- Tuncay Aktosun, Department of Mathematics, North Dakota State University
- Paolo Albano, Department of Mathematics, University of Bologna
- Giovanni Alessandrini, Dipartimento di Scienze Matematiche, University of Trieste
- Habib Ammari, Center of Applied Mathematics, CNRS
- Kais Ammari, Department of Mathematics, University of Nancy-I
- Douglas N. Arnold, Institute for Mathematics Applications
- George Avalos, Department of Mathematics, University of Nebraska
- Werner Ballmann, Mathematisches Institut, Rheinische Friedrich-Wilhelms-Universitaet Bonn
- Bartolome Barcelo, Departamento de Matematicas, Universidad Autunoma de Madrid
- Mikhail Belishev, Steklov Institute of Mathematics
- James G. Berryman, Department of Geophysics, Lawrence Livermore National Laboratory
- Gerard Besson, Institut Fourier, Laboratoire de Mathematiques, CNRS
- Russell Brown, Department of Mathematics, University of Kentucky
- Dmitri Burago, Mathematics Department, Penn State University
- John Cagnol, University Leonard de Vinci
- Jianguo Cao, Department of Mathematics, University of Notre Dame
- Yves Capdeboscq, Department of Mathematics, Rutgers University
- Jamylle Carter, Institute for Mathematics and its Applications
- Eduardo Chappa, Department of Mathematics, University of Washington
- Christopher B. Croke, Department of Mathematics, University of Pennsylvania
- Michael Dobranski, Department of Mathematics, University of Kentucky
- Fred Dulles, Institute for Mathematics and its Applications
- Patrick Eberlein, Department of Mathematics, University of North Carolina
- Matthias Eller, Department of Mathematics, Georgetown University

- Michael Galbraith, School of Mathematics, University of Minnesota
- Peter C. Gibson, Applied Mechanics, Technische Universitaet Darmstadt
- Leon Green, School of Mathematics, University of Minnesota
- Allan Greenleaf, Department of Mathematics, University of Rochester
- Robert Gulliver, School of Mathematics, University of Minnesota
- Scott Hansen, Department of Mathematics, Iowa State University
- Soenke Hansen, FB 17 Mathematik-Informatik, University of Paderborn
- Mary Ann Horn, Department of Mathematics, Vanderbilt University
- L. Steven Hous, Department of Mathematics, Iowa State University
- Kai Huang, Department of Mathematics, Michigan State University
- Victor Isakov, Department of Mathematics and Statistics, Wichita State University
- Chris Judge, Indiana University
- Hyeonbae Kang, Department of Mathematics, Seoul National University
- Chang-Wan Kim, Department of Mathematics, Korea Advanced Institute of Science and Technology
- Hyejoo Kim, Department of Statistics, Seoul National University
- Jeong-Hoon Kim, Mathematics Department, Brain21 Korea/Yonsei University
- Jinseog Kim, Department of Statistics, Seoul National University
- Jumi Kim, Mathematics Department, Brain 21 Korea/Yonsei University
- Seongjai Kim, Department of Mathematics, University of Kentucky
- Bruce Kleiner, Department of Mathematics, University of Michigan
- Gerhard Knieper, Fakultaet fuer Mathematik, Ruhr-Universitaet Bochum
- Kim Knudsen, Department of Mathematical Sciences, Aalborg University
- Vladimer V Kryzhniy, Applied Mathematics, Kuban State University of Technology
- Yaroslav Kurylev, Department of Mathematical Sciences, Loughborough University
- Irena Lasiecka, Department of Mathematics, University of Virginia
- Matti Juhani Lassas, Rolf Nevanlinna Institute, University of Helsinki

- Catherine Lebiedzik, Department of Mathematics, University of Virginia
- E. B. Lee, Department of Electrical and Computer Engineering, University of Minnesota
- Guang-Tsai Lei, Physiology and bio-physics, Mayo Clinic/Foundation
- Debra Lewis, Department of Mathematics, University of California, Santa Cruz
- Hyeona Lim, Department of Mathematics, Michigan State University
- Mi Kyoung Lim, Department of Mathematics, Seoul National University
- Walter Littman, School of Mathematics, University of Minnesota
- Patrick Martinez, UFR Mathematiques Informatique et Gestion, Universite Paul Sabatier
- Stephen McDowell, Department of Mathematics, University of Rochester
- Willard Miller, Institute for Mathematics and its Applications
- Steen Moeller, Department of Mathematical Sciences, Aalborg University
- Shari Moscow, Department of Mathematics, University of Florida
- Marwan Moubachir, Applied Mathematics, Laboratoire Central des Ponts et Chaussees
- Gen Nakamura, Department of Mathematics Graduate School of Science, Hokkaido University
- Nilima Nigam, Department of Mathematics and Statistics, McGill University
- Leonid N. Pestov, Sobolev Institute of Mathematics
- Jean-Pierre Puel, Laboratoire de Mathematiques Appliquees, Universite de Versailles St. Quentin
- Lizabeth Rachele, Department of Mathematics, Tufts University
- Rakesh, Department of Mathematical Sciences, University of Delaware
- Jan Marthedal Rasmussen, Informatics and Mathematical Modeling, Technical University of Denmark
- Paul Sacks, Department of Mathematics, Iowa State University
- Jin Keun Seo, Department of Mathematics, Yonsei University
- Sashirekha Shanmugavelu, Department of Computer Engineering, University of Minnesota
- Vladimir Sharafutdinov, Sobolev Institute of Mathematics
- Jaemin Shin, Department of Mathematics, Brain 21 Korea /Yonsei University
- Plamen D. Stefanov, Department of Mathematics, Purdue University

- Lev Steinberg, Department of Mathematics, University of Puerto Rico
- John Sylvester, Department of Mathematics, University of Washington
- Takashi Takiguchi, Department of Mathematics, National Defense Academy
- Alexandru Tamasan, Department of Mathematics, University of Washington
- Michael Taylor, Mathematics Department, University of North Carolina
- Roberto Triggiani, Department of Mathematics, University of Virginia
- Gunther Uhlmann, Department of Mathematics, University of Washington
- Nicolas Valdivia, Department of Mathematics, Wichita State University
- Judith Vancostenoble, Laboratoire de Mathematiques, MIP, Universite Paul Sabatier
- Michael S. Vogelius, Department of Mathematics, Rutgers University
- Jenn-Nan Wang, Mathematics, National Cheng Kung University
- Xiaoming Wang, Department of Mathematics, Iowa State University
- Anna Wienhard, Mathematisches Institut, Rheinische Friedrich-Wilhelms Universitaet Bonn
- Masahiro Yamamoto, Department of Mathematical Sciences, The University of Tokyo
- Stephen Zelditch, Department of Mathematics, Johns Hopkins University

1999–2000	Reactive Flows and Transport Phenomena
2000–2001	Mathematics in Multimedia
2001–2002	Mathematics in the Geosciences
2002–2003	Optimization
2003–2004	Probability and Statistics in Complex Systems: Genomics, Networks, and Financial Engineering
2004–2005	Mathematics of Materials and Macromolecules: Multiple Scales, Disorder, and Singularities
2005-2006	Imaging

IMA SUMMER PROGRAMS

1987	Robotics
1988	Signal Processing
1989	Robust Statistics and Diagnostics
1990	Radar and Sonar (June 18–29) New Directions in Time Series Analysis (July 2–27)
1991	Semiconductors
1992	Environmental Studies: Mathematical, Computational, and Statistical Analysis
1993	Modeling, Mesh Generation, and Adaptive Numerical Methods for Partial Differential Equations
1994	Molecular Biology
1995	Large Scale Optimizations with Applications to Inverse Problems, Optimal Control and Design, and Molecular and Structural Optimization
1996	Emerging Applications of Number Theory (July 15–26) Theory of Random Sets (August 22–24)
1997	Statistics in the Health Sciences
1998	Coding and Cryptography (July 6–18) Mathematical Modeling in Industry (July 22–31)
1999	Codes, Systems, and Graphical Models (August 2–13, 1999)
2000	Mathematical Modeling in Industry: A Workshop for Graduate Students (July 19–28)
2001	Geometric Methods in Inverse Problems and PDE Control (July 16–27)
2002	Special Functions in the Digital Age (July 22–August 2)
2003	Probability and Partial Differential Equations in Modern Applied Mathematics (July 21–August 1)
2004	n-Categories: Foundations and Applications (June 7–18)

IMA "HOT TOPICS" WORKSHOPS

- Challenges and Opportunities in Genomics: Production, Storage, Mining and Use, April 24–27, 1999

- Decision Making Under Uncertainty: Energy and Environmental Models, July 20–24, 1999
- Analysis and Modeling of Optical Devices, September 9–10, 1999
- Decision Making under Uncertainty: Assessment of the Reliability of Mathematical Models, September 16–17, 1999
- Scaling Phenomena in Communication Networks, October 22–24, 1999
- Text Mining, April 17–18, 2000
- Mathematical Challenges in Global Positioning Systems (GPS), August 16–18, 2000
- Modeling and Analysis of Noise in Integrated Circuits and Systems, August 29–30, 2000
- Mathematics of the Internet: E-Auction and Markets, December 3–5, 2000
- Analysis and Modeling of Industrial Jetting Processes, January 10–13, 2001
- Special Workshop: Mathematical Opportunities in Large-Scale Network Dynamics, August 6–7, 2001
- Wireless Networks, August 8–10 2001
- Numerical Relativity, June 24–29, 2002
- Operational Modeling and Biodefense: Problems, Techniques, and Opportunities, September 28, 2002
- Data-driven Control and Optimization, December 4–6, 2002
- Compatible Spatial Discretizations for PDE, November 3–7, 2003

SPRINGER LECTURE NOTES FROM THE IMA:

The Mathematics and Physics of Disordered Media
Editors: Barry Hughes and Barry Ninham
(Lecture Notes in Math., Volume 1035, 1983)

Orienting Polymers
Editor: J.L. Ericksen
(Lecture Notes in Math., Volume 1063, 1984)

New Perspectives in Thermodynamics
Editor: James Serrin
(Springer-Verlag, 1986)

Models of Economic Dynamics
Editor: Hugo Sonnenschein
(Lecture Notes in Econ., Volume 264, 1986)

Forthcoming Volumes:

Geometric Methods in Inverse Problems and PDE Control

Mathematical Foundations of Speech and Language Processing

Time Series Analysis and Applications to Geophysical Systems